PHYLUM PLATYHELMINT
CLASS TURBELLARIA

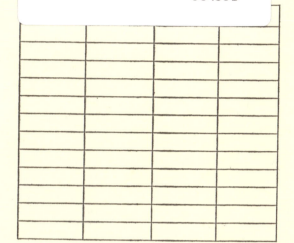

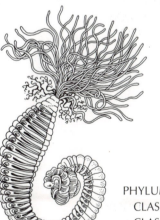

PHYLUM ANNELIDA
CLASS POLYCHAETA
CLASS OLIGOCHAETA
CLASS HIRUDINEA

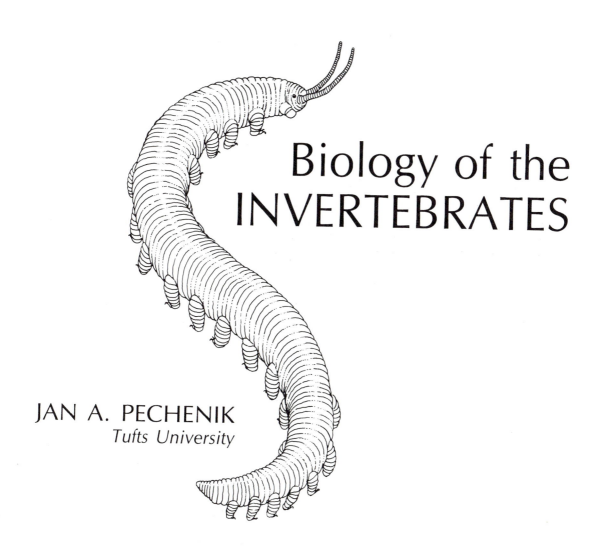

Biology of the
INVERTEBRATES

JAN A. PECHENIK
Tufts University

PRINDLE, WEBER & SCHMIDT, BOSTON

PWS PUBLISHERS

Prindle, Weber & Schmidt • ♣ • Duxbury Press • ♠ • PWS Engineering • △
Statler Office Building • 20 Park Plaza • Boston, Massachusetts 02116

PWS Publishers is a division of Wadsworth, Inc.

Library of Congress Cataloging in Publication Data

Pechenik, Jan A.
   Biology of the invertebrates.

   Bibliography: p.
   Includes index.
   1. Invertebrates.   I. Title.
QL362.P43   1985   592   84-25896
ISBN 0-87150-450-2

ISBN 0-87150-450-2

Front cover photos: (top) a soft coral , copyright © 1984 by Bruce A. Iverson, all rights reserved; (bottom) a flatworm, *Pseudoceros* sp., photo by Scott Johnson. Back cover photos: (top) a garden spider, photo by Richard H. Gross, Motlow State Community College; (middle) a candy cane sea star, copyright © Jeffrey L. Rotman; (bottom) an *Amphitrite ornata*, photo by William H. Lang.

Sponsoring editor: Jean-François Vilain. Production coordinators: David Chelton and David Foss. Text designer: Jeanne Ray Juster. Cover designer: Trisha Hanlon. Artists: Art Ciccone, Parry Clark, Wayne Clark, Peter Loewer, Cindy Moor, Deborah Schneck, and Eric Vincent. Composed in Optima and Electra by Interactive Composition Corp. Text printer and binder: Halliday Lithograph. Cover printer: New England Book Components. Printed in the United States of America.

85 86 87 88 89 — 10 9 8 7 6 5 4 3 2

TO MY PARENTS

PREFACE

Invertebrate Zoology is an enormous field. More than 98% of all known animal species are invertebrates, and that proportion will undoubtedly increase with time as more species are described. Invertebrates are distributed among more than 35 phyla and a mind-boggling number of classes, subclasses, orders, and families. The degree of morphological and functional diversity found within some groups, even within one order, can overwhelm the beginning student. The sheer enormity of the field and the large range of potential approaches to the subject matter make invertebrate biology challenging to teach and to learn. In preparing this book, I have endeavored to make the tasks of both teaching and learning easier and more interesting.

Learning invertebrate biology is like learning a foreign language. There is a considerable vocabulary to be acquired, but mastering the vocabulary need not be an end in itself. Here I present the basic vocabulary and grammar that will help open the door to the great literature of invertebrate biology and allow instructors to embellish and expand on the text to suit the focus of their particular course. New terms are printed in boldface and are defined in the text where they first occur. Frequently used terms are also defined in a glossary at the end of the book.

I have generally avoided drawing attention to particular species in the text, in keeping with my emphasis on unifying principles. In the selection of illustrations, on the other hand, I have deliberately chosen species that the student is especially likely to encounter in the lab, the field, or in reading the scientific literature. Although all phyla are covered, I have aimed for conciseness, not exhaustiveness; I have written a basic textbook, not an encyclopedia. The book is intended to be a functional part of the teaching process in the standard introductory course.

Although the book can be used as a supplement to lectures, it is designed especially to be read in preparation for lecture and lab. If the book has a bias, it tends to be towards functional morphology. Such treatment seems to bring the animals to life for the students and also prepares them for the careful observation of living animals in the laboratory or the field. An added benefit is that a discussion of function effectively reinforces knowledge of structure. To keep the book manageable in size and scope, I have also chosen to emphasize the features that set each group apart from all others, rather than to emphasize the diversity of form and function found within each group. Such diversity is best encountered in the laboratory or the field, after basic body plans and the associated vocabulary have been mastered.

Although I have chosen to concentrate on the manageable basics of invertebrate biology, I have given a glimpse of some of the major research areas in which invertebrates have played and are playing important and exciting roles. Most chapters close with a section entitled "Topics for Further Discussion and Investigation." I have spent considerable effort formulating questions and suggestions that encapsulate some of the major topics that have been and are being addressed for the animals covered in each chapter. I have selected references that are likely to be found in the typical college or university library and that should be intellectually accessible to an interested beginning student. The topics that I have chosen, along with the references that follow them, could be used as a basis for class discussion, term papers, or short essays, or simply as a convenient way to ease students into the original literature by having them investigate a topic that excites their curiosity. I expect that these references will help to make the book a valuable reference source long after the students have completed their introductory course.

I have made the book flexible enough to fit the various emphases of different instructors. Although the chapters are arranged in a traditional and logical order, they need not be read in that sequence. The chapters are self-contained and can be assigned in the order that best suits the organization of the particular course, once the introductory chapters have been covered.

I believe that all animal groups are inherently fascinating and worthy of study; accordingly, no groups are labeled "lesser" or "lower," and most phyla are treated in individual chapters, facilitating the assignment of readings. Within each chapter, the material has been arranged in manageable, readable chunks for the convenience of both students and instructors. For example, a section entitled "General Characteristics" might be assigned prior to a lecture on a particular group of organisms, while a section on "Feeding and Digestion" might be best assigned prior to lab or omitted entirely, depending upon the focus of the course. The separate chapter on hydrostatic skeletons (Chapter 8) can be assigned whenever the instructor feels it is most appropriate. The final chapter, concerning general principles of invertebrate reproduction and development, serves to bring together all of the major phyla in a consideration of problems shared by the members of all groups covered in the text.

I have prepared a comprehensive instructor's manual, including transparency masters selected from the illustrations in the text. This manual offers a number of specific aids for using the text to best advantage.

It is a pleasure to thank the many people who have helped in the preparation of this book. The following colleagues read one or more drafts of all or portions of the manuscript; I thank them for the care and thoroughness with which they read and commented. The entire manuscript was read by Bob Bullock (University of Rhode Island), Linda S. Eyster (Northeastern University and Harvard Medical School), John Hitt (Wittenberg University), Edward Ruppert (Clemson University), and Tom Wolcott (North Carolina State University).

Specific sections of the manuscript were read

by John Corliss, University of Maryland (Protozoa); Bruce Coull, University of South Carolina (Nematoda, Arthropoda, Tardigrada, Onychophora, Pentastomida); C. Bradford Calloway, Harvard University (Gastropoda, Bivalvia); Gordon Hendler, Smithsonian Institution (Echinodermata); G. Richard Harbison, Woods Hole Oceanographic Institution (Urochordata, Ctenophora); Edward S. Hodgson, Tufts University (Cnidaria); William M. Layton, Jr. (Dartmouth Medical School); Nancy Milburn, Tufts University (Arthropoda); John F. Pilger, Agnes Scott College (Cnidaria); Gary Polis, Vanderbilt University (Arthropoda); Pamela Roe, California State University, Stanislaus (Rhynchocoela); Terry Snell, University of Tampa (Rotifera); and Russell L. Zimmer, University of Southern California (Phoronida, Brachiopoda, Bryozoa). I accept responsibility for any errors and ambiguities that remain, and welcome constructive criticism from all readers.

I am also grateful to the many investigators and publishers who permitted reproduction of their drawings and photographs. Art Ciccone, Parry Clark, Wayne Clark, Peter Loewer, Cindy Moor, Debbie Schneck, and Eric Vincent succeeded in translating my sketches and instructions into splendid illustrations, which add much to the usefulness of the book. Special thanks go to my friend Rick Casabona who, just when I thought that all was lost, successfully transferred essential text files from an alien computer to one that I could talk to, saving me many pages of retyping.

This book might never have happened had it not been for the recruiting prowess and enthusiasm of Jean-François Vilain, of PWS Publishers (one can do so much with a French accent), and for the hard work of his associates, including David Chelton, David Foss, and Tyrel Holston. Their professionalism and attention to detail was most impressive and appreciated. My wife, Lindy Eyster, has been my firmest critic, major source of moral support, and best friend throughout the project; and all for only 10% of the royalties. Love is blind!

Jan A. Pechenik
Medford, Mass.

CONTENTS

1

ENVIRONMENTAL
CONSIDERATIONS

The organisms considered in this book are grouped into 37 phyla. The members of almost half of these phyla are entirely marine, and the members of the remaining phyla are found primarily in marine and, to a lesser extent, in freshwater habitats. Excluding the arthropods, invertebrates generally have been far less successful in invading terrestrial environments. Even those invertebrate species that are terrestrial as adults often have aquatic developmental stages. It is important, therefore, to consider some of the physical properties of salt water, fresh water, and air as potential habitats before proceeding to consider the animals that live and/or develop in these different environments. These physical properties play major roles in determining the structural, physiological, and behavioral characteristics required by animals for life in various habitats.

Air is dry, whereas water is wet. As trivial as this statement may seem, the repercussions with respect to morphology, respiratory physiology, nitrogen metabolism, and reproductive biology are tremendous, as seen in Table 1.1.

Diffusion of gases occurs only across moist surfaces. Since aquatic organisms are in no danger of drying out, gas exchange can be accomplished across the general body surface. Thus,

TABLE 1.1 Summary of the Different Life-styles Possible in the Two Major Environments, Aquatic and Terrestrial, as They Reflect Differences in the Physical Properties of Water and Air.

PROPERTY	WATER	AIR
Humidity	High: exposed respiratory surfaces; external fertilization*; external development; excretion of ammonia	Low: internalized respiratory surfaces; internal fertilization; protected development; excretion of urea and uric acid
Density	High: rigid skeletal supports unnecessary; filter-feeding life-style possible; external fertilization; dispersing developmental stages*	Low: rigid skeletal supports necessary; must move to find food; internal fertilization; sedentary developmental stages
Compressibility	Low: transmits pressure changes uniformly and effectively	High: less effective at transmitting pressure changes
Specific heat	High: temperature stability	Low: wide fluctuations in ambient temperature
Oxygen solubility	Low: 5–6 ml O_2/liter of water	High: 210 ml O_2/liter of air
Nutrient content	High: salts and nutrients available through absorption directly from water for all life stages*; adults may make minimal nutrient investment per egg*	Low: no nutrients available via direct absorption from air; adults supply eggs with all nutrients and salts needed for development
Light-extinction coefficient	High: animals may be far removed from sites of surface-water primary production	Low: animals never far from sites of primary production

*Signifies features that are especially characteristic of marine invertebrates and uncommon among freshwater invertebrates.

the body walls of aquatic invertebrates are generally thin and water-permeable, and any specialized respiratory structures that exist may be in direct contact with the surrounding medium. Gills, which can be structurally quite complex, are simply vascularized extensions of the outer body wall. Gills, therefore, increase the surface area available for gas exchange and, if they are especially thin-walled, may also increase the efficiency of respiration (measured as the volume of gas exchanged per unit time per unit area).

In contrast to the minimal complexity

required for aquatic respiratory systems, terrestrial systems must cope with potential desiccation (dehydration). Terrestrial species that rely on simple diffusion of gases through unspecialized body surfaces must have some means of maintaining a moist outer body surface, as by the secretion of mucus in earthworms. Truly terrestrial invertebrates generally have a water-impermeable outer body covering that prevents rapid dehydration. Gas exchange in such species must be accomplished through specialized, internal respiratory structures.

The union of sperm and egg, and the subsequent development of a zygote, can be achieved far more simply by aquatic invertebrates than by terrestrial species. Marine organisms, in particular, may shed sperm and eggs freely into the environment. Since the gametes and embryos of marine species are not subject to dehydration or to osmotic stress, fertilization and development can be completed entirely in the water. On the other hand, fertilization in the terrestrial environment must be internal to avoid dehydration of gametes; therefore, terrestrial invertebrates require more complex reproductive systems than do their marine counterparts. Successful fertilization of terrestrial eggs often involves complex reproductive behaviors as well.

Ammonia is the basic end product of amino acid metabolism in all organisms, regardless of habitat. Generally, ammonia is highly toxic, largely through its effects on cellular respiration. Even a small accumulation of ammonia in the tissues and blood is detrimental to individuals of most species. However, few terrestrial organisms can afford the luxury of constantly eliminating

ammonia as it is produced, since the water required to flush out the ammonia is in short supply. As an adaptation to life on land, terrestrial organisms usually incorporate ammonia into less toxic compounds (urea and uric acid), which can then be excreted in a smaller amount of water. This detoxification of ammonia requires additional biochemical pathways and an increased expenditure of energy. Aquatic invertebrates, on the other hand, simply use the surrounding water to dilute away metabolic ammonia as it is produced. Moreover, because water is wet, ammonia excretion may occur by simple diffusion across the general body surface of many aquatic invertebrates. In contrast, all terrestrial animals require complex excretory systems.

Water is a remarkably versatile solvent. The benefits to aquatic invertebrates are both direct and indirect. First of all, aquatic animals potentially can take up nutrients (including amino acids, carbohydrates, and salts) directly from the surrounding water by diffusion or by active uptake. In particular, dissolved salts and other water-soluble nutrients may be taken up directly from water by developing embryos and larvae. Embryos of terrestrial organisms must be supplied (by their parents) with all food and salts needed for development and must be protected from desiccation as well. As an indirect benefit to aquatic invertebrates, suspension in a nutrient-containing, wet medium permits primary producers to take the form of small, suspended, single-celled organisms (**phytoplankton**); roots are not mandatory. Phytoplankton cells can attain high concentrations in water, and they can be easily harvested and ingested by many filter-feeding aquatic herbivores and the developmental stages of these herbivores.

Water is far denser than air, a fact that has profound consequences for invertebrates. For one thing, a rigid skeletal support system is not required in water, since the medium itself is supportive. For the same reason, animals may often move with greater efficiency in water than in air, expending less energy to progress a given distance. Indeed, many aquatic invertebrate species expend virtually no energy at all for movement—they simply don't move. How do such animals feed without the ability to move? Because water is wet and dense, small, free-floating plants and animals (phytoplankton and **zooplankton,** respectively) live in suspension; this enables many other aquatic animals to make their livings "sitting down," filtering food particles directly from the medium as it flows past the stationary animal. Often, some energy must be expended to move water past the feeding structures of the animal, but the animal need not use energy in a search for food. Such a filter-feeding or suspension-feeding existence, quite commonly encountered in aquatic environments, seems to have been exploited only by web-building spiders in the terrestrial habitat. Potential food particles simply do not occur in high concentrations in the dry, unsupportive air.

External fertilization and external development of embryos, so commonly encountered among marine invertebrates, are made possible as much by the high density of water as by its wetness; the water supports both sperm and egg, and the embryo itself as it develops. In many groups of marine invertebrates, external fertilization and/or external larval development is the rule rather

than the exception. Because little energy may be required to remain afloat in the aquatic medium, developmental stages of aquatic invertebrates often serve as the dispersal stages for sedentary adults—exactly the opposite of the situation encountered among most terrestrial animals.

One additional advantage of water as a biological environment should be mentioned: its relatively high temperature stability with respect to air. Water has a high specific heat; that is, the number of calories required to heat one gram of water 1°C is considerably greater than that required to raise the temperature of one gram of most other substances by the same 1°C. Because of its high specific heat, water is slow to cool and slow to heat up; water temperature is relatively insensitive to short-term fluctuations in air temperature. Over a 24-hour period, air temperatures at mid-latitudes may vary by 20°C or more. For reasonably large volumes of water, local surface temperatures will probably not vary by more than 1–2°C over the same time interval.

Differences in seasonal temperature fluctuations are even more striking. Near Cape Cod, Massachusetts, for example, local seawater temperature may vary between approximately 5°C during the winter and approximately 20°C during the summer: a seasonal range of about 15°C. Air temperatures, on the other hand, fluctuate between approximately −25°C and 40°C, a seasonal range of 65°C. The range of water temperatures encountered during a year, even in small lakes and ponds in the same geographical areas, is much smaller than this. Because the rates of all chemical reactions, including those associated with organismal metabolism, are altered by temperature, wide fluctuations in temperature (especially those occurring over short time intervals) are highly stressful to most invertebrates. Invertebrates living in thermally variable environments require biochemical, physiological, and/or behavioral adaptations not required by organisms living in more stable habitats.

Life in water does pose a few problems. Light is extinguished over a shorter distance in water than in air, so that most aquatic **primary production** (net carbon fixation by photosynthesizing plants) is limited to the upper 20–50 meters or so. Moreover, the oxygen-carrying capacity of water, volume for volume, is about 2.5% of the oxygen-carrying capacity of air. An additional problem for aquatic organisms is the fact that the time required for a given molecule to diffuse across a given distance in water is much, much greater than the time required for the same molecule to diffuse across the same distance in air. An organism sitting completely still in motionless water would have a very definite gas-exchange problem once the fluid immediately in contact with the respiratory surface had given up all available oxygen (and/or had become saturated with carbon dioxide). On the other hand, even the slightest movement of the water surrounding the respiratory surface of an animal enhances gas exchange significantly. **Sessile** (nonmotile) organisms living in areas of significant water current velocity thus benefit in terms of gas exchange as well as nutrient replenishment. Sessile animals living in still water invariably have some means of creating water flow over their respiratory surfaces.

Organisms living in fresh water obtain many of the benefits available to marine organisms but face several difficulties unique to the freshwater environment. Marine invertebrates are approximately in **osmotic equilibrium** with the medium in which they live; that is, the concentration of solutes in their body fluids matches that of the surrounding seawater. In contrast, the internal body fluids of freshwater organisms are always higher in osmotic concentration than is the surrounding medium. That is, freshwater organisms are **hyperosmotic** to their surroundings, and water tends to diffuse inward along the osmotic concentration gradient. Some freshwater animals have reduced surface permeability to water, reducing the magnitude of this inflow. Complete impermeability to water is not possible, however, because respiratory surfaces must remain permeable for gas exchange to occur. Thus, all freshwater animals must be capable of constantly expelling large volumes of incoming fresh water. Also, because salts are relatively rare in the freshwater medium (by definition of freshwater), most of the salts necessary for embryonic development must be supplied to the egg by the mother. By contrast, all salts required for the differentiation and growth of marine embryos are readily available in the surrounding medium.

The relative paucity of salts in fresh water has additional ramifications for animals living in that environment. For example, freshwater organisms, which must constantly pump out incoming water, often possess sophisticated physiological mechanisms for reclaiming precious salts from the urine before the urine leaves the body; they also must possess mechanisms for making good any salt loss that does occur.

It should be noted that most bodies of fresh water are ultimately ephemeral, with smaller ponds and lakes being subject to drying up at yearly or even more frequent intervals. Most marine invertebrates are not faced with such a high degree of habitat unreliability. Finally, the pH of salt water is far less variable, both spatially and temporally, than is the pH of fresh water. The pH of salt water is maintained near 8.1 by the tremendous buffering capacity of the bicarbonate ion. Most freshwater environments lack such a high buffering capacity, and the pH of the water is therefore far more sensitive to local fluctuation of acid and base content.

From such considerations of the properties of air, salt water, and fresh water, it is easy to understand why life must have originated in the ocean. The specialized physiological and/or morphological adaptations essential for existence on land or in fresh water are not required for the relatively simple and generally less stressful existence possible in the marine environment. Once life arose, various **pre-adaptations** could eventually evolve, which made a transition from saltwater environments to other habitats possible. Such pre-adaptations apparently arose rarely in many groups of animals and not at all in others. Not surprisingly, most phyla are still best represented in the ocean, both in terms of species numbers and in terms of the diversity of body plans and life-styles.

2

INVERTEBRATE CLASSIFICATION

INTRODUCTION

Presently, more than one million animal species exist on Earth. At one time, there were presumably no animal species. The marvelous variety of animal life forms represented today evolved gradually, beginning several billion years ago. All of today's major animal groups are present in the earliest fossil record, which extends back only about 600 million years, so we will never know the precise evolutionary history of the animals presently populating the planet. Nevertheless, we know that all animals must have ancestral forms in common, and we can infer evolutionary relationships, with varying degrees of certainty, on the basis of morphological, developmental, physiological, and biochemical similarities and differences among groups of animals.

Before we can consider the evolutionary interrelationships among different animal groups, we must first determine the degrees of similarity and difference among and within the groups. Indeed, this must be done before we can make generalizations about any animal group or even decide the group to which an animal belongs. For hundreds of years, then, scientists have attempted to place organisms into categories that reflect structural, functional, developmental, and evolutionary cohesiveness. It is important to remember that we impose these classifications upon the diversity of life forms around us for practical reasons, in an attempt to introduce order. As we will see at intervals throughout this book, many organisms do not fit cleanly into any one group; it is relatively simple to decide upon the categories to be used but often far less simple to determine the category to which a given organism belongs. In this chapter, we will consider some of the schemes that have been developed to sort animals into groups.

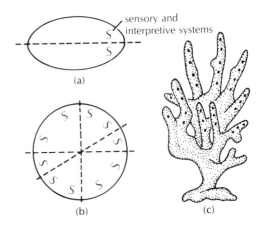

FIGURE 2.1. *Various types of body symmetry. (a) Bilateral symmetry. (b) Radial symmetry. (c) Asymmetrical body plan of a marine sponge.*

CLASSIFICATION BY CELL NUMBER AND BODY SYMMETRY

Animals have been categorized in several ways. The most basic division is based upon the number of cells composing any given individual. Most of the animals with which we are familiar are **multicellular** and are referred to collectively as the Metazoa, or as **metazoans.** Other animals are not multicellular and are considered to be either **unicellular** (single-celled) or **acellular** (without cells), a distinction that will be discussed further in the next chapter (The Protozoa). As we will see in the next several chapters, the point at which an association of cells can be viewed as composing a multicellular organism is not always clear-cut.

Animals may also be classified according to their general body form. Most metazoans show one of two types of body symmetry (Fig. 2.1). Animals like ourselves are **bilaterally symmetrical,** possessing right and left sides that are approximate mirror images of each other. Bilateral symmetry is highly correlated with **cephalization,** which is the concentration of nervous and sensory tissues and organs at one end of an animal, resulting in distinct anterior and posterior ends. For an animal that shows cephalization, two mirror images can be produced only when a slice is made parallel to the animal's long (anterior–posterior) axis, with the cut passing down the midline. Any cut perpendicular to this midline, even when passing through the center of the animal, creates two dissimilar pieces. This is not so for a **radially symmetric** animal organism. Such an animal can be divided into two reasonably equal halves by any cut that passes through the center of the organism. Thus, most animals belong to either the Radiata or the Bilateria, groupings that are often convenient for discussion but which have no formal taxonomic significance. Rarely, we will encounter invertebrates that are asymmetrical, i.e., there is no ordered pattern to their gross morphology.

Once again, what seems to be straightforward on the surface is never quite so simple when dealing with actual animals. Many species whose external appearances are the epitome of uncontroversial, radial symmetry have asymmetric internal anatomies. All knife cuts are thus created equal only with respect to superficials. Perhaps it would have been better to group animals based on degree of cephalization rather than on the basis of body symmetry. I bow, however, to historical precedent.

CLASSIFICATION BY DEVELOPMENTAL PATTERN

Multicellular invertebrates generally can be classified into two groups based upon the number of distinguishable germ layers formed during embryogenesis. **Germ layers** may be defined as groups of cells behaving as a unit during the early stages of embryonic development and giving rise to distinctly different tissues and/or organ systems in the adult. In **diploblastic** animals (*diplo*=G: double), only two distinct germ layers form during or following the movement of cells into the interior of the embryo. The outermost layer of cells is called the **ectoderm** (*ecto* = G: outer; *derm* = G: skin), and the innermost layer of cells is called the **endoderm** (*endo* = G: inner). Members of only two, or possibly three, phyla are considered to be diploblastic (see Fig. 2.8). Instead, the majority of metazoans are **triploblastic** (*triplo* = G: triple). During the ontogeny of triploblastic animals, cells of either the ectoderm or, more usually, the endoderm, give rise to a third germ layer, the **mesoderm** (*meso* = G: middle). The mesodermal layer of tissue always lies between the outer ectodermal tissue and the inner endodermal tissue.

Note that the absence of a distinct, embryonic third tissue layer does not imply that the adult of a diploblastic species will lack the derivatives of this layer that are found in the adult of a triploblastic species. Muscular elements, for example, derive from the mesodermal layer in triploblastic animals. Musculature is also a component of the diploblastic adult, despite the absence of a morphologically or behaviorally distinct group of cells that can be termed "mesoderm" in the early embryo.

Triploblastic animals can be further classified into three basic plans of body construction, based upon whether or not an organism has an internal body cavity independent of the gut, and based upon how this cavity, if present, is formed during embryogenesis. The most **primitive** animals (i.e., those believed to resemble most closely the ancestral metazoans) lack an internal body cavity. These are the **acoelomates** (*a* = G: without; *coelom* = G: a hollow space). Characteristically, the area lying between the outer body wall and the gut of acoelomates is solid, being occupied by mesoderm.

In a second group of animals, this area between the outer body wall and the gut is a fluid-filled space derived from the **blastocoel,** an internal space that develops in the embryo prior to gastrulation. The juvenile and adult organisms therefore possess an internal body cavity lying between the endoderm of the gut and the mesodermal musculature of the body wall (Fig. 2.2). Selective pressures that might account for the evolution of such a body cavity from an acoelomate ancestral form are not difficult to imagine. For example, the gut is now somewhat independent from muscular, locomotory activities of the body wall. Also, the animal now has internal space into which can bulge digestive organs, gonads, and developing embryos. Perhaps most significantly, the animal gains the mechanical advantages associated with a more effective locomotory system, to be discussed in Chapter 8 (Introduction to the Hydrostatic Skeleton).

This fluid-filled cavity derived from the embryonic blastocoel is termed a **pseudocoel,** and the organism housing it is said to be **pseudocoelomate.** The name is a bit misleading. The "pseudo" is not intended to disparage the "coel"; the body cavity is genuine. The "pseudo" prefix merely draws attention to the fact that this body cavity is not to be confused with a true coelom, which, as you will see, is a precisely defined internal cavity formed through quite specific processes.

This brings us to the third group of triploblastic animals, those with a true **coelom;** that is, an internal, fluid-filled body cavity lying between the gut and the outer body wall musculature and lined with mesoderm. The animals possessing such a body cavity are **coelomates** (or **eucoelomates;** *eu* = G: real, true). Coelom formation may occur by either of two quite dissimilar mechanisms; mode of coelom formation is a major characteristic that is used to assign coelomates to

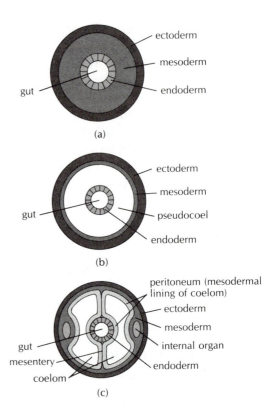

FIGURE 2.2. *(a) Diagrammatic cross section through the body of an acoelomate. Note that the space between the gut and the outer body wall musculature is completely filled with tissue derived from embryonic mesoderm. (b) Cross section through the body of a pseudocoelomate. Note that the gut derives entirely from endoderm and is therefore not lined with mesoderm. (c) Cross section through the body of a coelomate. Note that the entire coelomic space is bordered by tissue derived from embryonic mesoderm.*

one of two major subgroups: protostomes or deuterostomes. In the **protostomes,** coelom formation occurs by gradual enlargement of a split in the mesoderm (Fig. 2.3). This process is termed **schizocoely** *(schizo* = G:

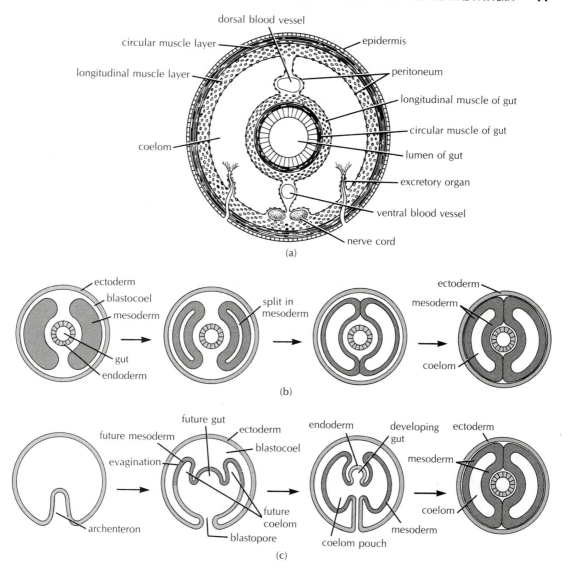

FIGURE 2.3. *(a) Detailed cross section through the body of a coelomate. The tissues bordering the coelomic space include the musculature of the gut; the mesenteries, which suspend the various organs in the coelom; and the peritoneum, lining the coelomic cavity. (From Hyman, 1951. The Invertebrates, Vol. II, McGraw-Hill. Reproduced by permission.) (b) Coelom formation by schizocoely; i.e., by an actual split, or schism, in the mesodermal tissue. (c) Coelom formation by enterocoely, in which the archenteron evaginates into the embryonic blastocoel.*

split). In the **deuterostomes,** on the other hand, the coelom forms through evagination of the archenteron into the blastocoel of the embryo (Figs. 2.3 and 2.4). Because the coelom of deuterostomes is formed from what eventually becomes the gut, coelom formation in this group of animals is termed **enterocoely** (*entero* = G: gut).

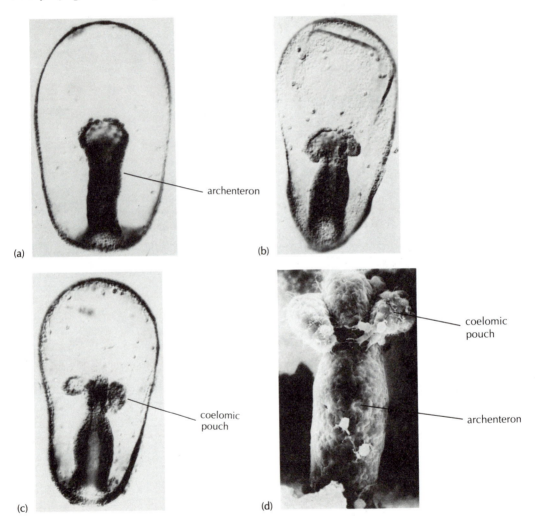

(a)   (b)

archenteron

(c)   (d)

coelomic pouch

coelomic pouch

archenteron

FIGURE 2.4.   (*a–c*) *Enterocoely in a sea urchin, showing the coelomic pouches forming from the sides of the archenteron and splitting off.* (d) *The coelomic pouches are clearly shown in this scanning electron micrograph. The pouches gradually enlarge to form the coelom.* (Courtesy of B. J. Crawford, from Crawford and Chia, 1978. J. Morphol., 157: 99.)

It is important to note that, regardless of whether coelom formation occurs by schizocoely or enterocoely, the end result is similar. The organism is left with a fluid-filled internal body cavity lying between the gut and the outer body wall musculature, and unlike the cavity of pseudocoelomates, this cavity is lined entirely by mesoderm. The fact that internal cavities develop by any of three distinctly different mechanisms (enterocoely, schizocoely, or persistence of blastocoel) suggests that such cavities have been independently evolved at least three times. Clearly, the selective pressures favoring the evolution of internal body cavities must have been substantial.

To summarize, triploblastic animals can be divided into acoelomates, pseudocoelomates, and coelomates, depending upon whether or not they possess an internal, fluid-filled body cavity, and depending on whether or not this body cavity is lined entirely by mesoderm. Also, coelomates can be further classified into protostomes and deuterostomes, depending, in part, upon how the coelom is created during development.

Actually, making this distinction between protostomes and deuterostomes can be difficult. Flatworms (Phylum Platyhelminthes) and ribbon worms (Phylum Rhynchocoela) are often grouped with the protostomes, despite the fact that these worms lack a coelom or even a pseudocoelom—they are acoelomates. Flatworms and ribbon worms are nevertheless often lumped together with the protostomes because the mode of coelom formation is only one of several characteristics distinguishing protostomes from deuterostomes, and in these other characteristics, flatworms and ribbon worms develop as perfectly good protostomes. For our purposes, it seems best to view the flatworms as being neither protostomes nor deuterostomes, but to simply recognize that their development can be characterized as being protostome-like. To recapitulate, the distinction between protostomes and deuterostomes in this text will refer only to coelomate animals.

The terms "protostome" and "deuterostome" were actually coined to reflect differences in the origin of the mouth (*stoma* = G: mouth). Among the protostomes, the mouth is formed from or near the blastopore (the opening from the outside into the archenteron). Thus the term "proto-stome," meaning "first mouth"; the mouth forms from the first opening that appears during development of the embryo. Among the deuterostomes, the mouth never develops from the blastopore. Indeed, the blastopore often gives rise to the anus, the mouth always forming as a second, novel opening elsewhere on the embryo. Thus, "deuterostome"; that is, "second mouth."

What other characteristics distinguish protostomes from deuterostomes? In addition to differences in the mode of coelom formation and differences in the embryological origin of the mouth, protostomes and deuterostomes differ with respect to how the mesoderm originates; the manner in which the openings to the digestive tract are formed; and the orientation of the spindle axes of the cells during cleavage.

Cleavage is usually either radial or spiral, depending on the orientation of the mitotic spindles relative to the egg axis. Generally, yolk is asymmetrically distributed within an egg, and the nucleus occurs in, or moves to,

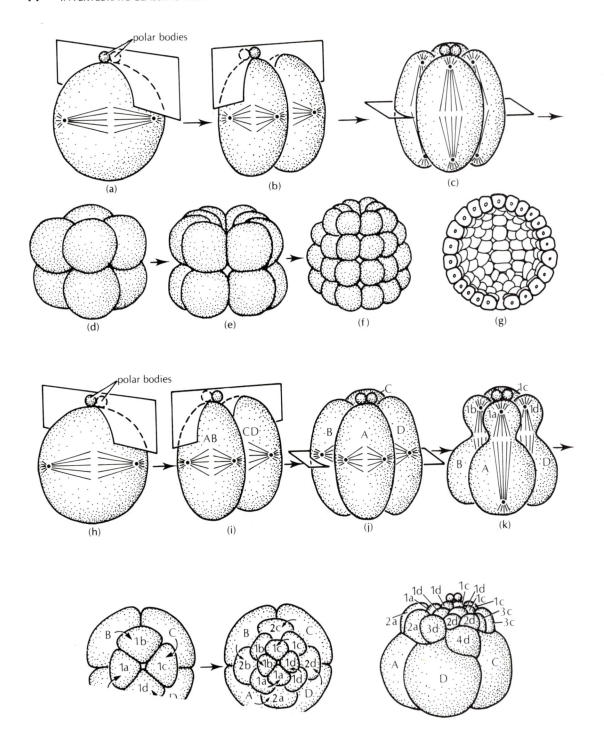

the region of lowest yolk density. This is the **animal pole,** and it is here that the polar bodies are given off during meiosis. The opposite end of the egg is termed the **vegetal** (not vegetable!) **pole.**

In **radial cleavage** (deuterostomes), the spindles of a given cell, and thus the cleavage planes, are oriented either parallel or perpendicular to the animal–vegetal axis. Thus, daughter cells derived from a division in which the cleavage plane is parallel to the animal–vegetal axis end up lying in the same plane as the original mother cell (Fig. 2.5). The two daughter cells resulting from a division perpendicular to the animal–vegetal axis come to lie directly one atop the other, with the center of the upper cell lying directly over the center of the underlying cell.

In contrast, the spindle axes of cells undergoing **spiral cleavage** (protostomes) are oriented (after the first two cleavages) at 45° angles to the animal–vegetal axis. Moreover, the division line may not pass through the center of the dividing cell. As a result, by the 8-cell stage we often see a group of smaller cells (**micromeres**) lying in the spaces between the underlying larger cells (**macromeres**). Cell division continues in this fashion, with the cleavage planes always oblique to the polar axis of the embryo.

A further difference between the two groups of coelomates concerns the source of mesoderm. Among protostomes, all mesodermal tissue derives from a single cell of the 64-cell embryo. This is not true of deuterostomes.

During their first one or two cleavages, some protostomes form **polar lobes** (not to be confused with polar bodies, which arise during maturation of the egg). A polar lobe is

FIGURE 2.5. *(a–g) Radial cleavage, as seen in the sea cucumber* Synapta digitata. *In frame (g), part of the embryo has been removed to reveal the blastocoel. (h–n) Spiral cleavage. The first two cleavages (h–i) are identical with those seen in radially cleaving embryos, forming four large blastomeres (j). The cleavage plane during the next cleavage, however, is oblique to the animal–vegetal axis of the embryo and does not pass through the center of a given cell (k). This produces a ring of smaller cells (micromeres) lying in between the underlying larger cells (macromeres), as shown in (l). The lettering system illustrated was devised by the embryologist E.B. Wilson in the late 1800s to make possible a discussion of particular cell origins and fates. The number preceding a letter indicates when a particular micromere was formed. Capital letters refer to macromeres, while lower-case letters refer to micromeres. With each subsequent cleavage, the macromeres divide to form one daughter macromere and one daughter micromere, while the micromeres divide to form two daughter micromeres. The 32-cell-stage embryo of the marine snail,* Crepidula fornicata, *is shown in (n). Note the 4d cell, from which all of the mesodermal tissue of protostomes will ultimately derive. (After Richards.)*

a conspicuous bulge of cytoplasm that forms prior to cell division. The lobe contains no nuclear material. After cell division is complete, the bulge is resorbed into the single daughter cell to which it is still attached (Fig. 2.6). Although the functional significance of this phenomenon for the embryo is still not fully understood, polar lobe formation has provided developmental biologists with an intriguing system through which to study the role of cytoplasmic factors in determining cell fate. In the basic experiment, the fully formed polar lobe is detached from an embryo, and the development of the lobeless embryo is subsequently followed. Polar lobe formation is characteristic of only some protostome species (some annelids and some molluscs) but is never encountered among deuterostomes.

The developmental features distinguishing protostomes from deuterostomes are summarized in Table 2.1.

Classification schemes in biology are always tidy. Unfortunately, it is often far simpler for biologists to construct a logical classification system than to neatly distribute animals within it. I have already hinted at some of the difficulties. Many animals show other developmental departures from the characteristics outlined above. In particular, some triploblastic invertebrate species show characteristics that are part deuterostome and part protostome, while others display certain features that are characteristic of neither group. Because all animal groups have had ancestral forms in common at some time during their evolution; because evolution is an ongoing process; and because embryos as well as adults are subject to the modifying forces of natural selection, it should not be too upsetting to encounter some species whose developmental characteristics fall outside the mainstream. In any event, those species exhibiting entirely or primarily protostome characteristics are most certainly more closely related to each other than to those species exhibiting purely deuterostome characteristics. The affinities of the "misfits" must remain uncertain, at least until additional embryological studies can be completed.

TABLE 2.1 Summary of the Basic Characteristics of Protostomous and Deuterostomous Coelomates

DEVELOPMENTAL CHARACTERISTIC	PROTOSTOMES	DEUTEROSTOMES
Mouth origin	from blastopore	never from blastopore
Coelom formation	schizocoely	enterocoely
Mesoderm origin	4d cell	other
Cleavage pattern	spiral, determinate	radial, indeterminate
Polar lobe formation	present in some species	not present in any species

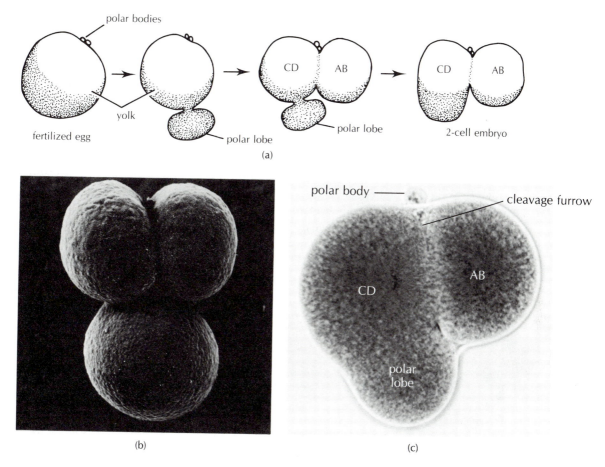

FIGURE 2.6. *(a) Polar lobe formation during the development of a protostome. Following resorption of the polar lobe, the two blastomeres are clearly unequal in size because the cytoplasm held within the polar lobe did not participate directly in the process of cleavage. © 1927; Columbia University Press. Reprinted by permission) (b) Scanning electron micrograph of the two-celled embryo of a marine snail,* Nassarius reticulatus. *The polar lobe (at bottom) is nearly equal in size to the blastomere into which it will be resorbed. (Courtesy of M. R. Dohmen.) (c) Polar lobe formation during first cleavage in the blue mussel,* Mytilus edulis, *as seen with light microscopy. The newly-formed cleavage furrow is visible between the AB and CD blastomeres. The polar lobe is clearly affiliated with one of the new cells (the CD blastomere). A polar body (a product of meiotic division prior to cleavage) can be seen at the animal pole of the embryo. (Photograph by Carolina Biological Supply Company.)*

CLASSIFICATION BY HABITAT AND LIFE-STYLE

Animals may also be categorized on the basis of habitat or life-style. For example, one group of animals may be **terrestrial**, living on land, while another is **marine**, living in the ocean. Marine animals, in turn, may be **intertidal** (living between the physical limits of high and low tides, and thus exposed to air periodically); **subtidal** (living below the low-tide line, and thus exposed to air only under extreme conditions, if ever); or **open ocean** creatures. In addition, animals may or may not be capable of locomotion, in which case they are either **free-living** (mobile); **sessile** (immobile); or, perhaps, **sedentary** (exhibiting only limited locomotory capabilities). Some organisms may be able to move, but their locomotory powers are negligible with respect to the movement of the medium in which they live; such individuals are said to be **planktonic** (from the Greek, "forced to drift or wander").

Animals are often categorized as to feeding type. For example, some species are **herbivores** (plant eaters), while others are **carnivores** (flesh eaters). Some species remove small food particles from the surrounding medium (**suspension-feeders**), while others

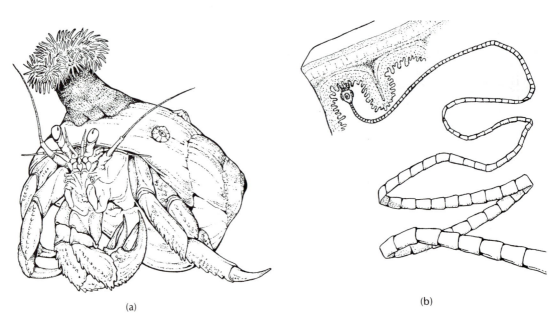

(a)

(b)

FIGURE 2.7. (a) A symbiotic relationship between a sea anemone, Calliactis parasitica, and a hermit crab, Eupagurus bernhardus. The crab deliberately places the anemones on its shell. (After Hardy.)
(b) A tapeworm, Taenia solium, shown attached to the intestinal wall of its vertebrate host. (After Villee.)

ingest sediment, digesting the organic component as the sediment moves through the digestive tract (**deposit-feeders**).

Members of one species frequently live in intimate association with those of another species. These **symbiotic associations,** or **symbioses,** frequently relate to the feeding biology of one or both of the participants (i.e., **symbionts**) in the association (Fig. 2.7). **Ectosymbionts** live near or on the body of the other participant, while **endosymbionts** live within the body of the other participant. When both symbionts benefit, the relationship is said to be **mutualistic,** or an example of **mutualism.** When the benefit accrues to only one of the symbionts and the other is neither benefitted nor harmed, the relationship is one of **commensalism,** and the benefitting member is the **commensal.** Lastly, some animals are **parasites;** i.e., they are utterly dependent upon their **host** for continuation of the species, generally subsisting on either the blood or the tissues of the host. A parasite may or may not substantially impair the activities of the host. The essence of parasitism is that the parasite is metabolically dependent upon the host, and that the association is obligate for the parasite.

The boundaries between parasitism, mutualism, commensalism, and predation are not always distinct. For example, a parasite that eventually kills its host essentially becomes a predator. A parasite that produces a metabolic end product from which the host benefits borders on being mutualistic. Indeed, it should not be surprising to see transitional forms in the process of evolving from one type of relationship to another. Such transitional forms once again make tidy categorization of animals into human-made schemes difficult; definitions of the above terms have been modified by various workers in an attempt to improve the fit, but every rule seems to have an exception.

CLASSIFICATION BY EVOLUTIONARY RELATIONSHIP

The last classification scheme we will consider is probably the most familiar: the taxonomic framework established several hundred years ago (1758) by Carolus Linnaeus. The system is hierarchical; that is, one taxon contains groups of lesser taxa, which in turn contain still more groups of lesser taxa, and so on. Thus:

Kingdom
Phylum
Class
Order
Family
Genus
Species

The members of a given phylum show a high degree of morphological and developmental similarity and are presumed to be more closely related to each other than to the members of any other phylum. Indeed, all the members of any particular phylum are presumed to have evolved from a single ancestral form. The evolutionary implications are similar for the other taxonomic categories as well.

The category of "species" has additional biological significance. Theoretically, the members of one species are reproductively isolated from members of all other species. The species, therefore, forms a pool of genetic material which only members of that species have access to, and which is isolated from the gene pool of all other species.

The scientific name of a species has two parts: the **generic name** and the **specific name;** that is, the name is a binomial. The generic and specific names (i.e., the **species name**) are usually italicized in print and underlined in writing. The generic name begins with a capital letter, but the specific name does not. For example, the proper scientific name for one of the common shallow-water snails found off Cape Cod, Massachusetts, is *Crepidula fornicata*. Related species are *Crepidula plana* and *Crepidula convexa*. Once the generic name is spelled out, it may be abbreviated when used subsequently. Thus, *Crepidula fornicata*, *C. plana*, and *C. convexa* are common shallow-water gastropods found near Woods Hole, Massachusetts. They all belong to the phylum Mollusca and are contained within the class Gastropoda, family Calyptraeidae. The family Calyptraeidae contains other genera besides *Crepidula*; the class Gastropoda contains other families besides the Calyptraeidae; and the phylum Mollusca contains other classes besides the Gastropoda. The taxonomic classification system is indeed hierarchical.

One last point about species names must be made. We often find an additional name following the generic and specific names of an organism. This additional name is capitalized but is not italicized and may be contained in parentheses. This is the name of the person who first described the organism. If the animal was named by Linnaeus, his name is often abbreviated as "L," since Linnaeus is associated with the descriptions of so many species. If the organism was originally described as being in a different genus than the one in which it is currently placed, the describer's name is enclosed within parentheses. Thus, the snail *Ilyanassa obsoleta* (Say) was described by a man named Say, who originally assigned the species to another genus (the genus *Nassa*). It was subsequently determined that this snail was sufficiently dissimilar from other members of this genus to warrant its removal. Occasionally you will see a person's name followed by a date, identifying the year in which the species was first described. For example, the shrimplike animal known as *Euphausia superba* Dana 1858 was first described by Mr. Dana in 1858, and it has remained in the genus *Euphausia* since it was originally named.

A listing of all groups containing invertebrates is presented in Fig. 2.8, showing where each phylum fits into the classification systems discussed in this section. The number following each listing gives the page on which the group is first discussed. The distribution of described species among the various animal phyla is summarized in Fig. 2.9. Note the rather small percentage of species contained within the Phylum Chordata (only about 5% of all described species) and that this phylum contains invertebrates as well as vertebrates.

Ideally, a taxonomic classification scheme should reflect degrees of phylogenetic relatedness, i.e., the evolutionary history of the various animal groups. All members of a given taxonomic group should be derived from a single ancestral form. We can often

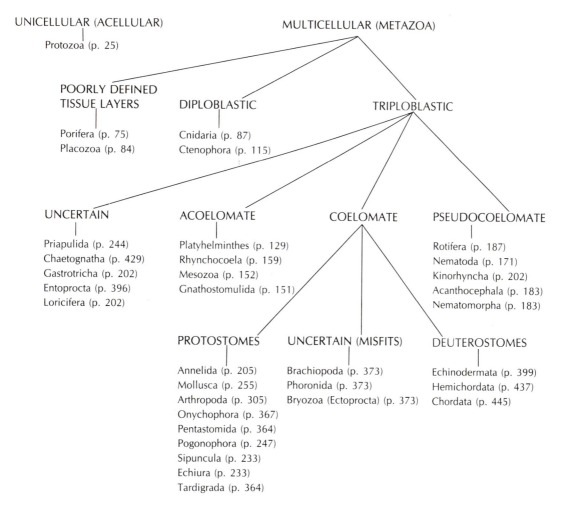

UNICELLULAR (ACELLULAR)

Protozoa (p. 25)

MULTICELLULAR (METAZOA)

POORLY DEFINED
TISSUE LAYERS

Porifera (p. 75)
Placozoa (p. 84)

DIPLOBLASTIC

Cnidaria (p. 87)
Ctenophora (p. 115)

TRIPLOBLASTIC

UNCERTAIN

Priapulida (p. 244)
Chaetognatha (p. 429)
Gastrotricha (p. 202)
Entoprocta (p. 396)
Loricifera (p. 202)

ACOELOMATE

Platyhelminthes (p. 129)
Rhynchocoela (p. 159)
Mesozoa (p. 152)
Gnathostomulida (p. 151)

COELOMATE

PSEUDOCOELOMATE

Rotifera (p. 187)
Nematoda (p. 171)
Kinorhyncha (p. 202)
Acanthocephala (p. 183)
Nematomorpha (p. 183)

PROTOSTOMES

Annelida (p. 205)
Mollusca (p. 255)
Arthropoda (p. 305)
Onychophora (p. 367)
Pentastomida (p. 364)
Pogonophora (p. 247)
Sipuncula (p. 233)
Echiura (p. 233)
Tardigrada (p. 364)

UNCERTAIN (MISFITS)

Brachiopoda (p. 373)
Phoronida (p. 373)
Bryozoa (Ectoprocta) (p. 373)

DEUTEROSTOMES

Echinodermata (p. 399)
Hemichordata (p. 437)
Chordata (p. 445)

FIGURE 2.8. *Arrangement of the major groups according to the factors discussed in this section. Note that the placement of some groups within this framework is uncertain at present.*

make reasonable guesses about the origins of various animal groups, based upon studies of developmental patterns, studies of morphological and biochemical characteristics, and careful examination of animals preserved in the fossil record. Nevertheless, the evolutionary history of the different animal groups will never be known with certainty, and ferreting out probable relationships is no easy task. Indeed, there is no universally accepted procedure for deducing evolutionary relationships; considerable disagreement on

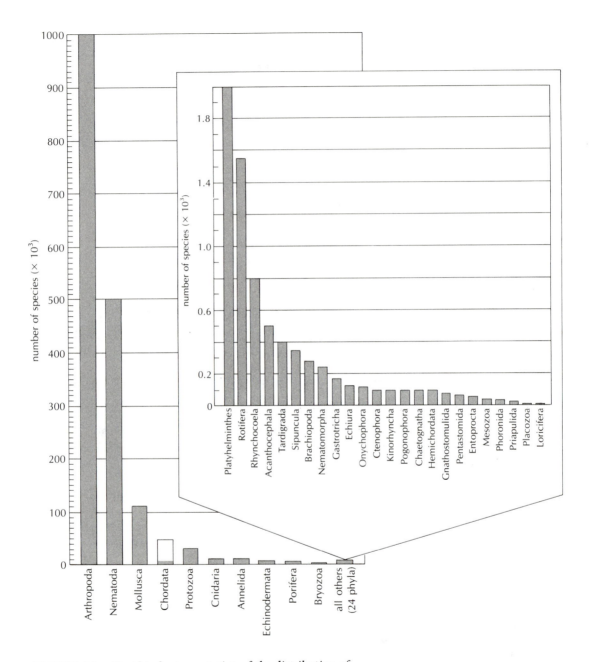

FIGURE 2.9. *Graphical representation of the distribution of described species among the 34 major groups of invertebrates. Phyla containing fewer than 2,000 described species are presented in the inset. (Note the different scale on the Y-axis of the inset.) The open (unshaded) area of the bar labeled "Chordata" represents vertebrate species. All other species in all other phyla are invertebrates.*

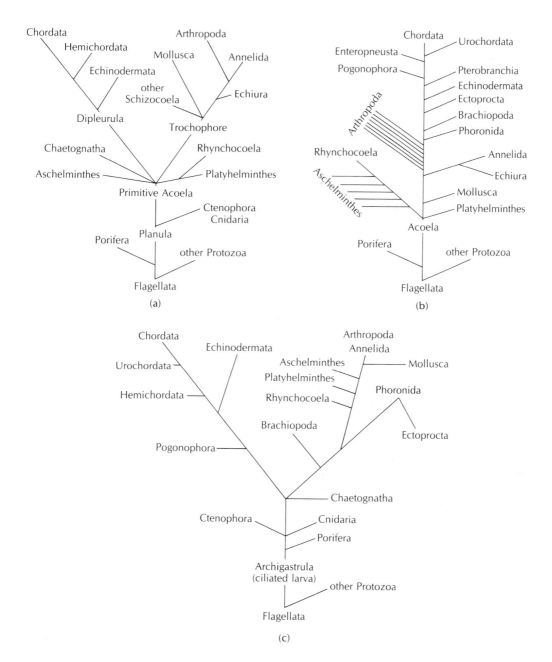

FIGURE 2.10. *Three schemes that have been proposed to illustrate presumed phylogenetic relationships between animals. (a) according to Hyman, 1940; (b) according to Hadži, 1953; (c) according to Marcus, 1958. (From Clark, 1964. Dynamics in Metazoan Evolution. Oxford University Press.)*

this point is well reflected in the appropriate literature of the past 20–30 years.[1] Basically, the present controversy seems to focus on the relative importance of phenotypic similarities among taxa, phenotypic differences among taxa, and the degree to which one is willing to admit (and deal with the fact) that phenotype may be a very misleading indicator of underlying genetic similarities and differences—through the process of **convergence,** distantly related animals may come to resemble each other rather closely.

A variety of phylogenetic trees have been proposed. Several of these are illustrated in Fig. 2.10. At least some of the differences among the various schemes may be attributed to presently inadequate data. As additional information about the various groups is gradually obtained, the evidence in favor of one scheme over some others may become more compelling. Classification schemes are by no means static, and the assignment of a given animal or group of animals to a particular position within the taxonomic hierarchy is not an irrevocable event.

TOPICS FOR FURTHER DISCUSSION AND INVESTIGATION

1. There are presently three major theories of animal classification: the theory of **phenetics** (based entirely on degree of overall anatomical and biochemical similarity); the theory of **cladistics** (based entirely upon inferred recency of common descent); and the theory of **evolutionary classification** (which attempts to consider both ancestry and the degree to which organisms have subsequently diverged from the ancestral form). Discuss the advantages and disadvantages inherent in any two of these three approaches to the inferring of phylogenetic relationships.

Bock, W., 1965. Review of Hennig, *Phylogenetic Systematics. Evolution*, 22: 646.

Farris, J., 1967. The meaning of relationship and taxonomic procedure. *Syst. Zool.*, 16: 44.

Mayr, E., 1965. Numerical phenetics and taxonomic theory. *Syst. Zool.*, 14: 73.

Mayr, E., 1974. Cladistic analysis or cladistic classification? *Zool. Syst. Evol.-forsch.*, 12: 94.

(Reprinted in E. Mayr, 1976. *Evolution and the Diversity of Life—Selected Essays.* Cambridge, Mass.: Harvard University Press, pp. 433–476.)

2. What are the characteristics of the "ideal" classification system, and why is this ideal so difficult to attain?

Archie, J.W., 1984. A new look at the predictive value of numerical classifications. *Syst. Zool.*, 33: 30.

Farris, J., 1967. The meaning of relationship and taxonomic procedure. *Syst. Zool.*, 16: 44.

Farris, J., 1982. Simplicity and informativeness in systematics and phylogeny. *Syst. Zool.*, 31: 413.

Mayr, E., 1974. Cladistic analysis or cladistic classification? *Zool. Syst. Evol.-forsch.*, 12: 94. (Reprinted in E. Mayr, 1976. *Evolution and the Diversity of Life—Selected Essays.* Cambridge, Mass.: Harvard University Press, pp. 433–476.)

---

1 See Topics for Discussion.

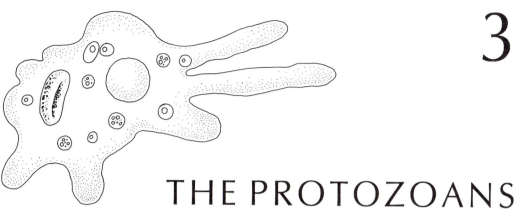

3

THE PROTOZOANS

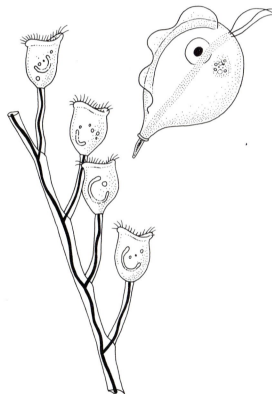

Subkingdom Proto•zoa
(G: FIRST ANIMALS)

INTRODUCTION

The Protozoa, composed of approximately 37,000 described species, can occur wherever there is moisture. Free-living protozoans are found in both marine and freshwater habitats, and in soil. In addition to the many free-living species, a great many protozoans live in close association both with animals (including other protozoans) and with plants, either as commensals or as parasites. Indeed, every major group of protozoans contains at least some parasitic species, and the members of two major groups are exclusively parasitic. All protozoans are small, usually 5–250 micrometers (μm) in length, and are therefore difficult to work with. Probably, most extant protozoans have yet to be described.

Although most people never see protozoans, few other groups of organisms rival them in economic and scientific importance. Protozoans play major roles both in primary production and in decomposition, and may serve as a major food source for many invertebrates and, indirectly, for many vertebrates as well. In addition, protozoans cause a number of human ailments, including malaria, African sleeping sickness, and dysentery, and a variety of devastating diseases of poultry, sheep, cattle, cabbage, and other human food sources. Protozoans have also provided biologists with outstanding material for genetic, physiological, developmental, and ecological studies. Nevertheless, these ubiquitous and important organisms have been known for only about 300 years. Their discovery awaited the invention of the microscope and the patience of the Dutch draper and amateur scientist Antony van Leeuwenhoek, who discovered the Protozoa in 1674.

Protozoans, more than any other group of organisms, absolutely defy a tidy categorization. Individual protozoans are not composed of individual cells; in essence, each protozoan *is* an individual cell. This is the one major characteristic that sets the Protozoa apart from all other animals, and is one of the few characteristics applicable to nearly all protozoans. In addition, most protozoans lack specialized circulatory, respiratory, and excretory (waste removal) structures. The surface area of their bodies is high relative to body volume, so that exchange of gases and removal of soluble wastes can occur by simple diffusion across the entire exposed body surface. The cytoplasm of a few species has recently been shown to contain hemoglobin, although its role in gas exchange has not yet been demonstrated for these animals. The relatively high surface area of protozoans may also facilitate active uptake of dissolved nutrients from the surrounding fluids.

It is difficult to make any other generalizations about this group. Indeed, in many respects the diversity of form and function encountered among the Protozoa rivals that encountered among all other invertebrates combined. Discussing all of the protozoans within one chapter is somewhat analogous to attempting a single discussion of the Annelida, Mollusca, Arthropoda, Echinodermata, and several other groups as well.

Protozoans are remarkably diverse in terms of size, morphology, mode of nutrition, locomotory mechanism, and reproductive biology. Indeed, scientists now commonly believe that the Protozoa did not arise from a single ancestral form, but rather constitute a collection of organisms that descended from several independent ancestral forms. That is, protozoans are probably **polyphyletic** (many-tribed, i.e., descended from many different ancestors) rather than **monophyletic** (single-tribed).

Precise evolutionary relationships are difficult to determine for any group of organisms, but this is particularly true for the Protozoa. The fossil record for protozoans is sparse, and what does exist is not particularly helpful in deducing relationships; members of some rather sophisticated protozoan groups possessing hard parts are encountered among the earliest fossils of about 600 million years ago, and have forms that are virtually identical to those found in some genera today. The small size of most individuals makes structural studies slow and difficult (but not insurmountably so). And, as with all organisms, structural similarities need not imply close evolutionary relationships; similar structures often arise independently in different, unrelated groups of organisms in response to similar selective pressures. (This phenomenon is known as **convergent evolution**.) As a result of these factors, the classification of protozoans has been a source of particular difficulty and controversy. By definition, a phylum must be monophyletic. Thus, it has been argued that the "Phylum Protozoa" should be elevated to a higher taxonomic level, such as subkingdom. This view is adopted here. Ah, but in what Kingdom is the subkingdom Protozoa to be placed?

Protozoans fit rather uncomfortably within the Kingdom Animalia. In a single drop of pond or ocean water, we can find some unicellular organisms that are capable of photosynthesis (i.e., plant-like organisms); some that feed by ingesting solid food particles (i.e., animal-like organisms); some that live amidst decaying plant and animal matter and "feed" by taking up dissolved organic material across their body surface (i.e., saprobic organisms [*sapro* = G: rotten]); and some that are capable of two or even all three modes of nutrition, either simultaneously or at different times. To further complicate matters, most of the photosynthesizing species are capable of active movement, while many of the carnivorous species are not. Many, but not all, of the plant-like species lack a major botanical characteristic: cellulose. It seems absurd to include all of these forms within a single group, the Protozoa, and yet there are no simple alternatives. The distinction between plant and animal is just not easily made here.

One solution to the dilemma, suggested by the great German scientist Ernst Haeckel about 120 years ago, is to place all these organisms within a separate Kingdom, the Protista, setting them apart from those organisms that are clearly plants and those that are clearly animals. More recently—within the past 20 years or so—it has been proposed that the Kingdom Protista contain all eukaryotic, unicellular organisms that lack cellulose, regardless of nutritional mode. The eukaryotic, nonmulticellular, cellulose-containing photosynthesizing organisms would then be placed within the Kingdom Plantae, along with the multicellular organisms that we all recognize as plants.

This arrangement for the Protista acknowledges what we now presume to be a close evolutionary relationship between the plant-like and animal-like organisms. However, we are still left with the difficulty of determining whether a given protist is more animal-like (i.e., "protozoan") or plant-like (i.e., "protophytan") in nature. Several major protist groups have been claimed by both botanists and zoologists. The present treatment deals primarily with organisms that ingest solid food, but points out plant-like relatives where they are known to exist within the group under discussion.

To summarize, we assume protozoans represent a potpourri of several distantly related groups of single-celled organisms exhibiting a variety of life-styles and lacking cellulose. Protozoans will be considered as a Subkingdom of the Kingdom Protista. The classification scheme followed here is basically that suggested recently by an international committee of protozoologists.[1] Additional modifications to the present scheme of protozoan classification are continually

---

1 N.D. Levine et al., 1980. *J. Protozool.*, 27: 37.

being proposed as new ultrastructural studies are published. Classification of the parasitic phyla has been simplified in the present treatment.

GENERAL CHARACTERISTICS

Protozoans are clearly among the most primitive of life forms. As described above, even the distinction between plant and animal is not clear in this group. Nevertheless, in many respects protozoans are as complex as any multicellular organism (**metazoan**), and a great deal of protozoan biology remains poorly understood despite much sophisticated research. The entire creature is bounded by a **plasmalemma** (cell membrane) that is structurally and chemically identical to that of multicellular organisms. The cytoplasm bounded by the plasmalemma resembles that of other animal cells, except that it is often differentiated into a clear, gelatinous outer region, the **ectoplasm**, and an inner, more fluid region, the **endoplasm**. Within the cytoplasm are organelles and other components typical of metazoan cells, including nuclei, nucleoli, chromosomes, Golgi bodies, endoplasmic reticulum (with and without ribosomes), lysosomes, centrioles, mitochondria, and, in some individuals, chloroplasts. In a sense, then, we may regard the typical protozoan as being a single-celled organism. Yet, the protozoan is functionally more versatile than any single cell found within any multicellular animal. The cells of metazoans are specialized for particular structural and/or physiological functions. In contrast, each protozoan is a complete entity, capable of feeding, digesting, respiring, excreting, sensing, moving, and reproducing.

In addition to the organelles encountered typically in multicellular animals, many protozoans contain organelles not generally found among metazoans. These organelles are, most notably, contractile vacuoles, trichocysts, and toxicysts. Many protozoans are also characterized by highly complex arrays of microtubules and microfilaments.

FIGURE 3.1. (a,b) Diagrammatic illustration of two types of contractile vacuole systems encountered among protozoans. The two differ largely in the complexity of the spongiome and in other aspects of the fluid collection system. Other contractile vacuole complexes (not illustrated) may lack ampullae, permanent pores, and the associated bundles of microtubules. (c,d) The behavior of two types of contractile vacuole complexes during filling and emptying; arrows represent the flow of liquid out of the emptying vacuole. The vacuole is viewed from above in the upper series (discharge emerges perpendicular to the plane of the page) and from the side in the lower series. In (c), a permanent pore is lacking, and the vacuole forms through the fusion of many small fluid-filled vesicles; the vacuole disappears completely following discharge. In type (d) the contractile vacuole is filled by conspicuous ampullae and discharges through a permanent pore. The ampullae may begin to refill before the vacuole discharges its fluid to the outside. (Modified from Patterson, 1980. Biol. Rev., 55: 1.)

Still other, highly specialized organelles are restricted to various small groups of protozoans.

Contractile vacuoles are organelles involved in the expulsion of excess water from the cytoplasm.[2] Apparently, fluid is collected

from the cytoplasm by a system of membranous vesicles and tubules called the **spongiome**. The collected fluid is transferred to a contractile vacuole and is subsequently discharged to the outside through a pore in the plasmalemma (Fig. 3.1). Contractile vacuoles are most commonly seen among freshwater protozoan species, because concentrations of dissolved solutes in the cytoplasm

2 See Topics for Discussion, No. 4.

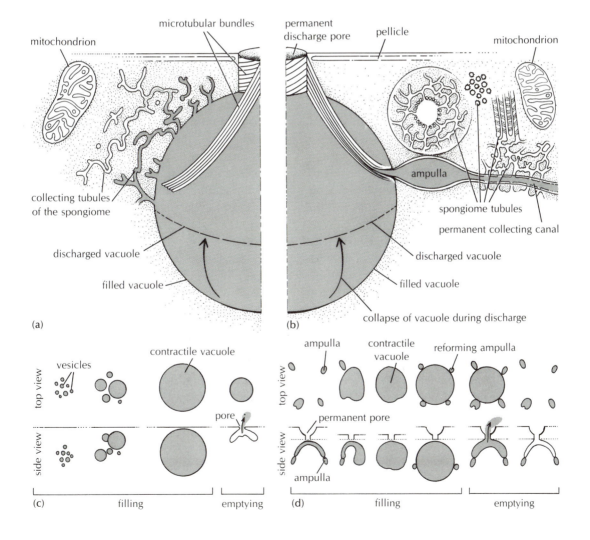

of freshwater protozoa are much higher than solute concentrations in the surrounding medium. Thus, water continuously diffuses into the cytoplasm across the plasmalemma, and does so in proportion to the magnitude of the solute concentration gradient (i.e., the **osmotic gradient**). Without some form of compensating mechanism, water would continue to flow into the animal until either the osmotic gradient across the plasmalemma was reduced to zero, or the animal burst. Moreover, cells can function over only a narrow range of internal solute concentrations, and protozoans are no exception; even a small dilution of the cytoplasmic solute concentration could be disabling.

From this discussion, we see that the contractile vacuole must function not only to prevent swelling (by ridding the body of fluid as fast as water crosses the plasmalemma), but also to maintain a physiologically acceptable solute concentration within the cell (Fig. 3.2). In other words the contractile vacuole must function both in **volume regulation** (maintaining a constant body volume), and in **osmotic regulation** (maintaining a constant intracellular solute concentration).[3] The contractile vacuole achieves the latter result by pumping out of the cell a fluid that is dilute relative to the surrounding cytoplasm in the cell. That is the solute is essentially separated from the water and retained within the cell before the remaining fluid is expelled. The mechanism by which water is separated from the cytoplasm is not known, nor is the mechanism by which the vacuole fluid is discharged to the

outside, although researchers are making some progress in these studies. Contractile vacuoles typically fill and discharge several to many times per minute, and a single animal may possess several contractile vacuoles.

Trichocysts develop within membrane-bound vesicles in the cytoplasm and eventually come to lie along the periphery of the protozoan. The trichocysts themselves are elongated capsules that can be triggered by a variety of mechanical and/or chemical stimuli to discharge a long, thin filament (Fig. 3.3). This discharge occurs within several thousandths of a second and is thought to be initiated through an osmotic mechanism involving a rapid influx of water. The adaptive significance of trichocysts is unknown; likely possibilities are that they function in protection or in anchoring the animal during feeding. Related structures called **toxicysts** are clearly involved in predation; filaments discharged from toxicysts paralyze prey and initiate digestion. At least ten other types of **extrusomes** (organelles capable of ejectability) have so far been described from the Protozoa.

As single-celled organisms, protozoans necessarily lack gonads. Although sexual reproduction is known for many groups, protozoan reproduction is typically asexual (*a* = G: without). Asexual reproduction is commonly encountered among all groups of protozoans and is the only form of reproduction reported for many species. By definition, asexual reproduction does not generate new genotypes. Protozoans reproduce asexually through **fission**, a controlled mitotic replication of chromosomes and splitting of the parent into two or more parts. Indeed, asexual reproduction among

---

3 See Topics for Discussion, No. 5.

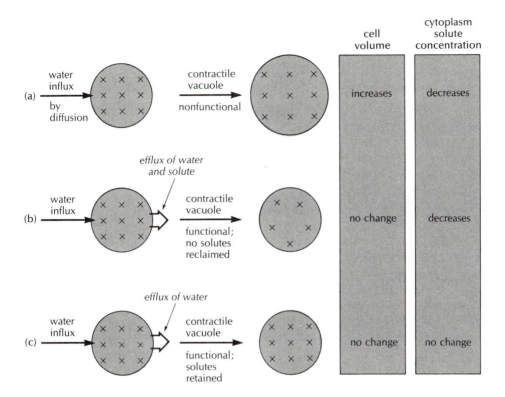

FIGURE 3.2. *Diagrammatic illustration of the functional significance of contractile vacuoles in freshwater protozoans. The vacuole system functions both to maintain body volume and to maintain proper solute concentrations within the cell. (× represents solute) (a) No functioning contractile vacuole; water diffuses inward, swelling the cell and diluting internal solute. (b) The contractile vacuole pumps out cytoplasmic fluid, including dissolved solute; cell volume is maintained, but solute is continuously lost. (c) The contractile vacuole pumps out fluid containing little solute, maintaining both cell volume and osmotic concentration.*

protozoans has long been exploited by biologists as a general model for the study of mitosis. **Binary fission** occurs when the protozoan splits into two individuals (Fig. 3.4). In **multiple fission**, a great many nuclear divisions precede the rapid differentiation of the cytoplasm into a great many distinct individuals. In **budding**, a portion of the parent breaks off and differentiates to form a new, complete individual. In some multinucleate species, the parent simply divides in two in the absence of any mitotic division, the orig-

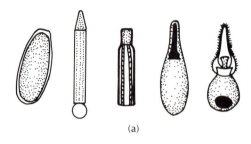

(a)

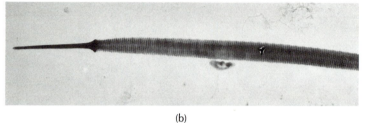

(b)

FIGURE 3.3. (a) Undischarged extrusomes of different types. At least twelve morphologically distinct types of extrusome have been described. Some expel mucus (left), others eject filaments of varying length. Paralytic toxins are injected by one class of extrusome (toxicysts, not shown). Others (far right) are used primarily for adhering to prey during food capture. (From Corliss, 1979. Amer. Zool., 19: 573.) (b) Transmission electron micrograph of a trichocyst that has been discharged from Paramecium sp. (Courtesy of M.A. Jakus, National Institutes of Health.)

inal nuclei being distributed between the two daughter cells. This process is termed **plasmotomy** (*tomy* = G: TO CUT).

In keeping with the widespread occurrence of asexual reproductive capabilities, many protozoans possess great capacity for regeneration. One example of this capacity is the phenomenon of encystment and excystment exhibited by many freshwater and parasitic species. During **encystment**, substantial dedifferentiation of the animal takes place. The individual loses its distinctive surface features, including cilia and flagella, and becomes rounded. The contractile vacuole(s) pump out all excess water from the cytoplasm, and a gelatinous covering (or some other type of covering) is secreted. This covering soon hardens to form a protective **cyst**. In this form, the quiescent animal can withstand long periods (several to many years in some species) of exposure to what would otherwise be intolerable environmental con-

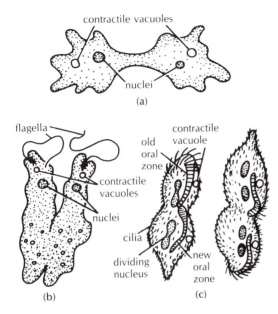

contractile vacuoles

nuclei

(a)

flagella

old oral zone

contractile vacuole

contractile vacuoles

nuclei

cilia

dividing nucleus

new oral zone

(b)

(c)

FIGURE 3.4. *Binary fission among the Proto-zoa: (a) in an amoeba; (b) in a flagellate; (c) in a ciliate. [(a) after Pennak. (b,c) after Hyman; after Gregory.]*

ditions of acidity, dryness, thermal stress, and food or oxygen deprivation.[4] Considerable dispersal by winds, animals, and other vectors may occur while individuals are in the encysted state. Once conditions improve, **excystment** quickly ensues, with the regeneration of all former internal and external structures.

Patterns of sexual reproduction will be discussed as appropriate on a group-by-group basis, as few generalizations are possible.

Food particles are digested internally among protozoans; since they are single-celled organisms, digestion is, of necessity,

entirely intracellular. Ingested food particles generally become surrounded by membrane (similar to the plasmalemma), forming a distinct **food vacuole**. These vacuoles move about in the fluid cytoplasm of the body as the vacuole contents are digested enzymatically. By feeding protozoans food stained with pH-sensitive chemicals, researchers have determined that the contents of the vacuoles first become quite acidic and later become strongly basic. As in other animals, including humans, digestion requires exposure of the food to a series of enzymes, each with a specific role to play and each with a narrowly defined pH optimum. The controlled changes of pH that occur within the food vacuoles of protozoans allow for the sequential disassembly of foods by a series of different enzymes, despite the absence of a digestive tract *per se*. Once solubilized, nutrients move across the vacuole wall and into the endoplasm of the cell. Indigestible solid wastes are commonly discharged to the outside through an opening in the plasma membrane.

Clearly, the members of the Protozoa are unlike other animals in many features of their biology. Protozoans are problematic, and intriguing, in another important respect as well: even the distinction between unicellular and multicellular is not always easily made. Although most protozoan species occur as single, one-celled individuals, many other species are **colonial**; that is, a single individual divides asexually to form a colony of attached, genetically identical individuals. Usually, the individuals of a protozoan colony are morphologically and functionally identical. However, the individuals comprising the colonies of several species show a

_____

4 See Topics for Discussion, No. 7.

degree of structural and functional differentiation that is very reminiscent of that encountered within primitive groups of metazoans. Protozoans thus bridge the gap not only between plants and animals, but between unicellular and multicellular life forms as well. As the protozoologist John Farmer has said, such "transitional forms play havoc with attempts to keep taxonomy tidy." But they certainly do add interest to studies of invertebrate zoology and evolution.

PHYLUM CILIOPHORA

Ciliates show the highest degree of subcellular specialization encountered among the Protozoa. Nevertheless, I begin with the ciliates because they are predominately free-living and, compared with other protozoan groups, are relatively uniform in basic body plan. The phylum contains over 7,500 species. Several features are unique to the ciliates, and several other features are characteristic of this group above all others.

First among the unique features of this phylum is the presence of external ciliation in at least some stage of the life cycle. The ultrastructure of cilia is remarkably uniform through the Ciliophora and, in fact, throughout the animal (and plant) kingdom(s) as well, although some modifications to the basic pattern do occur sporadically.

> *Phylum Cilio•phora*
>
> (G: CILIA BEARING)

The following description of ciliary structure and function is generally applicable to ciliary organelles encountered among animals in all other phyla.

Structure and Function of Cilia

Each cilium is cylindrical and arises from a **basal body (kinetosome)**. Within the cilium are found a number of long rods called **microtubules**, composed of a protein known as **tubulin**. Tubulin is extremely similar to the actin of metazoan striated muscles. A cross section of the kinetosome of a cilium in the region beneath the outer body surface shows a ring of nine groups of microtubules, with three microtubules to a group (Fig. 3.5). The "A," or innermost microtubule of each group, is physically connected to the "C," or outermost microtubule of an adjacent group, via a thin filament. Additional filaments connect the "A" microtubule to a central tubule, like the spokes of a wheel. This configuration of microtubules changes somewhat near the distal end of the cilium.

A cross section made through a cilium external to the body surface shows a ring of nine groups of microtubules, with only two microtubules per group; the "C" microtubule is not found (Fig. 3.5). Instead, a pair of **dynein arms** projects outward from each "A" microtubule toward the "B" tubule of the neighboring pair of microtubules. The primary protein component of these arms (**dynein**) is similar in some respects to muscle myosin. Like myosin, dynein possesses the ability to cleave ATP, releasing chemical energy. A pair of single microtubules is also located centrally within the cilium. These two mi-

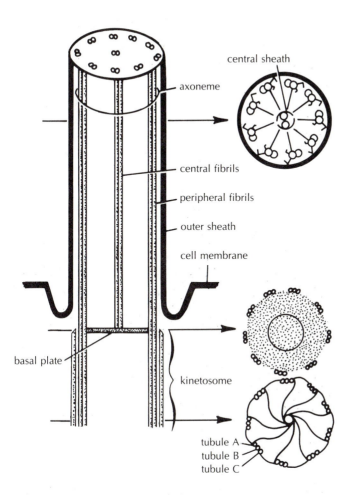

central sheath

axoneme

central fibrils

peripheral fibrils

outer sheath

cell membrane

basal plate

kinetosome

tubule A
tubule B
tubule C

FIGURE 3.5. *Ciliary ultrastructure. The appearance of the cilium in cross section changes along its length as indicated. (After Sherman and Sherman; and after Wells.)*

crotubules form the central shaft of the cilium, and they are often surrounded by a membrane, constituting a **central sheath**. Distinct filaments extend from the "A" microtubule of each outer doublet in toward this central sheath. The entire microtubular complex, consisting of the nine doublet mi-

crotubules and the inner pair of single microtubules, is termed the **axoneme**.

Although many details of ciliary operation remain to be elucidated, present evidence indicates that ciliary bending is achieved through the differential sliding of some groups of adjacent microtubules relative to others within the same cilium. The energy for this sliding appears to derive from the ATPase activity of the dynein arms, and the sliding itself appears to involve an interaction between tubulin and dynein, in a manner highly reminiscent of the interaction between the actin and myosin filaments of metazoan striated muscle. As we will see later in this chapter, microtubules play a variety of roles, both direct and indirect, in the locomotion of most cells—even in those lacking cilia.

The rhythmic and coordinated bending and recovery of the cilia provide the means of locomotion and often aid in food collection for members of this phylum. In addition ciliary activity is probably important in removing wastes from the immediate vicinity of the animal and in continually bringing oxygenated water into contact with the general body surface. Ciliates are the fastest-moving of all protozoans, achieving speeds of up to two millimeters per second.

To be effective, i.e., to create unidirectional movement of the animal in water or unidirectional flow of water past the animal, the **power stroke** and the **recovery stroke** of a cilium must not be identical in form. By analogy, if you are performing the breast stroke while swimming, and move your arm forward to the front of your head by the same path and with the same force that you used to bring the arm back to your side, you would move backward by about the same distance that you moved forward during your power stroke. The cilium faces a similar dilemma. During ciliary locomotion, the power stroke, as the name implies, does work against the environment. The cilium is outstretched for maximum resistance as it

bends downward toward the body (Fig. 3.6). During the recovery stroke, the cilium bends in such a manner as to considerably reduce resistance, doing less work against the environment. Thus, the recovery stroke does not undo the work of the power stroke. In addition, the cilium moves fastest in the power stroke, increasing the force of the power stroke relative to that of the recovery stroke. By analogy, more force is required to move your arm through the water quickly than slowly.

Generally, a large part of the body surface of ciliates is covered by distinct rows of cilia, and movement is achieved by coordinated, **metachronal** beating of the cilia in each row (Fig. 3.7). In metachronal beating, the power and recovery strokes of a cilium are begun immediately following the initiation of the comparable strokes of an adjacent cilium. The direction of metachronal beating may be quickly reserved in response to chemical,

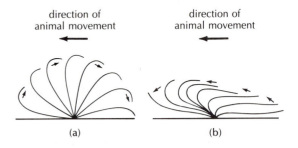

FIGURE 3.6. (a) Power and (b) recovery strokes in isolated cilia. Arrows indicate the direction of motion of the cilium. Note that more of the ciliary surface area is involved in pushing against the water in (a) than in (b). Thus, the recovery stroke does less work and does not undo all the work of the power stroke.

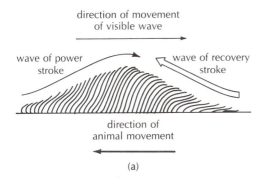

direction of movement
of visible wave

wave of power
stroke

wave of recovery
stroke

direction of
animal movement

(a)

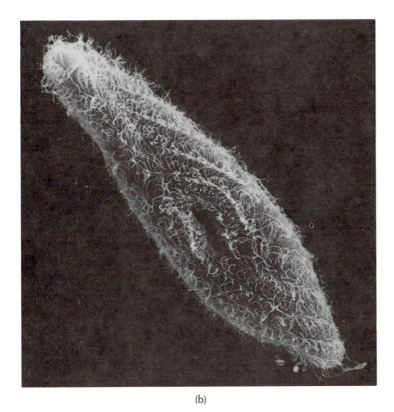

(b)

FIGURE 3.7.   (a) *A metachronal wave passing along a row of cilia. (After Wells; after Sleigh.) (b) A ciliate, (*Paramecium sonneborni,*) showing metachronal waves of ciliary activity. Note the oral opening near the middle of the body. The animal is 40 μm in length. (Courtesy of K.J. Aufderheide. From Aufderheide et al., 1983. J. Protozool., 30: 128.)*

mechanical, or other stimuli, permitting mobile animals to quickly reverse direction and perhaps escape undesirable situations.

Patterns of Ciliation

Individual cilia, seen external to the cell body, are associated with each other through a complex **infraciliature** below the body surface (Fig. 3.8). A striated fibril, called a **kinetodesmos**, extends from each kinetosome (basal body) in the direction of an adjacent cilium of the same row. Thus, running along the right side of each row of basal bodies is a cord of fibers, termed the **kinetodesmata**. Direct microtubular connections between adjacent kinetosomes have rarely been demonstrated. Each of these kinetosomes has its own array of microtubular, microfibrillar, and other organelles, giving ciliate bodies a more complex cytoarchitecture than is found in many other protozoans, and in any metazoan.

This infraciliature is found only within the Ciliophora, and is encountered in the adults of all ciliate species, even in those lacking external ciliation as adults. The structure of this infraciliature is one of the primary tools used to distinguish the different ciliate species and to assess the degree to which different species are related (Fig. 3.9). The functional significance of the complex structure of the infraciliature has not yet been conclusively demonstrated.[5]

Cilia cover virtually the entire body of some species, but are reduced or otherwise modified to varying degrees in others. In some species, groups of cilia are functionally associated in such a way as to form discrete organelles. One such organelle is the so-called **undulating membrane**, a flattened sheet of cilia that moves as a single unit (Fig. 3.10a). A second commonly encountered ciliary organelle is termed a **membranelle**; here, a smaller number of cilia in several adjacent rows appear to lean toward each other, forming, in effect, a two-dimensional triangular tooth (Fig. 3.10a,b). In addition, cilia may form a discrete bundle (**cirrus**), which tapers to a point toward the tip (Figs. 3.10a,b; 3.11). The cilia comprising such organelles are structurally identical with those that function as individuals, and no permanent physical attachments between the cilia comprising undulatory membranes, membranelles, or cirri have been observed. The mechanism by which their activities are so closely coordinated remains uncertain.

Besides the infraciliature and general placement and pattern of body ciliation, another characteristic of major taxonomic significance is the position, ultrastructure,

---

5 See Topics for Discussion, No. 3.

FIGURE 3.8. *The complex infraciliature of the ciliates* (a) Conchophthirus *sp., and* (b) Tetrahymena pyriformis. *These ultrastructural details were unknown before the advent of the electron microscope. [(a) From Corliss, 1979. Amer. Zool., 19: 573. (b) From Allen, 1967. J. Protozool. 14: 553.]*

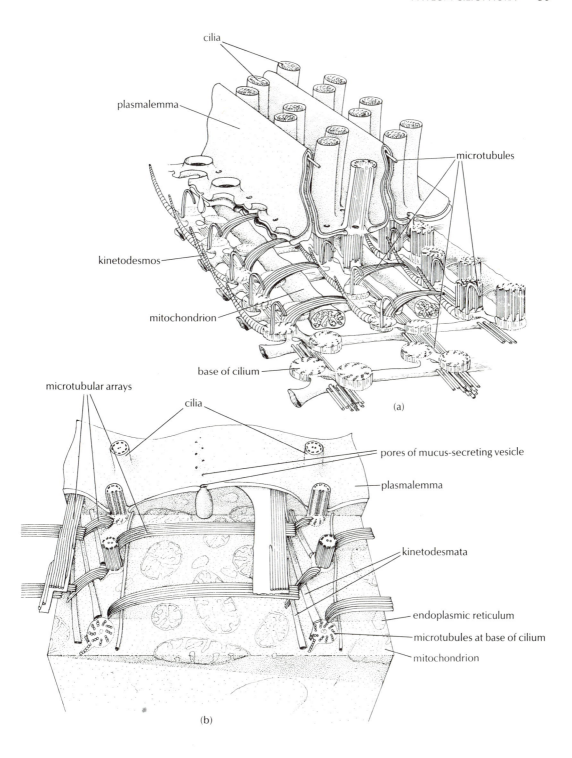

cilia

plasmalemma

microtubules

kinetodesmos

mitochondrion

base of cilium

(a)

microtubular arrays

cilia

pores of mucus-secreting vesicle

plasmalemma

kinetodesmata

endoplasmic reticulum

microtubules at base of cilium

mitochondrion

(b)

(a)   (b)

FIGURE 3.9. (a) *Climacostomum sp., stained to reveal the infraciliature. (b) Detail of three rows* (kineties) *of kinetodesmata. (Courtesy of C.F. Dubochet, from Dubochet et al., 1979. J. Protozool., 26: 218.)*

and pattern of ciliation of the oral region. The mouth opening, called the **cytostome** (*cyto* = G: cell; *stoma* = G: mouth), may be located anteriorly, laterally, or ventrally on the body. Often the cytostome is preceded by one or more pre-oral chambers, whose complexity varies greatly among major groups of ciliate species.

Reproduction Characteristics

In addition to the presence of cilia and an infraciliature, a second unique characteristic of all ciliates is the possession by each individual of two types of nuclei. That is, the nuclei of every ciliate are **dimorphic** (of two kinds). The ciliates are thus **heterokaryotic**

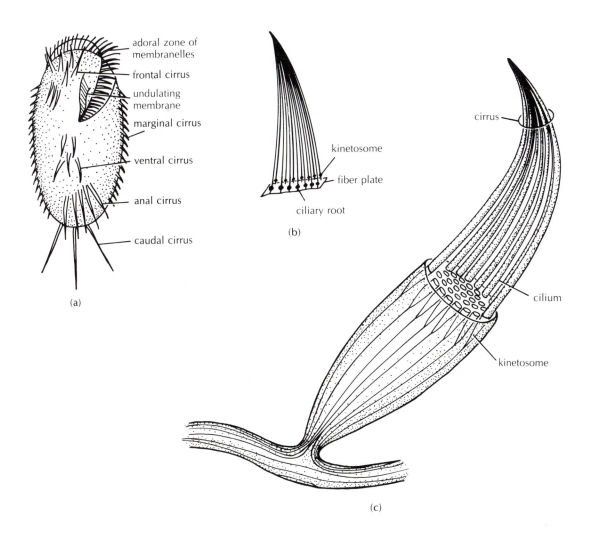

adoral zone of
membranelles

frontal cirrus

undulating
membrane

marginal cirrus

ventral cirrus

anal cirrus

caudal cirrus

(a)

kinetosome

fiber plate

ciliary root

(b)

cirrus

cilium

kinetosome

(c)

FIGURE 3.10. (a) Stylonychia *sp., showing several ciliary
organelles: undulating membranes, membranelles, and cirri. (From
Schmidt and Roberts, 1981. Foundations of Parasitology, 2nd ed.
C.V. Mosby; after Kudo.) (b) Diagrammatic illustration of
membranelle structure. (From Schmidt and Roberts; after Kudo.)
(c) Diagrammatic representation of the structure of a single cirrus.
Cirri typically contain between 24 and 36 individual cilia.
(Reprinted with permission of Macmillan Publishing Company
from The Invertebrates: Function and Form, 2nd ed., by Irwin W.
Sherman and Vilia G. Sherman. Copyright © 1976 by Irwin W.
Sherman and Vilia G. Sherman.)*

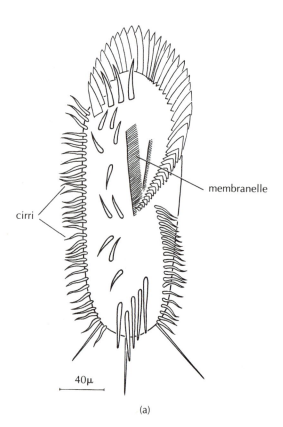

cirri

membranelle

40μ

(a)

50μ

(b)

FIGURE 3.11. *(a) Illustration of* Stylonychia lemnae. *(b) The same organism, seen with the scanning electron microscope. (Courtesy of D. Ammermann, from Ammermann and Schlegel, 1983.* J. Protozool. 30: 290.)

(*hetero* = G: different; *karyo* = G: nucleus), while all other cells, including other protozoans, are **monomorphic** (one kind) or **homokaryotic** (*homo* = G: alike). Every individual contains one or more large nuclei, termed **macronuclei**, and one or more smaller nuclei, termed **micronuclei**. Micronuclei are often more abundant than macronuclei within a given individual; some ciliate species possess more than 80 micronuclei per individual.

The macronucleus is polyploid, contains both DNA and RNA, and is involved both in the day-to-day operations of the protozoan and in differentiation and regeneration. The animals cannot live without their macronuclei, but can live without their micronuclei. However, micronuclei are essential for sexual reproduction. In contrast, the macronucleus plays no direct role in the sexual activities of ciliates.

Sexuality among ciliates never involves the formation of gametes. Rather, their primary sexual activity invariably involves a process called **conjugation**, a temporary physical association between two "consenting" individuals during which genetic material is exchanged. The exchange takes place through a tube connecting the cytoplasm of the two individuals. (Fig. 3.12).

During conjugation, the macronuclei disintegrate and the micronuclei, which are diploid, divide by meiosis and mitosis so that four haploid **pronuclei** are formed from each micronucleus. Typically all but one of these pronuclei degenerate, and the remaining pronucleus undergoes mitosis to form two identical haploid pronuclei. One of these two pronuclei somehow migrates through the cytoplasmic tube into the other individual. Exchange of pronuclei is reciprocal; that is, each individual winds up with one migrant pronucleus. Each migrant micronucleus subsequently fuses with the stay-at-home pronucleus, restoring the diploid condition through the formation of a **synkaryon** (i.e., a nucleus formed by the fusion of the pronucleus from one individual with its partner's pronucleus). Once the two conjugants separate, following the transfer and fusion of micronuclei, the synkaryon of each ex-conjugant divides mitotically from one to several times. Some of the products form micronuclei, while others give rise to macronuclei. Cytoplasmic divisions (i.e., actual reproduction of individuals) may follow, resulting in several individual offspring that are probably genetically dissimilar, both with respect to each other and to the parental conjugants.

Although sexual dimorphism *per se* appears to be lacking in ciliates, the species that have been well-studied do exhibit different **mating types**. These mating types, in turn, belong to separate varieties called **syngens**.

For example, *Paramecium aurelia* has 16–18 known syngens, each containing several different mating types. Conjugation occurs only between individuals of different mating types and only between individuals within one syngen. In the case of **P. aurelia**, the various syngens have recently been formally recognized as separate taxonomic species, each with its own pair of mating types.

All ciliates are capable of asexual reproduction, as are all other protozoans. Among ciliates, asexual reproduction takes the form of transverse binary fission. The animal becomes bisected perpendicular to its long axis, so that complete individuals result from the anterior and posterior halves of the parent (Fig. 3.13). In contrast, binary fission in other groups of protozoa is longitudinal (i.e., parallel to the long axis), producing unicellular offspring that are mirror images of each other. By convention, these offspring are called "daughters"; this should not be taken as a reference to sexuality.

During binary fission in ciliates, the micronuclei divide mitotically and redistribute throughout the cytoplasm. The macronuclei elongate, but do not undergo mitosis. In some cases, all macronuclei fuse prior to elongation, forming one very large macronucleus. A cleavage furrow gradually forms, dividing the macronucleus, and the body itself, into anterior and posterior halves. The detached, posterior half of the body must then regenerate all external and internal structures that have been appropriated by the anterior half, and vice versa. In many species, much of this differentiation is completed prior to cell division. Commonly, all parental organelles are eventually re-

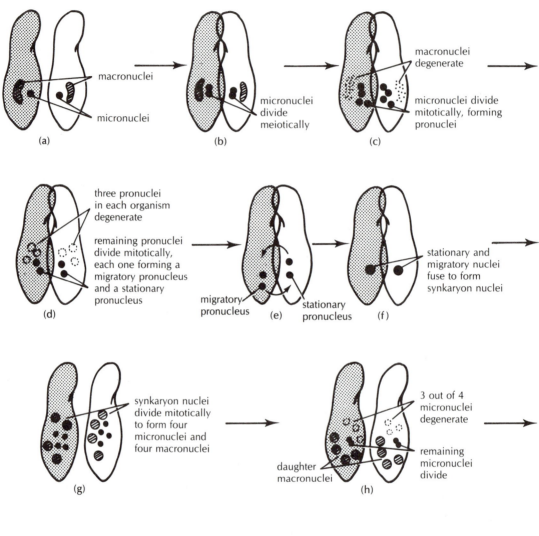

macronuclei

micronuclei

(a)

micronuclei divide meiotically

(b)

macronuclei degenerate

micronuclei divide mitotically, forming pronuclei

(c)

three pronuclei in each organism degenerate

remaining pronuclei divide mitotically, each one forming a migratory pronucleus and a stationary pronucleus

(d)

migratory pronucleus

stationary pronucleus

(e)

stationary and migratory nuclei fuse to form synkaryon nuclei

(f)

synkaryon nuclei divide mitotically to form four micronuclei and four macronuclei

(g)

3 out of 4 micronuclei degenerate

daughter macronuclei

remaining micronuclei divide

(h)

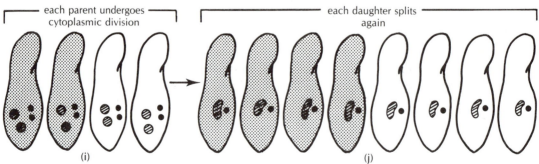

each parent undergoes cytoplasmic division

each daughter splits again

(i)

(j)

FIGURE 3.12.   *Conjugation in* Paramecium caudatum. *(a) Two individuals of different mating types come together. (b) A tubular cytoplasmic bridge forms between the two conjugants, and the micronucleus divides meiotically. (c) The macronucleus degenerates and the micronuclei divide mitotically, forming four pronuclei in each individual. (d) Three pronuclei degenerate, while the remaining pronucleus undergoes another mitotic division, forming one migratory pronucleus and one stationary pronucleus. (e) The migratory pronuclei and exchanged through the cytoplasmic bridge, and fuse with the stationary pronuclei (f). (g,h) The exconjugants separate, while a series of nuclear divisions and degenerations produce four macronuclei and a pair of micronuclei. (i,j) Cytoplasmic divisions produce four daughter individuals from each parent.*

placed in both daughters. Such differentiation is, as always, under the control of the macronucleus.

One other form of nuclear reorganization commonly encountered among ciliates remains to be discussed. In this process, called **autogamy**, a form of sexual activity takes place with the involvement of only a single individual! In some respects, events are reminiscent of the preliminaries to conjugation. The macronucleus, or macronuclei, degenerate while the micronucleus, or micronuclei, undergo meiosis so that two pronuclei form from each micronucleus (Fig.

FIGURE 3.13.   *Binary fission in* Stentor coeruleus. *(Courtesy of D.R. Diener, from Diener et al., 1983. J. Protozool., 30: 83.)*

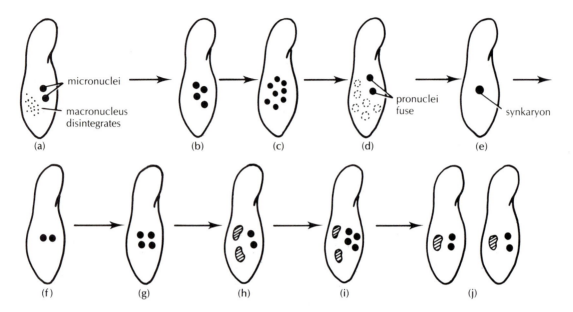

FIGURE 3.14 *Diagrammatic illustration of autogamy.*
(a) Degeneration of macronucleus. (b–d) Micronuclei replicate
meiotically and then mitotoically; selective degeneration of haploid
pronuclei follows. (e) Two pronuclei fuse to form a synkaryon.
(f–i) A series of mitotic divisions produce macronuclei and
micronuclei. (j) Cytoplasmic division produces two individuals,
each with the proper number of micro- and macronuclei. (Based on
Raikov in Chen, 1972. Research in Protozool., 4: 193. Pergamon
Press.)

3.14). Meiosis is followed by several mitotic divisions of the pronuclei. Two of these pronuclei fuse together, forming a zygote nucleus (i.e., a **synkaryon**), while the remaining division products disintegrate. Subsequent events resemble those of conjugation following separation of the conjugants. Through this process, new genotypes may be formed, and a new macronucleus is generated, as in conjugation.

Periodic renewal of the macronucleus appears to be essential for a number of protozoan species. Cultures will eventually die out after many generations of asexual reproduction if some form of nuclear reorganization is prevented from occurring periodically. Regeneration of the macronucleus thus appears to have a rejuvenating effect on the animals, which would otherwise apparently exhibit senescence and, ultimately, death.

Other Features

One other morphological feature particularly characteristic of ciliates is the covering of the body by an often complex series of membranes, forming a **pellicle** (Fig. 3.15). The inner membranes, lying beneath the single plasmalemma enveloping the body, form a series of elongated, flattened vesicles called **alveoli** (*alveolus* = L: a cavity). Cilia project to the outside between adjacent alveoli. The pellicle may be rigid or highly flexible, depending upon how the membranes are organized, and may serve a protective function in some species. Trichocysts are characteristically associated with the pellicle of ciliates.

In keeping with the unusually high level of structural complexity characteristic of ciliates, only the members of this group have permanently established excretory pores associated with their contractile vacuoles. The openings are maintained by arrays of microtubules. The system for the collection of cytoplasmic fluid and its transfer to the contractile vacuole is also particularly complex among ciliates.

A variety of life-styles are exhibited within the Phylum Ciliophora. About 65% of ciliate species are free-living, and the majority of these are motile. Some other species form temporary attachments to living or nonliving substrates for feeding purposes, while others are permanently attached (i.e., **sessile**) and

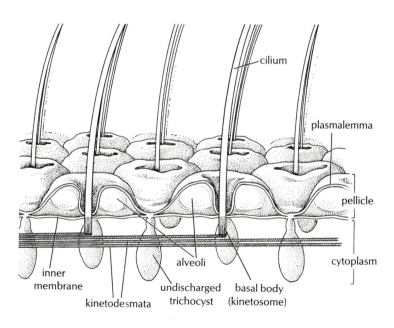

FIGURE 3.15.   *Illustration of pellicle, showing alveoli. (After Corliss; and after Sherman and Sherman.)*

may form colonies. Although all ciliates have a distinct pellicle, some sessile species also produce a rigid, protective encasement termed a **test** or **lorica** (*lorica* = L: armor). Probably the best-known examples of such test-forming species are encountered among two groups of ciliates known as tintinnids and folliculinids (Fig. 3.16). Folliculinids are quite remarkable animals in that, although they are permanently attached to solid substrate as adults, they are capable of dedifferentiating to a "larval" free-living form, abandoning their encasement, relocating at a distance from the original site, secreting a new case, and resuming the adult life-style. In fact, some species seem to dedifferentiate and then undergo fission, with the posterior half of the animal remaining behind to redifferentiate at the original location while the anterior half of the animal leaves to establish itself elsewhere!

About one-third of all ciliate species are parasitic in or on a variety of other invertebrates (including crustaceans, molluscs, bryozoans, and annelids) and vertebrates (including the digestive tracts of humans, producing an ulcerative condition of the intestine, and the skin of fishes, causing the disease known to freshwater fish afficionados as "ick"). Some species are commensals, attaching to the outer surface of organisms such as crabs, or within the body of hosts such as cattle, horses, sheep, frogs, and cockroaches.

Most free-living ciliates are **holozoic**; that is, they ingest particulate foods. Some species may be **raptorial**, i.e., hunting and ingesting animal prey. Still others may be passive suspension-feeders, primarily ingesting bacteria, unicellular algae, other protozoans, or small metazoans.

The mouths of some raptorial species can expand to body width or greater, permitting the ingestion of prey that are large with respect to the ingestor. One species, *Didinium nastum*, about 125 μm in length, can ingest prey (exclusively members of the ciliate genus *Paramecium*) two to three times larger than itself (Fig. 3.17)!

Suspension-feeders, in contrast, are relatively passive. Food particles are carried to the mouth by water currents generated by ciliary activity in the mouth region. Commonly, the suspension-feeding species have stalks, as in members of the genus *Vorticella*. The stalks of such species may contain a **spasmoneme**, i.e., a coiled, membrane-bound bundle of contractile fibers (Fig. 3.18). Contraction of the spasmoneme fibers results in a very rapid shortening of the stalk, and constitutes the animal's only escape response. Other sessile species, such as *Stentor* spp., lack stalks, but contractile fibers in the body proper often permit extensive shape changes to occur (Fig. 3.18a).

FIGURE 3.16.   *(a–c) Folliculinids, animals and loricas. [(a) after Jahn, 1949. (b,c) Courtesy of M. Mulisch, from Mulisch and Hausmann, 1983. J. Protozool., 30: 97.) (d) A tintinnid,* Stenosemella sp. *(From Corliss, 1979. Amer. Zool., 19: 573.) (e) Lorica of a tintinnid,* Tintinnopsis platensis, *as seen with a scanning electron microscope. (f)* T. parva *without its test. [(e,f) Courtesy of K. Gold, from Gold, 1979. J. Protozool., 26: 415.]*

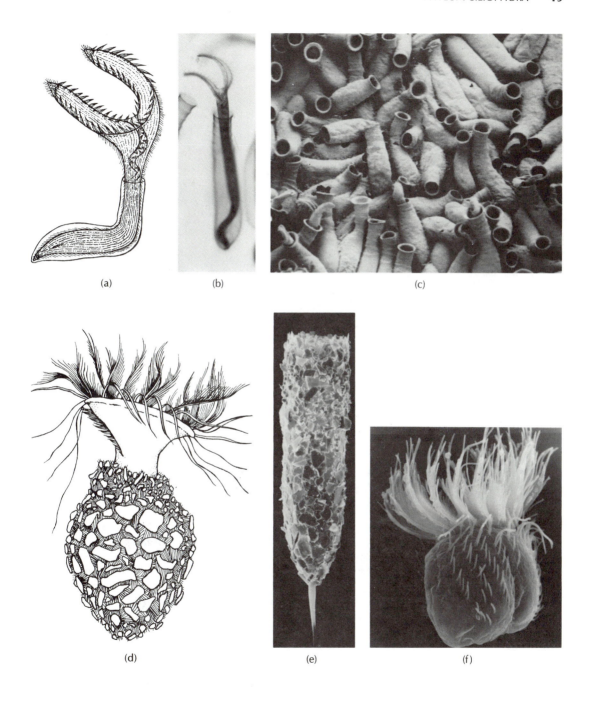

(a)

(b)

(c)

(d)

(e)

(f)

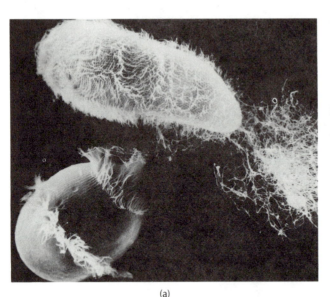

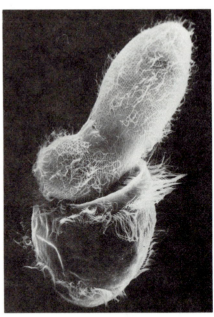

(a)

(b)

FIGURE 3.17. *(a) This* Paramecium multimicronucleatum *has discharged numerous trichocysts in response to attack by another ciliate,* Didinium nastum. *(b) The persistent D. nastum begins to ingest its prey. (Courtesy of G.A. Antipa, San Francisco State University.)*

FIGURE 3.18. *Ciliate diversity. (a) Stentor. (After Sherman and Sherman.) (b)* Tracheloraphis kahli. *(From McConnaughey, Bayard H., and Zottoli, Robert, 1983. Introduction to Marine Biology, 4th ed. St. Louis, MO: The C. V. Mosby Co.; redrawn from Fenchel, T., 1969. Ecology of Marine Microbenthos IV. Ophelia 6: 1, July. (c)* Diophyrs scutum. *(From McConnaughey and Zottoli; after Fenchel.) (d)* Didinium nastum. *(From Hyman; after Blochmann.) (e) A colony of* Epistylis *sp. (After Hyman.) (f)* Vorticella. *(After Sherman and Sherman.) (g) Vorticellids covering the shell of a larval marine bivalve. Note the coiled, spring-like stalks. The larval shell is approximately 200 µm in length. (Courtesy of C. Bradford Calloway and R.D. Turner.) (h) A suctorian,* Tokophyra quadripartita. *The body is approximately 100–175 µm long. (From Hyman; after Kent.) (i) Prey cytoplasm being moved through a hollow tentacle of the suctorian predator. Haptocysts aid in maintaining contact between prey and predator. (After Jahn.)*

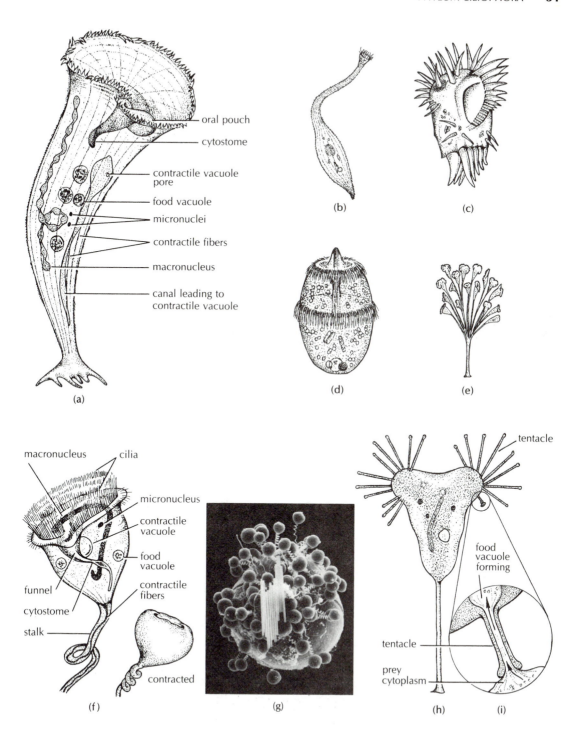

(a)

oral pouch

cytostome

contractile vacuole pore

food vacuole

micronuclei

contractile fibers

macronucleus

canal leading to contractile vacuole

(b)

(c)

(d)

(e)

macronucleus   cilia

micronucleus

contractile vacuole

food vacuole

contractile fibers

funnel

cytostome

stalk

contracted

(f)

(g)

tentacle

food vacuole forming

tentacle

prey cytoplasm

(h)   (i)

One group of usually sessile ciliates, the suctorians, merit special attention in that they, of all free-living ciliates, seem to have diverged farthest from the basic body plan. The body may attach to a substrate directly or by means of a long stalk (Fig. 3.18 h). This stalk is never contractile. Cilia are totally absent in adult suctorians, although the immature stages are ciliated and free-swimming. The systems of kinetodesmata of the immature dispersive stage are retained even after external ciliation is lost during development. Cytostomes are never found, even in the immature stage. Besides the infraciliature, the only clear indication that adult suctorians are indeed ciliates is the presence of both macro- and micronuclei. Unlike other ciliates, asexual reproduction occurs not by binary fission, but instead by simple budding off of small parts of the adult. These buds differentiate into ciliated swimming forms, which disperse, attach elsewhere, and differentiate to adulthood.

Since suctorians lack both cytostomes and cilia, food collection and ingestion must clearly take place by methods atypical of ciliates in general. Food particles are captured by tentacles, which stud the general body surface (Fig. 3.18h). Some of these tentacles are hollow and bear knobs at their ends. Numerous membrane-bound organelles called **haptocysts** stud these knobs. When an appropriate prey organism contacts the tentacles, it triggers discharge of the haptocysts. The haptocysts then penetrate the body surface of the prey, presumably by enzymatic secretion, and help maintain contact between prey and suctorian tentacle. Prey cytoplasm is then drawn through the lumen of the hollow tentacles into the suctorian body, by an unknown mechanism (Fig 3.18i). As the name of the group implies, these are ingenious little suckers indeed.

PHYLUM SARCOMASTIGOPHORA

This phylum contains the majority of protozoan species, well in excess of 18,000 species. These species are found free-living in marine and freshwater environments, in soil, and as parasites of invertebrates, vertebrates, and plants. The members of this group have in common the absence of cilia, the absence of infraciliature, and the absence of conjugation; where sexual reproduction is known to occur, gamete formation is generally involved. In addition, most sarcomastigophorans are homokaryotic, containing a single type of nucleus. Some species are multinucleate, but the nuclei are always monomorphic, in contrast to the distinctly different macronuclei and micronuclei of the ciliates. In general, members of this great phylum are considered to be less highly organized than are members of the Ciliophora, particularly with regard to locomotory systems and feeding structures. A distinct, permanent cytostome is found in some species, but it never attains the complexity of the ciliate cytostome. Most species lack a per-

Phylum Sarco•mastigo•phora
(G: FLESH WHIP BEARING)

manent mouth opening, and some never ingest particulate food at all. Similarly, a pellicle is found in some sarcomastigophorans, but it is never as complex as that of ciliates. Contractile vacuoles, when present, lack associated canal systems.

The diversity of form and function found within the Sarcomastigophora vastly exceeds that found within the Ciliophora, particularly with respect to body form, nutritional mode, and mode of locomotion. Shapes range from completely amorphous and everchanging, to highly structured, with elaborate, rigid skeletal supports. Diversity of nutritional mode is equally remarkable. Several species are plant-like: they contain chlorophyll, obtain energy from light, and fix carbon dioxide as their primary carbon source. Other species feed on small particulate matter exclusively, while still others subsist on dissolved organic matter. Some species are capable of switching nutritional modes as environmental conditions change. Locomotion in some species is accomplished by means of elongated, cilia-like organelles called **flagella**. However, other species locomote by means of relatively amorphous blobs of cytoplasm called **pseudopodia**.

Based on the dramatic differences in locomotory mechanisms encountered within this phylum, the Sarcomastigophora is subdivided into two distinct subphyla: the Sarcodina and the Mastigophora. These two subphyla are linked by numerous transitional forms that show characteristics of both groups, either simultaneously or at different stages in the life cycle. Similarly, the Mastigophora can be subdivided further into phytomastigophorans (*phyto* = G: plant) and

zoomastigophorans (*zoo* = G: animal), based upon differences in nutritional mode. Transitional forms betray a close evolutionary relationship between the phyto- and zoomastigophorans. Indeed, mastigophoran-like organisms were very likely the ancestors of both multicellular plants and multicellular animals.

Subphylum Sarcodina

Members of this group lack a pellicle, and, at least as adults, lack flagella. Flagellated stages appear during the life cycles of some species, however. Internal organelles (e.g., contractile vacuoles in freshwater species) are not fixed in position within the cell, as they are in ciliates, but rather move freely within the cytoplasm. As in the Ciliophora, specialized sensory organelles are absent. The typical sarcodinid is characterized above all by cytoplasmic extensions of the body called **pseudopodia**, or pseudopods (*pseudo* = G: false; *pod* = G: a foot).

Pseudopodia take several forms, as illustrated in Fig. 3.19. **Lobopodia** are typical of the common amoeba. These pseudopodia are broad with rounded tips, like fingers (*lobo* = G: a lobe), and bear a distinctly clear ectoplasmic area, called the **hyaline cap**, near each tip. In contrast, all other forms of pseudopod lack a hyaline cap and are very slender. **Filopodia** (*filo* = L: a thread) often branch, while **reticulopodia** always branch extensively, forming dense pseudopodial networks (*reticul* = L: a network). Species with any of these three types of root-like

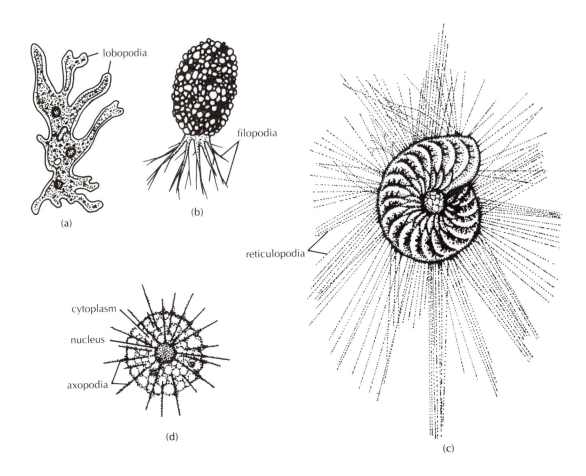

FIGURE 3.19.   *(a) A naked sarcodinid,* Amoeba proteus, *with lobopodia. (From Hyman, 1940.* The Invertebrates, Vol. I. McGraw-Hill. *Reproduced by permission.) (b)* Pseudodifflugia *sp., with filopodia. (After Pennak.) (c) A foraminiferan,* Polystomella strigillata, *with reticulopodia. (After Kingsley; after Whiteley.) (d)* Actinophrys sol, *with axopodia. (After Beck and Braithwaite.)*

pseudopodia are placed within the Superclass Rhizopodea (*Rhizo* = G: root; *pod* = G: a foot). The remaining sarcodinids possess pseudopodia borne on thin, radiating, skeletal support filaments. These species are placed within a separate superclass, the Actinopodea (*actino* = G: a ray), and their pseudopodia are called **axopodia** (*axo* = G: an axle). This last type of pseudopodium is typically stiffened by the presence of microtubules, which form a rather complex inner core.

SUPERCLASS RHIZOPODEA

Many rhizopod species use their pseudo-podia, especially lobopodia and filopodia, to move.[6] Typically, the organism moves by flowing into the advancing pseudopodia, a process called **cytoplasmic streaming**. The mechanism by which this is accomplished is not certain, although it seems clear that movement involves a controlled transition of cytoplasm between the gelatinous, ecto-plasmic form (**gel**) and the more fluid, endo-plasmic form (**sol**). The factors coordinating this transformation in different parts of the body are not known. The transformation it-self may reflect selective polymerization and depolymerization of actin microfilaments, which are abundant in the cytoplasm. My-osin is also present in the cytoplasm, so that a model of sliding actin filaments has also been proposed for the sol-gel transitions. In this model, the interaction between actin and myosin would resemble that observed in the muscle tissue of multicellular animals. Pseudopodial locomotion is extremely slow, usually less than five micrometers per sec-ond.

In **naked** species, i.e., those in which indi-viduals are surrounded only by a plasma membrane, pseudopodia can generally form at practically any point on the body surface. In such species the body is truly formless, lacking permanent anterior, posterior, or lat-eral surfaces (Fig. 3.20). In many other spe-cies, the body is at least partially surrounded by a protective covering, called a **test** or

FIGURE 3.20. Chaos *sp.*, *a naked sarcodinid that can attain lengths of up to 5 millimeters. Note the numerous pseudopodia extending in various directions about the periphery of the organism. Also note the ingested* Paramecium, *to the left of the contractile vacuole. (Photograph by Carolina Biological Supply Company.)*

shell. The test may be a secretion product of the cell itself or may consist of a structured agglutination of fine sand particles or detri-tus. In such cases, the body is given form, and the pseudopodia project through one or more openings in the test (Fig. 3.19).

Pseudopodia are often used to capture food as well as for locomotion; in many spe-cies, pseudopodia are used exclusively for food capture.[7] All free-living sarcodinids are particle feeders (i.e., holozoic). Commonly, ingestion involves **phagocytosis** (*phago* = G: to eat). In this process, lobopodia or filopodia advance on both sides of, and often on top of, the intended prey, forming a **food cup**.

---

6 See Topics for Discussion, No. 2.

7 See Topics for Discussion, No. 8.

The tips of these pseudopodia soon come together and fuse, thereby internalizing the prey item. Sarcodinids typically feed on bacteria, other protozoans, and unicellular algae (e.g., diatoms). In **pinocytosis** (*pino* = G: to drink), much smaller pseudopods are formed, capturing extremely small particulates or fluids rich in dissolved organic matter. Reticulopodia function as a sticky, food-catching net, as well as in locomotion, whereas axopodia are used almost exclusively for capturing food particles. Species with axopodia are generally incapable of active locomotion, most species being free-floating in fresh or salt water.

More than 40% of all free-living sarcodinids are foraminiferans (*foramin* = L: an opening; *fera* = L: to bear). These animals are primarily marine, and all species are free-living (i.e., nonparasitic). Foraminiferans secrete multi-chambered tests, largely of calcium carbonate (Fig. 3.21). The individual chambers of these tests are often demarcated from each other by perforated septa. Sticky reticulopodia project from minute openings in the test, and function primarily in food capture. Repeated extension and shortening of the pseudopodia also permits slow crawling over the ocean bottom.

The tests of foraminiferans are well represented in the fossil record. Although tests of present-day species rarely exceed 9 mm in diameter, with most species being less than 1 mm, fossilized tests of up to 150 mm in diameter have been found—quite a remarkable size for a protozoan. The extensive fossilized remains of foraminiferans have taken on considerable economic importance; for example, cement and blackboard chalk are foraminiferan-containing products. In addition, geologists have become quite interested in certain foraminiferan fossils as fairly reliable indicators of likely places to drill for oil.

SUPERCLASS ACTINOPODEA

Radiolarians (Fig. 3.22) are also prominent in the fossil record, owing to the ornate siliceous (silica-containing) skeletons possessed by most species. The complexity and symmetry of this rigid infrastructure make many radiolarians exceedingly beautiful. In contrast to the mostly bottom-dwelling Foraminifera, radiolarians are exclusively planktonic animals, passively carried about by surface water currents in the ocean. The radiolarian body is generally spherical, and is divided into an **intracapsular zone** and an **extracapsular zone** by a perforated, spherical membrane or capsule. Food vacuole formation and digestion occur in the extracapsular region. The nucleus is contained in the intracapsular zone. A high degree of biochemical specialization of the two zones is also characteristic of at least some species.

The pseudopods of radiolarians generally take the form of axopods supported by an inner core of microtubules. The long, thin, glass-like projections of the siliceous endoskeletons of these organisms give a spiny, rayed appearance to the members of many species. Similar animals, called acantharians, have skeletons composed of strontium sulfate.

Another substantial group of sarcodinids with axopodia are the heliozoans (*helio* = G: the sun; i.e., the "sun animals"). (See Fig. 3.22c.) Like the radiolarians and acantharians, heliozoans are primarily floating

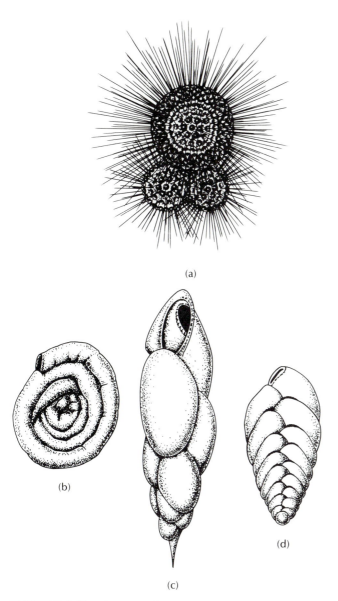

(a)

(b)

(c)

(d)

FIGURE 3.21.   *Foraminiferan tests. Probably 80% of all described foraminiferan species are extinct, and are known only from the fossil record. (a) Globigerina bulloides. (After Kingsley.) New York: Halstead Press; London and Basingstoke: Macmillan. Reprinted by permission. (b) Glomospira gordialis (×140). (c) Stainforthia concava (×200). (d) Brizalina spathulata (×100). [(b,c,d) from Haynes, 1981. Foraminifera.*

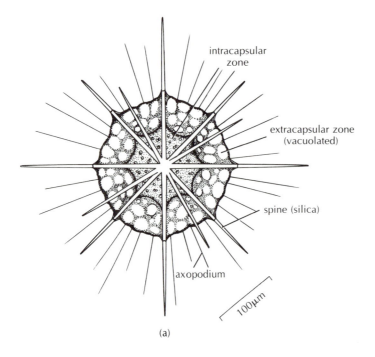

(a)

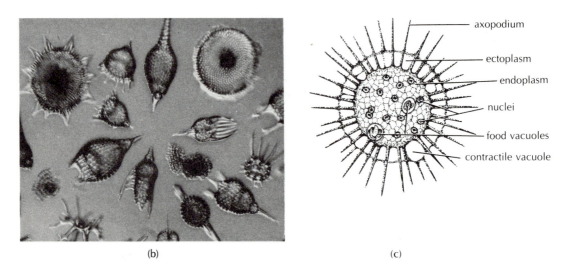

(b)   (c)

FIGURE 3.22. *(a)* A *radiolarian,* Acanthometra *elasticum. (After Farmer.) (b) Radiolarian skeletons, composed of silica. (Courtesy of Eric V. Gravé.) (c)* A *heliozoan,* Actinosphaerium *sp. (After Beck and Braithwaite.)*

animals, but are largely restricted to fresh-water habitats. As in the Radiolaria, helio-zoan bodies are demarcated into a frothy outer region of ectoplasm in which digestion occurs, and a less highly vacuolated inner region of endoplasm containing the nucleus. However, a distinct physical boundary (i.e., a capsular membrane) between the two regions is absent among heliozoans. The axopodia serve primarily to capture food items, although they function in locomotion in some species, too. In such cases, a coordinated retraction and re-extension of axopods permits individuals to roll slowly over a surface. Microtubules located in the cores of the axopods are apparently involved in mediating the changes in axopod length.

Other Sarcodinid Characteristics

The naked sarcodinids described to date have come primarily from fresh water and moist soil. Naked sarcodinids, as yet undescribed, are also found in the ocean. A small percentage of known species (about 2%) are parasites of both vertebrates and invertebrates, including other protozoans. Amoebic dysentery is probably the best known human ailment caused by a member of this group. Some sarcodinid species are parasites within the bodies of other parasites; i.e., they are **hyperparasites**.

All sarcodinid species reproduce asexually, chiefly by binary fission or multiple fission. In addition, sexual reproduction has been described for many species. In some cases, a single individual essentially becomes a gamete, which then fuses with another individual. In other cases, flagellated gametes are formed, pairs of which fuse to form zygotes. Encystment is common among the Sarcodina, especially in freshwater and parasitic species, providing a means of withstanding unfavorable environmental circumstances.

The presence of flagellated gametes in the amoeboid life cycle has intriguing phylogenetic implications. Before leaving the Subphylum Sarcodina for the true flagellates (Subphylum Mastigophora), we should briefly consider several of the sarcodinids that further bridge the gap between the two subphyla. Many examples could be given. One of these transitional forms is *Naegleria gruberi*. Under normal conditions in the laboratory, the animal is a completely convincing, amorphous, crawling amoeba. Lobopodia form anywhere on the cell surface for locomotion and phagocytosis, and the single contractile vacuole has no fixed position within the body. However, a variety of stimuli, including salinity changes or simply maintaining the animals in suspension, quickly induce a complete transformation of the amoebae into equally convincing flagellates.[8] A pellicle forms, the contractile vacuole becomes fixed in one position, and functional flagella appear. The entire transformation takes place within about one and one-half hours, and has been the subject of considerable study by developmental biologists (Fig. 3.23).

A second transitional form consists of a major group of organisms, the Mycetozoa ("fungus animals"; *myceto* = G: a fungus) or "slime molds." For reasons that will soon become apparent, the slime molds are often classified as plants or as fungi. Nevertheless,

---

8 See Topics for Discussion, No. 6.

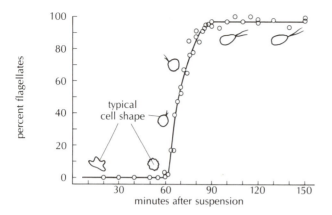

FIGURE 3.23. *The transformation of* Naegleria gruberi *from an amoeboid form to a flagellated form. The transformation was initiated at 25°C by moving the amoebae from the surfaces on which they had been living, into suspension in a test tube. The total number of cells in the culture did not change appreciably during the experiment. (From Fulton and Dingle, 1967. Devel. Biol., 15: 165.)*

for much of their lives, the slime molds exist as individual amoeboid cells, moving and feeding on particulate matter by means of lobopodia. Most species are harmless and are commonly found among decaying vegetation in terrestrial habitats. Some species are plant pathogens of considerable commercial importance. In all cases, the amoeboid individuals eventually aggregate. Among the true slime molds, the individual cell membranes then fuse together, forming a large multinuclear mass—a giant, multinucleate amoeba called a **plasmodium** (Fig. 3.24).

The plasmodium contains functional contractile vacuoles, and ingests particulate food in standard amoeboid fashion. If two plasmodia of the same species come into contact with each other, cell fusion may occur, cre-

ating an even larger plasmodium that continues to move and feed. A decrease in nutrient availability causes the plasmodium to develop fungus-like **sporangia**, or fruiting bodies. Highly resistant spores are released from these sporangia, and only appropriate environmental conditions will bring about germination of these spores. An amoeboid form or a flagellated form may emerge from a spore, depending upon the species. In some cases, the amoeboid individual subsequently transforms into a flagellated individual, which proceeds to eat and undergo repeated binary fission. Eventually, the flagellates give rise to amoebae, which ultimately aggregate to form a multinucleate plasmodium that then produces another round of resistant spores.

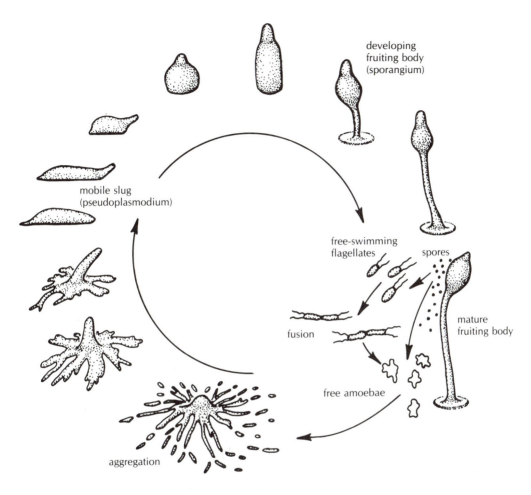

developing
fruiting body
(sporangium)

mobile slug
(pseudoplasmodium)

free-swimming
flagellates

spores

mature
fruiting body

fusion

free amoebae

aggregation

FIGURE 3.24. *The life cycle of a slime mold.* Dictyostelium *is perhaps one of the best-studied slime molds. Note that the spores may give rise to swimming flagellates, which eventually fuse and form amoebae, or may germinate amoebae directly, depending upon the species. (Adapted from* Biological Science, *Third Edition, by William T. Keeton, illustrated by Paula De Santo Bensadoun. Copyright © 1980, 1979, 1978, 1972, 1967 by W. W. Norton & Company, Inc. By permission of W. W. Norton & Company, Inc.)*

As you will see in the following section, flagellates are quite unlike sarcodinids, both in structure and function. Clearly, though, the dramatic morphological differences be- tween adult flagellates and adult amoebae are misleading; the evolutionary re- lationships between these two groups must be very close.

Subphylum Mastigophora (= Flagellata)

Mastigophorans, also called flagellates (depending upon whether your classical affinities are Greek or Latin, respectively), are characterized by the possession of a pellicle, giving the body a definite shape, and, most especially, by the possession of one or more flagella. Most species are free-living and motile. A cytostome is present in some species, but its morphology is never as complex as that encountered within the Ciliophora. Contractile vacuoles may be present, particularly in freshwater species; if present, their position is fixed within the cell cytoplasm, in contrast to the contractile vacuoles of sarcodinid species.

Flagella are structurally much like cilia. In cross section, flagella exhibit a characteristic arrangement of nine pairs of microtubules ringing a pair of central microtubules over most of their length, as in cilia. Like cilia, flagella are produced from basal bodies, and the movement of flagella is believed to involve the sliding of microtubules in relation to each other. The cross-sectional appearance of a flagellum in the region of its basal body is modified as previously described for cilia. Unlike cilia, however, flagella often bear numerous external hair-like projections (mastigonemes) along the length of the organelle. Presumably, these mastigonemes increase the effective surface area of the flagellum, thus increasing the power that it can generate when it is moved through the water. Also, flagella are usually longer than cilia, and the typical flagellate bears many fewer flagella than the typical ciliate has cilia. Unlike cilia, several waves of movement may be in progress simultaneously along a single flagellum and the wave may be initiated at the tip of the flagellum rather than at the base, pulling the organism forward. Flagellar locomotion can be fairly rapid, up to about 200 micrometers per second. This is only about one-tenth the speed attained by many ciliates, but is about forty times the speed attained by the fastest amoebae.

Subclass Phytomastigophora

Some mastigophorans contain chlorophyll, obtain their energy directly from sunlight, and rely exclusively on carbon dioxide as a carbon source. These species are said to be **photolithotrophic** (*photo* = G: light; *litho* = G: a stone [i.e., an inorganic substance]; *tropho* = G: food). Other photosynthesizing species use light as an energy source but require various dissolved organic compounds as well. These species are categorized as **photoorganotrophic**. In both cases, the organisms lack mouths and never form food vacuoles. Together, these two groups comprise the plant-like flagellates: the **phytomastigophorans** (Fig. 3.25). Both freshwater

---

Subphylum Mastigo•phora (= Flagellata)
(G: WHIP BEARING) (L: WHIP)

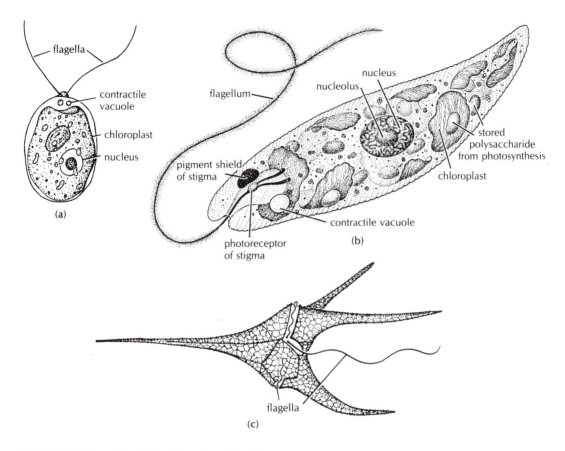

FIGURE 3.25. *Flagellate diversity: Phytoflagellates.*
(*a*) Chlamydomonas. (*After Pennak.*) (*b*) Euglena. (*From Purves
and Orians,* Life: The Science of Biology. *Sinauer
Associates/Willard Grant Press.*) (*c*) Ceratium hirundinella, *a
dinoflagellate. (After Pennak.)*

forms, such as *Euglena*, and marine forms,
including the dinoflagellates, are common.
Dinoflagellates are best known for commonly exhibiting **bioluminescence** (biochemical production of light), and for occasionally producing highly toxic "red tides."

Many phytoflagellates bear a red, cup-

shaped, photosensitive organelle called a
stigma. This is one of the most specific sensory organelles known among protozoans.
Both flagella and stigma are clearly adaptive
for phytomastigophorans to the extent that
they help individuals maintain themselves in
the narrow region of the water column where

sufficient light is available for net photosynthesis to occur.

Although some phytoflagellates are completely **autotrophic** (i.e., self-nourishing through photosynthesis), many species are capable of feeding on particulate foods if necessary. Some euglenids, for example, become holozoic if maintained in darkness for a sufficient time. These individuals then ingest solid food and form food vacuoles in perfectly good animal fashion.

Subclass Zoomastigophora

Most strictly animal-like flagellate species (i.e., the **zoomastigophorans**, (Fig. 3.26) are free-living in either fresh or salt water. One of the most interesting groups of free-living species are the choanoflagellates, a group found primarily in fresh water. Many choanoflagellate species are sessile, being permanently attached to a substrate (Fig. 3.26a–c). Each individual bears a single flagellum, which extends for part of its length through a cylindrical network (collar) of closely spaced protoplasmic strands. Flagellar movements create feeding currents; small food particles stick to the protoplasmic collar and are ingested. Commonly, individuals are stalked and/or embedded in a gelatinous secretion. Most species are colonial. Members of the genus *Proterospongia* form colonies of up to several hundred cells; these colonies may bear a striking resemblance to primitive sponges, as discussed in the next chapter. Whether this similarity reflects a true phylogenetic relationship between the unicellular flagellates and the multicellular sponges, or whether the similarities are a product of independent, convergent evolution, is not certain.

Another intriguing group of free-living flagellates is the Rhizomastigida (Fig. 3.26 d). The members of this group possess both locomotory flagella and pseudopodia, either simultaneously or in succession. These species are clearly at the crossroads of flagellate–amoeba organization, and provide further evidence of a close evolutionary relationship between the two forms. Rhizomastigophorans are most commonly encountered in freshwater environments, but marine and parasitic species are also known.

About 25% of zooflagellate species are parasitic or commensal with plants, invertebrates, and vertebrates, including humans. Species in these groups often exhibit levels of structural and functional complexity not observed in any other sarcomastigophoran. Life cycles are especially complex, including adaptations for perpetuating the species from host to host. **Intermediate hosts** commonly serve as **vectors**, which serve to transport developing individuals from one **definitive host** to another; sexual maturity of the protozoan parasite can be attained only in the definitive host (Fig. 3.27). The problems implied in maintaining a parasitic life-style—such as locating or attracting proper intermediate and definitive hosts, adapting to the physiological requirements of different hosts, and evading host immune responses—are common to parasites of all animal phyla; additional discussion of these issues is found in Chapter 7.

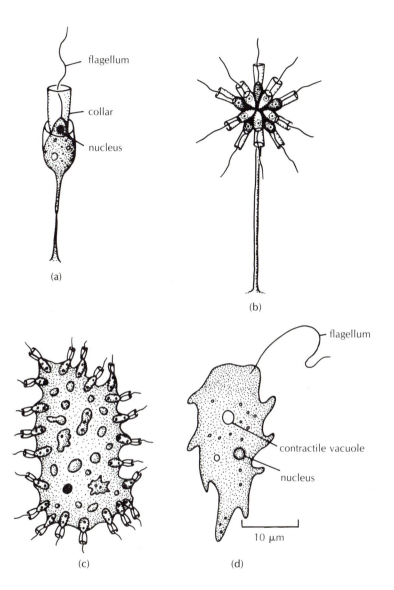

flagellum

collar

nucleus

(a)

(b)

flagellum

contractile vacuole

nucleus

10 μm

(c)   (d)

FIGURE 3.26. *Flagellate diversity: Zooflagellates.*
(a–c) Choanoflagellates: (a) Codosiga *sp. (From Hyman; after*
Lapage.) (b) Codosiga botrytis, *a colonial species. (After Kingsley.)*
(c) Proterospongia, *another colonial species, with the individuals*
embedded in a thick, gelatinous matrix. (From Hyman; after Kent.)
(d) A rhizomastigophoran, Mastigamoeba longifilum. *(After*
Farmer.)

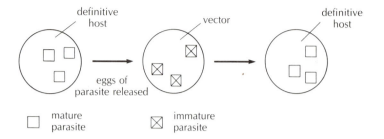

definitive host

vector

definitive host

eggs of parasite released

☐ mature parasite

⊠ immature parasite

FIGURE 3.27 *Diagram of a typical parasitic life cycle, involving an invertebrate vector in the transfer of the parasite from one definitive host to another. The parasite cannot attain sexual maturity within the body of the vector.*

Trypanosomes are parasites of especial interest to humans (Fig. 3.28). Various species are pathogens in flowering plants and in cattle, sheep, goats, horses, and other domesticated animals. Trypanosomes, with the tsetse fly as the vector, are responsible for the well-known African sleeping sickness, and for many other human ailments. *Leishmania donovani*, for example, causes extreme disfigurement and death (with a mortality rate approaching 95%) in many areas of the world. Sand flies serve as vectors for the transmission of leishmanial parasites from human to human. Dogs also serve as suitable definitive hosts, functioning, in a sense, as reservoirs for later human infestation. The human defense system recognizes these trypanosomes as foreign beings, and the trypanosomes are quickly phagocytozed by appropriate cells of the human reticuloendothelial system. Remarkably, however, *L. donovani* is not then digested within these cells, but instead succeeds in greatly increasing in numbers through repeated binary fission within the cells. How the host defense system is circumvented in this way is not known. Other trypanosomes are equally successful at evading the humoral immune response, either by changing the chemical composition of their antigenic surface coats frequently enough to prevent the host from producing antibody in effective concentrations, or by quickly invading specific tissues that are reasonably well isolated from circulating antibody.[9]

Assuming that natural selection ultimately promotes associations that do not kill the host (an obvious detriment to continued existence of the parasite population), the most highly evolved associations between zoomastigophorans and their hosts must be nonpathogenic in nature. The trichomonads represent one such group of primarily commensal flagellates. Nonpathogenic species generally attract less research attention (and funding) than do pathogenic species, but some trichomonads wreak enough havoc upon humans to have merited careful study. *Trichomonas vaginalis*, for example, is a typ-

9 See Topics for Discussion, No. 9.

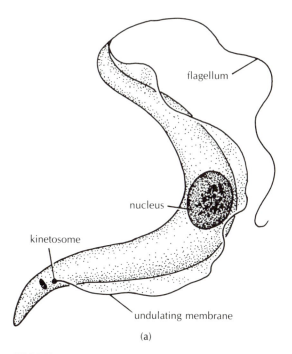

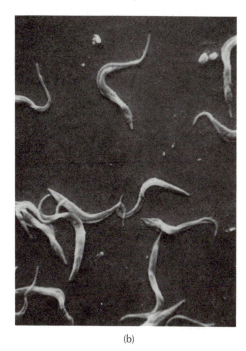

(a)

(b)

FIGURE 3.28. *Flagellate diversity: Trypanosomes.*
(a) Trypanosoma lewisi. *Note the undulating membrane (not to be confused with the ciliate structure of the same name).*
(Reprinted with permission of Macmillan Publishing Company from The Invertebrates: Function and Form, *2nd ed., by Irwin W. Sherman and Vilia G. Sherman. Copyright © 1976 by Irwin W. Sherman and Vilia G. Sherman.) (b) A scanning electron micrograph of* T. cruzi, *the agent of American trypanosoma. Magnified nearly 2000×. (Courtesy of J.P. Kreier, The Ohio State University.)*

ically small protozoan parasitizing the human vagina, prostate, and urethra. *T. vaginalis* generally causes little or no discomfort in men, but often produces considerable inflammation and irritation in women. The disease is readily transmittable, sexually and otherwise (through contact with toilet seats and towels, for example).

Probably the most morphologically complex flagellate species are found among the Hypermastigida, a group of protozoan commensals living in the guts of termites, cockroaches, and wood roaches (Fig. 3.29). A great many individuals can be easily obtained for study by squeezing out some of the fluid from the guts of termites. One species alone (*Trichonympha campanula*) may account for up to one-third of the biomass of an individual termite! The species in this group are generally large (for flagellates), typically attaining body lengths of several hundred micrometers. The bodies are commonly di-

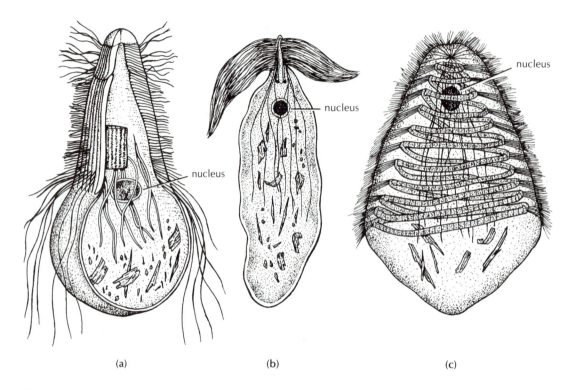

(a)   (b)   (c)

FIGURE 3.29. *Symbiotic flagellates, the most complex of all flagellate species. (a)* Trichonympha collaris *(about 150 μm), taken from the intestine of a termite. (From Hyman; after Kirby.) (b)* Rhynchonympha tarda, *from the gut of a wood roach. (From Hyman; after Cleveland.) (c)* Macrospironumpha xylopletha, *from the gut of a wood roach. (From Hyman; after Cleveland.)*

visible into several morphologically and functionally distinct regions, and typically bear numerous (up to several thousand!) flagella.

Perhaps the most intriguing feature of the biology of these animals is the mutual dependence between host and flagellate. Several species inhabiting termite guts are capable of digesting cellulose, something that most animals, including termites, can-

not do. If the termite is deprived of its flagellates, it soon dies, even though it continues to ingest wood. Similarly, wood-eating roaches soon die if deprived of their flagellate gut fauna. The flagellates, on the other hand, gain protection from desiccation and predation, but must rely upon the host for a source of cellulose to digest; cellulose serves as the primary carbon source for these protozoans.

Subphylum Opalinata

The members of this class are exclusively parasitic, or commensal, within the guts or recta of frogs, toads, and a few fishes. Their precise taxonomic position has been, and continues to be, uncertain. The bodies of opalinids bear numerous rows of cilia (with an associated infraciliature). However, opalinids lack a mouth, lack dimorphic nuclei, and exhibit sexual reproduction by gamete formation rather than by conjugation. On balance, then, they seem to be more flagellate than ciliate in nature . Once again, we see the difficulties of attempting to wedge living organisms into necessarily artificial schemes of categorization.

PHYLUM SPOROZOA

The members of this phylum are all endoparasites. The gregarines are parasites of invertebrates, while the coccidians are parasites of both vertebrates and invertebrates. Many species are blood-cell parasites of humans. Most conspicuous among these are members of the genus *Plasmodium*, the agent of malaria (not to be confused with the plasmodium stage of slime molds, page 60). Malaria has been called the most important disease in the world today; probably more than 1.5 billion people living in tropical and semi-tropical areas are at risk, and perhaps one-hundred million people die of malaria yearly. Mosquitos serve as vectors of the disease, so that the control and eradication of malaria requires effective manipulation of the mosquito population.

In typical parasite fashion, sporozoan life cycles are complex, including sexual reproduction by means of flagellated gametes; asexual reproduction through fission; and the production of highly resistant spores, which remain dormant until taken up by an appropriate host. The adults are highly specialized for existence as endoparasites, and completely lack locomotory organelles.

PHYLUM CNIDOSPORA

The members of this group are all endoparasites and superficially resemble the sporozoans in some respects, including life cycle and morphology. As in the sporozoans, the body lacks locomotory organelles, and the production of dispersive, resistant spores is a prominent feature of the life cycle.

However, members of the Cnidospora possess a distinctive organelle, the **polar filament**, which is encountered within no other group of animals. Polar filaments are commonly wound tightly within **polar capsules**, which, in turn, occupy much of the space within the spore stage of the life cycle (Fig. 3.30). If the spore is ingested by (or in some other way comes into contact with) a suitable host, the polar filaments discharge from their capsules and apparently serve to anchor the spore to the host tissues while further events take place. An amoeboid **sporoplasm** soon exits from the attached spore and invades the host tissues. The sporoplasm replicates itself many times by asexual and often complex means, so that the location of a suitable host by a single spore may result in the eventual production of thousands of new spores within that host.

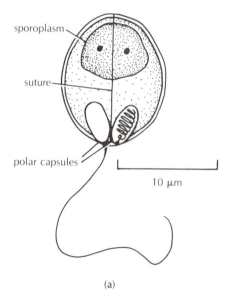

sporoplasm

suture

polar capsules

10 μm

(a)

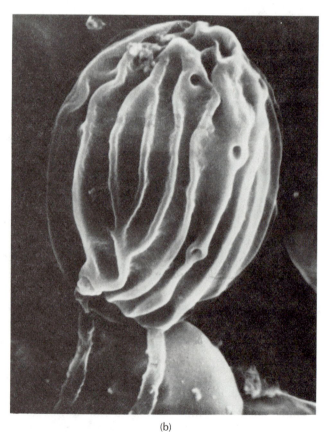

(b)

FIGURE 3.30. *(a) Spore with two polar capsules, one with an extruded filament. Infective sporozooites emerge from the spore following rupture along the suture. (From Farmer, John N., 1980. The Protozoa. St. Louis, MO: The C.V. Mosby Co.) (b) Scanning electron micrograph of a spore of* Chloromyxum catostomi. *(Courtesy of J.K. Listebarger, from Listebarger and Mitchell, 1980. J. Protozool., 27: 155.)*

T A X O N O M I C   S U M M A R Y

KINGDOM PROTISTA
 SUBKINGDOM PROTOZOA

  PHYLUM CILIOPHORA—THE CILIATES

  PHYLUM SARCOMASTIGOPHORA
   SUBPHYLUM SARCODINA
    SUPERCLASS RHIZOPODEA—AMOEBAE,
      FORAMINIFERANS
    SUPERCLASS ACTINOPODEA—RADIOLARIANS,
      ACANTHARIANS, HELIOZOANS
   SUBPHYLUM MASTIGOPHORA—THE
     FLAGELLATES
     SUBCLASS PHYTOMASTIGOPHORA
     SUBCLASS ZOOMASTIGOPHORA
   SUBPHYLUM OPALINATA

  PHYLUM SPOROZOA

  PHYLUM CNIDOSPORA

T O P I C S   F O R   F U R T H E R   D I S C U S S I O N   A N D   I N V E S T I G A T I O N

1. In the present chapter, two phyla of purely parasitic protozoans were described: the Sporozoa and the Cnidospora. Recent ultrastructural studies have revealed a high degree of morphological heterogeneity among the members of these two groups; species that were considered to be sporozoans and cnidosporans are now distributed among five phyla rather than only two. The agent of malaria (species in the genus *Plasmodium*) is now contained in the new phylum Apicomplexa. Discuss the life history of *Plasmodium*.

See any recent parasitology textbook, such as Noble, E.R., and G.A. Noble, 1982. *Parasitology*, 5th edition. Philadelphia: Lea and Febiger.

2. The locomotion of sarcodinids has received substantial attention in recent years. Discuss the locomotory mechanism of animals from either group A or group D, following, or compare the locomotion of group B or C animals with that of animals in either group A or D.

Group A: Heliozoans

Edds, K.T., 1975. Motility in *Echinosphaerium nucleofilum*. I. An analysis of particle motions in the axopodia and a direct test of the involvement of the axoneme. *J. Cell Biol.*, 66: 145.

Edds, K.T., 1975. Motility in *Echinosphaerium nucleofilum*. II. Cytoplasmic contractility and its molecular basis. *J. Cell Biol.* 66: 156.

Watters, C., 1968. Studies on the motility of the heliozoa. I. The locomotion of *Actinosphaerium eichhorni* and *Actinophrys* sp. *J. Cell Sci.*, 3: 231.

Group B: Test-bearing sarcodinids

Eckert, B.S., and S.M. McGee-Russell, 1973. The patterned organization of thick and thin microfilaments in the contracting pseudopod of *Difflugia*. *J. Cell Sci.*, 13: 727.

Group C: Foraminiferans

Travis, J.L., J.F.X. Kenealy, and R.D. Allen, 1983. Studies on the motility of the Foraminifera. II. The dynamic microtubular cytoskeleton of the reticulopodial network of *Allogromia laticollaris*. *J. Cell Biol.*, 97: 1668.

Group D: Amoebae

Allen, R.D., D. Francis, and R. Zeh, 1971. Direct test of the positive pressure gradient theory of pseudopod extension and retraction in amoebae. *Science*, 174: 1237.

Cullen, K.J., and R.D. Allen, 1980. A laser microbeam study of amoeboid movement. *Exp. Cell Res.*, 128: 353.

Pollard, T.D., and S. Ito, 1970. Cytoplasmic filaments of *Amoeba proteus*. I. The role of filaments in consistency changes and movement. *J. Cell Biol.*, 46: 267.

Taylor, D.L., J.S. Condeelis, P.L. Moore, and R.D. Allen, 1973. The contractile basis of amoeboid movement. I. The chemical control of motility in isolated cytoplasm. *J. Cell Biol.*, 59: 378.

3. Discuss the coordination of ciliary beating in free-living ciliates. What is the evidence that the infraciliature is not involved in this coordination?

Naitoh, Y., and R. Eckert, 1969. Ciliary orientation: controlled by cell membrane or by intracellular fibrils? *Science*, 166: 1633.

Tamm, S.L., 1972. Ciliary motion in *Paramecium*: a scanning electron microscopic study. *J. Cell Biol.*, 55: 250.

4. Discuss the present limits to our understanding of how the contractile vacuole fills and empties.

Ahmad, M., 1979. The contractile vacuole of *Amoeba proteus*. III. Effects of inhibitors. *Canadian J. Zool.*, 57: 2083.

Organ, A.E., E.C. Bovee, and T.L. Jahn, 1972. The mechanisms of the water expulsion vesicle of the ciliate *Tetrahymena pyriformis*. *J. Cell Biol.*, 55: 644.

Patterson, D.J., and M.A. Sleigh, 1976. Behavior of the contractile vacuole of *Tetrahymena pyriformis* W: a redescription with comments on the terminology. *J. Protozool.*, 23: 410.

Riddick, D.H., 1968. Contractile vacuole in the amoeba, *Pelomyxa carolinensis*. *Amer. J. Physiol.*, 215: 736.

Wigg, D., E.C. Bovee, and T.L. Jahn, 1967. The evacuation mechanism of the water expulsion vesicle ('contractile vacuole') of *Amoeba proteus*. *J. Protozool.*, 14: 104.

5. What is the evidence that the contractile vacuole plays a role in osmotic regulation?

Hampton, J.R., and J.R.L. Schwartz, 1976. Contractile vacuole function in *Pseudocohnilembus persalinus*: responses to variation in ion and total solute concentration. *Comp. Biochem. Physiol.*, 55A: 1.

Schmidt-Nielson, B., and C.R. Schrauger, 1963. *Amoeba proteus*: studying the contractile vacuole by micropuncture. *Science*, 139: 606.

Stoner, L.C., and P.B. Dunham, 1970. Regulation of cellular osmolarity and volume in *Tetrahymena*. *J. Exp. Biol.*, 53: 391.

6. Several protozoans can alter their morphology considerably, quickly transforming from a convincing member of one taxon to an equally convincing member of another. Describe the morphological changes involved and discuss the environmental factors that initiate these dramatic transformations.

Fulton, C., and A.D. Dingle, 1967. Appearance of the flagellate phenotype in populations of *Naegleria* amebae. *Devel. Biol.*, 15: 165.

Nelson, E.M., 1978. Transformation in *Tetrahymena thermophila*. Development of an inducible phenotype. *Devel. Biol.*, 66: 17.

Willmer, E.N., 1956. Factors which influence the acquisition of flagella by the amoeba, *Naegleria gruberi*. *J. Exp. Biol.*, 33: 583.

7. Discuss the adaptive significance of cyst formation by freeliving protozoans.

Corliss, J.O., and S.C. Esser, 1974. Comments on the role of the cyst in the life cycle and survival of free-living Protozoa. *Trans. Amer. Microsc. Soc.*, 93: 579.

Maguire, B., Jr., 1963. The exclusion of *Colpoda* (Ciliata) from superficially favorable habitats. *Ecology*, 44: 781.

8. Discuss feeding and digestion in naked sarcodinids.

Bowers, B., and T.E. Olszewski, 1983. *Acanthamoeba* discriminates internally between digestible and indigestible particles. *J. Cell Biol.*, 97: 317.

Jeon, K.W., and M.S. Jeon, 1983. Generation of mechanical forces in phagocytosing amoebae: light and electron microscopic study. *J. Protozool.*, 30: 536.

Mast, S.O., and F.M. Root, 1916. Observations on ameba feeding on rotifers, nematodes and ciliates, and their bearing on the surface tension theory. *J. Exp. Zool.*, 21: 33.

Salt, G.W., 1968. The feeding of *Amoeba proteus* on *Paramecium aurelia*. *J. Protozool.*, 15: 275.

9. How do trypanosomes evade the immune responses of their mammalian hosts?

Turner, M., 1980. How trypanosomes change coats. *Nature*, 284:13.

Vickerman, K., 1978. Antigenic variation in trypanosomes. *Nature*, 273:613.

4

THE PORIFERANS

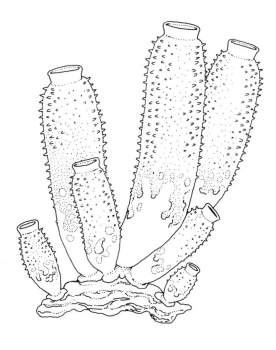

> **Phylum Pori • fera**
> (L: PORE BEARING)

INTRODUCTION

Sponges constitute the simplest of multicellular animals (metazoans). Less than 3% of all sponge species are found in fresh water, the remaining 97% of the species being marine. There are no terrestrial sponges. The approximately 5000 or more species comprising this group of animals are perhaps more remarkable for the characteristics they lack than for those they possess. Unlike any other metazoan, sponges lack a nervous system, and have no true musculature. Thus, locomotion is generally beyond them. No specialized reproductive, digestive, respiratory, sensory, or excretory organs are found in this group, either; indeed, no organs are found at all. Often, sponges are amorphous, asymmetric creatures, although there are some very beautiful exceptions to this generalization. No sponge has anything corresponding to "anterior," "posterior," or "oral" surfaces. Moreover, only a few different types of cells are encountered within any given individual. These cells are functionally independent to the extent that an entire sponge can be dissociated into its constituent cells in the laboratory, with no long-term impact. The cells dedifferentiate to an amoeboid form, reaggregate, and redifferentiate to reform the sponge.[1] Probably the most remarkable aspect of sponge biology is that they work so well with so little.

Considering the lack of organs and systems in sponges, and the lack of direct intercellular communication, one can wonder whether sponges are truly multicellular animals or whether they represent a highly evolved form of colonial living by individual cells. In fact, some freshwater species of colonial protozoans (the choanoflagellates) bear very definite morphological similarities to the simplest sponges, and, as recently as one hundred years ago, it was suggested that sponges be classified as colonial protozoans. In the opinion of several currently practicing zoologists, members of the Porifera are sufficiently dissimilar from all other animals to warrant placement in a separate kingdom.

GENERAL CHARACTERISTICS

A sponge is essentially a fairly rigid, perforated bag, whose inner surface is lined with flagellated cells. The empty space of this bag is called the **spongocoel**. The flagellated cells lining the spongocoel are called **choanocytes** (literally, "funnel-cells"), or **collar cells** in recognition of the cylindrical arrangement of

the cytoplasmic extensions ("collars") surrounding the proximal portion of each flagellum (Fig. 4.1). These collar cells:

(1) generate currents that help maintain circulation of sea water within and through the sponge;
(2) capture small food particles;
(3) capture incoming sperm for fertilization.

Adjacent to the choanocyte layer is a gelatinous, nonliving layer of material called the **mesohyl** layer (*meso* = G: middle; *hyl* = G: stuff, matter). Although the mesohyl is acellular and nonliving, it contains live cells. Amorphous, amoeboid cells called **archaeocytes** wander throughout the mesohyl by typical cytoplasmic streaming, which involves the formation of pseudopodia as in amoeboid protozoans. The archaeocytes are responsible for the digestion of food particles captured by the choanocytes, so that digestion is entirely intracellular. Some archaeocytes also store digested food material. In addition, archaeocytes may give rise to both sperm (which are flagellated) and eggs, although gametes also may arise through morphological modification of existing choanocytes. Archaeocytes may also play a role in transporting sperm cells from the choanocytes, which capture them, to the eggs. In at least some species, however, the choanocyte that captures incoming sperm dedifferentiates to amoeboid form and transports the sperm through the mesohyl to the egg directly. Finally, archaeocytes play a role in the elimination of wastes, and, in addition, secrete the supporting elements located in the mesohyl layer. These support elements may be calcareous or siliceous **spicules**, or

---

1 See Topics for Discussion, No. 6.

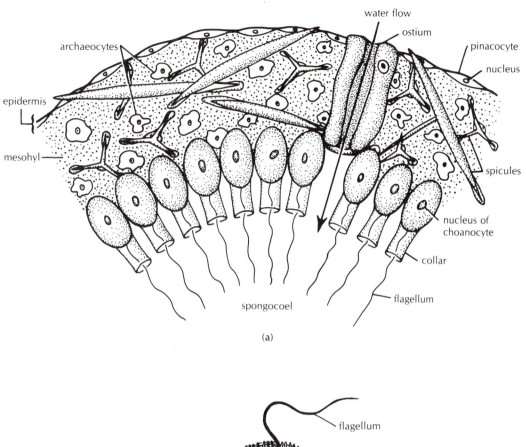

(a)

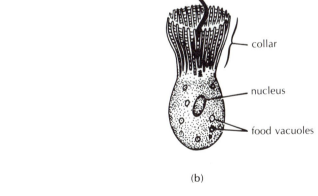

(b)

FIGURE 4.1. *(a) Diagrammatic illustration of the body wall of a sponge. (Modified from Hyman, 1940.* The Invertebrates, Vol. I. *McGraw-Hill. Reproduced by permission.) (b) Detail of a choanocyte. (After Rasmont.)*

may be fibers composed of a collagenous protein called **spongin**. The cells secreting spicules are termed **sclerocytes**, and those producing spongin fibers are termed **spongocytes**[2] (Fig. 4.2). Both of these cell types are derived from archaeocytes. Clearly, archaeocyte cells are quite versatile. The spicules and fibers secreted by the sclerocytes and spongocytes are of great importance (1) to systematists, as an indispensible factor in species identification, and (2) to sponges, which depend upon the support elements for maintaining shape and, probably, for discouraging predation.

At certain times of year, numbers of archaeocytes form clusters in many freshwater sponge species, and in some marine species as well. These clustered archaeocytes cooperatively secrete a single thick, complex covering; the entire result is called a **gemmule** (Fig. 4.3). Gemmules are far more resistant to desiccation and freezing than are the sponges that produce them. In fact, the gemmules of many species must spend several months at low temperature before the gemmules are capable of hatching, i.e., they require a period of **vernalization**. Under appropriate environmental conditions, the living cells leave the gemmule ("hatch") through a narrow opening, and differentiate to form a functional sponge.[3] Gemmule formation thus appears to be a mechanism through which sponges can withstand unfavorable environmental conditions, by entering a stage of developmental arrest, i.e., a **diapause**. In light of some of the differences

2 See Topics for Discussion, No. 2.
3 See Topics for Discussion, No. 3.

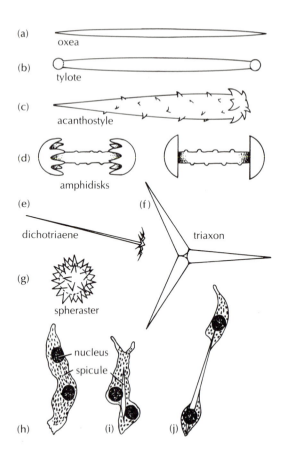

FIGURE 4.2. *(a–g) Representative sponge spicule morphologies. (From Beck and Braithwaite, 1968.* Invertebrate Zoology Laboratory Workbook, *3rd ed. Burgess Publishing Company, Minneapolis, Minnesota.) (h–j) The production of a sponge spicule by a binucleate sclerocyte. The spicule is initiated between the two nuclei. After the spicule is completed, the cells will wander off into the mesoglea. (From Hyman; after Woodland.)*

between freshwater and marine environments mentioned in Chapter One, the adaptive significance of gemmule formation by freshwater sponges should be particularly apparent.

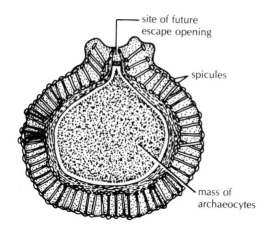

site of future
escape opening

spicules

mass of
archaeocytes

FIGURE 4.3. *The gemmule of a freshwater sponge,* Ephydatia, *as seen in cross section. (From Hyman; after Evans.)*

The outer layer of a sponge is composed of flattened contractile cells called **pinacocytes** (Fig. 4.4). These cells also line the incurrent canals and the spongocoel in places where choanocytes are lacking. Contraction of pinacocytes enables the sponge to undergo minor shape changes, and may also play a role in regulating water flow through the sponge by varying the diameter of the incurrent openings.

Because they lack muscles, nerves, and deformable bodies, sponges are utterly dependent upon water flow for food, gas exchange, dissemination and collection of sperm, and removal of wastes. Partly as a consequence of choanocyte activity and partly as a consequence of sponge architecture,[4] water flows into the spongocoel through narrow openings (**ostia**) and exits

4 See Topics for Discussion, Nos. 4, 5.

the spongocoel through larger openings (**oscula**). Ostia are always numerous on the body of a sponge, but there may be as few as one osculum present per individual.

PORIFERAN DIVERSITY

Advances in sponge morphology imply, over evolutionary time, selection for maximizing current flow through the spongocoel and increasing the amount of surface area available for food collection. There are three basic levels of sponge construction: asconoid, syconoid, and leuconoid, in order of increasing complexity. Each form simply reflects an increased degree of evagination of the choanocyte layer away from the spongocoel, increasing the extent of flagellated surface area enclosed by the sponge (Fig. 4.5). The majority of sponge species are of leuconoid construction.

Sponges are distributed among four classes, based largely upon the chemical composition and morphology of the support elements. Members of the class Calcarea bear spicules composed only of calcium carbonate ($CaCO_3$). Representatives of all three types of construction occur in this class. Indeed, the only living asconoid forms are found among the Calcarea. Members of the largest class (containing more than 4000 species), the Demospongiae, are uniformly of leuconoid construction. The supporting spicules and fibers of the Demospongiae may be composed of spongin and/or silica, but never of $CaCO_3$. All freshwater sponges (150 species) are found in this class. Interestingly, these freshwater species possess contractile vacuoles, which are organelles specialized

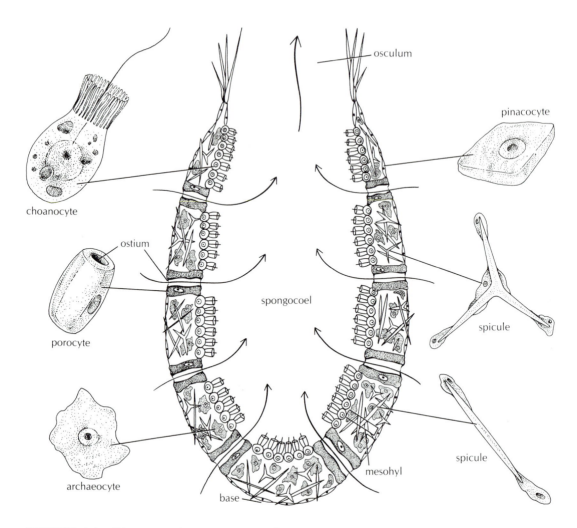

FIGURE 4.4. *Diagrammatic representation of a simple (asconoid) sponge, illustrating the various cellular and structural components of the animal. In asconoids, the incurrent canal is simply a tube passing through a modified pinacocyte, called a porocyte. Note that six cells are involved in producing a triradiate spicule. (Reprinted with permission of Macmillan Publishing Company from* The Invertebrates: Function and Form, *2nd ed., by Irwin W. Sherman and Vilia G. Sherman. Copyright © 1976 by Irwin W. Sherman and Vilia G. Sherman.)*

FIGURE 4.5. *Diagrammatic illustrations of the different levels of complexity of sponge architecture. Arrows indicate the direction of water flow: in at the ostia and out at the osculum (oscula). (a) Asconoid sponge; (b) syconoid sponge; (c) leuconoid sponge. (From Bayer and Owre, 1968.* The Free-Living Lower Invertebrates. *Macmillan.) (d) Detail of water circulation in a leuconoid sponge. (From Johnston and Hildemann, 1982. In Cohen and Sigel (eds.),* The Reticuloendothelial System, *Vol. 3. Plenum Publ. Corp., pp. 37–57.)*

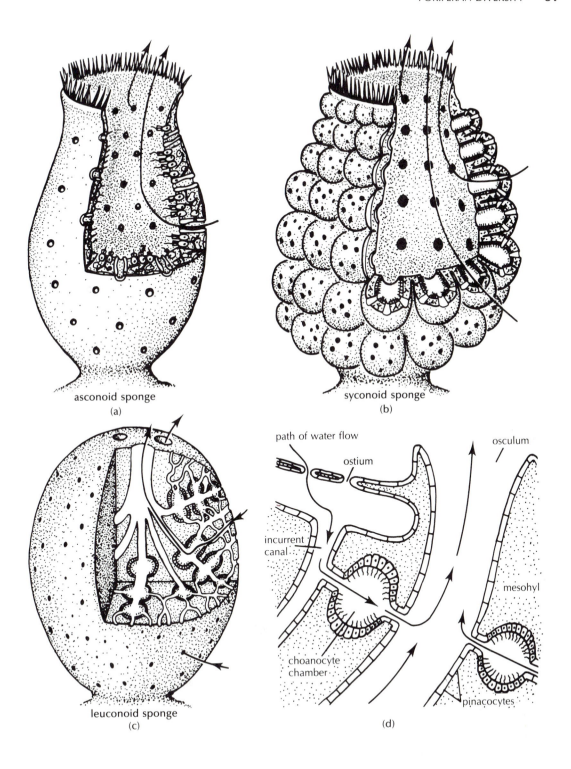

asconoid sponge
(a)

syconoid sponge
(b)

leuconoid sponge
(c)

path of water flow

ostium

incurrent
canal

choanocyte
chamber

osculum

mesohyl

pinacocytes

(d)

for the elimination of water from cytoplasm and are found elsewhere only among the Protozoa. A third class of sponges, the Sclerospongiae (*sclero* = G: hard), contains only a few species. Like members of the Demospongiae, all members of the Sclerospongiae are of leuconoid construction. However, the body is supported by all three of the skeletal materials encountered among sponges: CaCO$_3$, silica, and spongin. This class is known only from coral reefs. Finally, sponges whose bodies are supported entirely by six-rayed siliceous spicules are placed in the class Hexactinellida. These sponges, known as the glass sponges, are marvels of structural complexity and symmetry. Their canal systems may be either syconoid or leuconoid.

Representative sponges are illustrated in Fig. 4.6.

OTHER FEATURES OF PORIFERAN BIOLOGY

Reproduction and Development

Sponges reproduce asexually, through the production of either gemmules or of buds, or sexually, through the production of eggs and sperm. Most sponges are **hermaphroditic**, i.e., a single individual produces both types of gametes. Fertilization is internal, a surprising elaboration for what is generally considered to be a rather unsophisticated animal. Development of the fertilized egg is unlike the development of other metazoan eggs. Typically, upon attaining the 16-cell stage, one group of 8 cells continues to divide rapidly and to become flagellated at the anterior end of each cell. The 8 cells of the second group divide far more slowly, and remain unflagellated. A small internal cavity opens to the outside in the middle of this group of relatively large, slowly dividing cells. In most species, the embryo is said to turn inside out through this opening, so that the flagella of the rapidly dividing cells come to lie on the outer surface of the embryo, where they can propel the animal through the water once the larva is discharged from the parent. It is important to note that the embryo remains only one cell thick after turning inside out. This process, called **inversion**, is therefore not to be confused with the process of gastrulation.

The hollow, swimming larva, ciliated at one end only, is termed an **amphiblastula** (Fig. 4.7). Other sponge species form a solid **parenchymula** larva, with nearly the entire

FIGURE 4.6. *Sponge diversity. (a) A freshwater species encrusting a twig. (After Hyman.) (b) An encrusting marine sponge. (c) Neptune's Goblet sponge, Poterion neptuni. [(b–c) after Pimentel.] (d) A syconoid sponge, Sycon sp. (After MacGinitie and MacGinitie.) (e) A fingered sponge, Microciona sp. (After Pimentel.) (g) Euplectella sp., Venus's Flower Basket. This is a member of the Hexactinellida. The spicules are six-rayed and composed of silica, as shown in (f). [(f) from Barnes, 1980. Invertebrate Zoology, 4th ed. Saunders. (g) After Bayer and Owre.]*

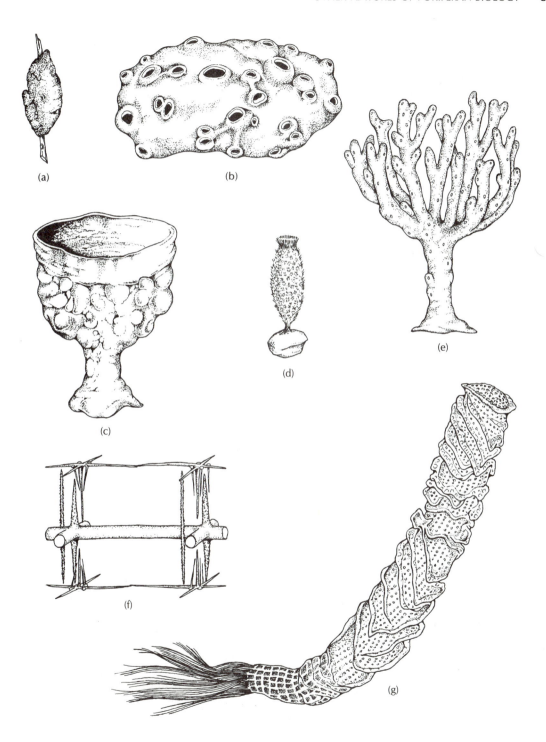

(a)

(b)

(c)

(d)

(e)

(f)

(g)

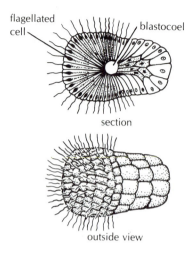

flagellated cell

blastocoel

section

outside view

FIGURE 4.7. *Typical amphiblastula larva. This free-swimming larva will attach to a substrate before undergoing further development. The large number of flagellated cells seen in the larva will eventually differentiate to form choanocytes. (From Hyman; after Hammer.)*

outer body surface covered by flagella. Before losing the ability to swim, sponge larvae attach to a substrate. During the ensuing process of **gastrulation**, in which the embryonic body wall becomes two cells thick, the layer of flagellated cells apparently becomes internalized once again. Cells from various parts of the embryo undergo extensive migrations, and begin to, or continue to, differentiate to form the future adult sponge.

With respect to their mode of reproduction, sponges are unusual in that both marine and freshwater species have a free-swimming larval stage in their development. Such a free-swimming dispersal stage is often suppressed in the life histories of other freshwater invertebrates, as noted in Chapter 1.

PHYLUM PORIFERA—THE SPONGES
  CLASS CALCAREA
  CLASS DEMOSPONGIAE
  CLASS SCLEROSPONGIAE
  CLASS HEXACTINELLIDA

◀ T A X O N O M I C   S U M M A R Y

T O P I C S   F O R   F U R T H E R   D I S C U S S I O N   A N D   I N V E S T I G A T I O N

1. Sponges are quite distinct from most other metazoans in their lack of symmetry and in their lack of true tissue layers and organs. In some respects, the most primitive sponges resemble certain colonial flagellated protozoans. Similarly troublesome animals are contained in a separate phylum, the Placozoa. Only a single placozoan species has been described, *Trichoplax adhaerens* (Fig. 4.8), and it is known only from the laboratory. What features do placozoans share with the

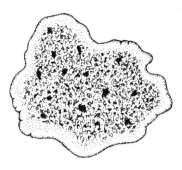

FIGURE 4.8. *The placozoan,* Trichoplax adhaerens. *The individual is approximately 0.5 mm in its longest dimension.*

sponges? What features do placozoans share with members of the Protozoa?

Margulis, L., and K.V. Schwartz, 1982. *Five Kingdoms.* San Francisco: W.H. Freeman Co., pp. 166–167.

Parker, S.P. (Ed.), 1982. Placozoa, in *Synopsis and Classification of Living Organisms.* New York: McGraw-Hill, p. 639.

2. How are sponge spicules secreted by sclerocytes?

Dendy, A., 1926. Origin, growth, and arrangement of sponge spicules. *Q.J. Microsc. Sci.,* 70: 1.

Elvin, D., 1971. Growth rates of the siliceous spicules of the fresh-water sponge *Ephydatia müelleri* (Lieberkuhn). *Trans. Amer. Microsc. Sci.,* 90: 219.

Ledger, P.W., and W.C. Jones, 1977. Spicule formation in the calcareous sponge *Sycon ciliatum. Cell Tissue Res.,* 181: 553.

3. Investigate the environmental control of hatching from sponge gemmules.

Benfey, T.J., and H.M. Reiswig, 1982. Temperature, pH, and photoperiod effects upon gemmule hatching in the freshwater sponge, *Ephydatia mülleri* (Porifera, Spongillidae). *J. Exp. Zool.,* 221: 13.

Fell, P.E., 1974. Diapause in the gemmules of the marine sponge, *Haliclona loosanoffi,* with a note on the gemmules of *Haliclona oculata. Biol. Bull.,* 147: 333.

4. How does sponge architecture contribute to increased flow of water through a sponge, and to increased feeding efficiency?

LaBarbera, M., and S. Vogel, 1982. The design of fluid transport systems in organisms. *American Scientist,* 70: 54.

Vogel, S., 1974. Current-induced flow through the sponge, *Halichondria. Biol. Bull.,* 147: 443.

Vogel, S., 1978. Organisms that capture currents. *Sci. Amer.,* 234: 108.

5. No one has ever demonstrated nerve tissue among the Porifera. Nevertheless, in some species there is evidence of cooperation among different areas of the sponge, resulting in the regulation of water flow through the animal. What is the evidence for such internal coordination, and through what mechanisms might it be accomplished in the absence of nerve cells?

Lawn, I.D., G.O. Mackie, and G. Silver, 1981. Conduction system in a sponge. *Science,* 211: 1169.

6. How do dissociated sponge cells recognize each other in the reaggregation process?

Galtsoff, P.S., 1925. Regeneration after dissociation (an experimental study on sponges). I. Behavior of dissociated cells of *Microciona prolifera* under normal and altered conditions. *J. Exp. Zool.,* 42: 183.

Hildemann, W.H., J.S. Johnston, and P.L. Jokiel, 1979. Immunocompetence in the lowest metazoan phylum: transplantation immunity in sponges. *Science,* 204: 420.

Humphreys, T., 1963. Chemical dissolution and in vitro reconstruction of sponge cell adhesions. I. Isolation of and functional demonstration of the components involved. *Devel. Biol.,* 8: 27.

McClay, D.R., 1974. Cell aggregation: properties of cell surface factors from five species of sponge. *J. Exp. Zool.*, 188: 89.

Spiegel, M., 1954. The role of specific surface antigens in cell adhesion. I. The reaggregation of sponge cells. *Biol. Bull.*, 107: 130.

Wilson, H.V., and J.T. Penney, 1930. The regeneration of sponges (*Microciona*) from dissociated cells. *J. Exp. Zool.*, 56: 73.

7.   Based upon your knowledge of sponge biology and of the properties of air and water, why are there no terrestrial sponges?

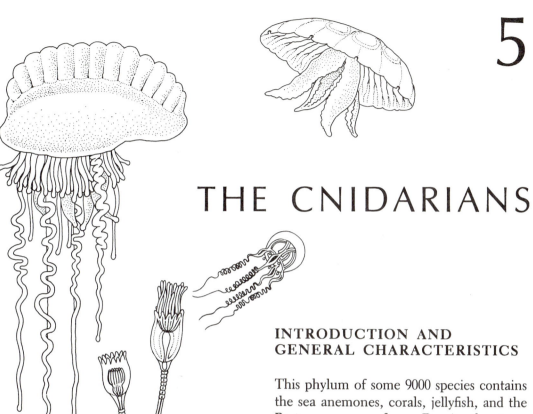

5

THE CNIDARIANS

INTRODUCTION AND GENERAL CHARACTERISTICS

This phylum of some 9000 species contains the sea anemones, corals, jellyfish, and the Portuguese man-of-war. Despite the structural and functional diversity represented by these different organisms, the members of this phylum are clearly related. All cnidarians have a basic radial symmetry and possess only two layers of living tissue (the epidermis and the gastrodermis). All cnidarians possess a gelatinous layer, the **mesoglea**, located between the epidermis and gastrodermis. Although the mesoglea is itself non-living, it may contain living cells. All

Phylum Cnidaria (= Coelenterata)
(G: A STINGING THREAD) (G: HOLLOW GUT)

cnidarians possess tentacles surrounding the mouth, and possess only a single opening to the digestive system. All cnidarians possess nematocysts.

Nematocysts (literally, "thread-bags"), unique to the members of this phylum, are organelles formed within cells called **cnidoblasts** (or **cnidocytes**). Each nematocyst consists of a rounded, proteinaceous capsule, with an opening at one end that is often occluded by a hinged operculum. Within the sac is a long, hollow, coiled tube. During nematocyst discharge, the hollow tube shoots out explosively from the sac, turning inside out as it goes (Fig. 5.1). Discharge results from a combination of chemical and tactile stimulation, perceived, presumably, through a modified cilium (the **cnidocil**) that projects from the cnidoblast.[1]

Many different morphological and functional types of nematocyst are found within the phylum, and even within an individual cnidarian. Within a given individual, nematocysts may be specialized for wrapping

---

1 See Topics for Discussion, Nos. 5–7.

around small objects, sticking to surfaces, penetrating surfaces, and/or secreting toxins. Nematocysts function in food collection, defense, and, to some extent, in locomotion. They are especially abundant on the feeding tentacles of all species, and within the digestive cavity of some species. The functional significance of nematocysts lies in their great number per square mm of body surface, rather than in their individual size; a nematocyst capsule rarely exceeds 50 μm in diameter. Nematocyst morphology is often important in making species identifications.

In contrast to members of the Porifera, cnidarians possess nerves and muscles, although a central nervous system is lacking. Instead, the nervous system of cnidarians consists of a network of nerve cells (neurons) and their processes (neurites), which generally synapse on one another repeatedly before terminating at a neuromuscular junction (Fig. 5.2). Although nerve impulses may cross certain synapses in one direction only, many synapses permit impulses to pass in both directions. Moreover, a given cell body may give rise to two or more neurites, radiating in different directions. Thus, a nerve

FIGURE 5.1. *(a) Discharge of a nematocyst, stimulated by a chemical and/or physical contact with the cnidocil. The mechanism of discharge remains unclear, although it is apparently mediated by an inflow of water along a concentration gradient. (From Wells, 1968, Lower Animals, McGraw-Hill. Reproduced with permission.) (b) Penetration of the nematocyst filament into another animal. Spines on the filament are exposed as the filament emerges [see part (a)], cutting through the prey tissues. (From Hardy, 1965. The Open Sea: Its Natural History, Houghton Mifflin.) (c) Medusa capturing prey. Nematocysts commonly inject toxins that paralyze the prey prior to ingestion. (From various sources.)*

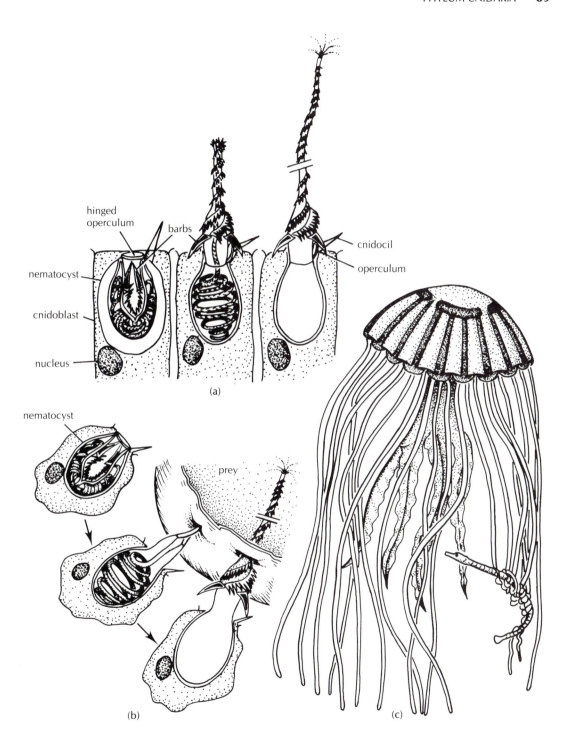

hinged
operculum

barbs

nematocyst

cnidoblast

nucleus

cnidocil

operculum

(a)

nematocyst

prey

(b)

(c)

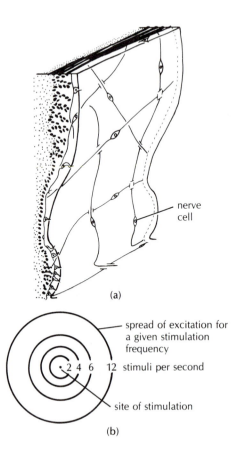

nerve cell

(a)

spread of excitation for a given stimulation frequency

2 4 6   12 stimuli per second

site of stimulation

(b)

FIGURE 5.2. *(a) Diagrammatic illustration of a cnidarian nerve net. The nerve cells synapse with each other repeatedly. Nerve impulses may cross the synapses in both direction. (After Bullock and Horridge.) (b) The cnidarian surface area affected by stimulation of a nerve cell varies directly with the frequency of stimulation. That is, the greater the frequency of stimulation, the greater the surface area affected.*

impulse received by one neuron may proceed in several directions at once.

With such a nerve network, stimulation of a given sensory cell in the epithelium results in an outward spread of excitation over the entire body of the animal (Fig. 5.2b). The amount of surface area of the cnidarian that is affected by stimulating a given nerve cell increases in proportion to the frequency of stimulation.

In addition to this slow-conducting nerve network, a second, fast-conducting nerve network generally underlies the epithelium. These cells are less branched (bipolar as opposed to multipolar) than are the cells in the slow-conducting network, so that signal transmission is more directed. Furthermore, neurites in the two nets differ in size: nerves in the fast-conducting network are of greater diameter, allowing more rapid conduction of nerve impulses.

For those contemplating reincarnation, a major drawback to life as a cnidarian would seem to be the absence of an anus. All undigested food material passes out through the same single opening through which the food entered: the mouth. This is not particularly appetizing, from the human point of view. Moreover, the shortcomings of life without an anus are not merely aesthetic. The sequential disassembly of food material that occurs in an open-ended tubular gut is not possible in the coelenterate digestive system. Moreover, movements of the coelenterate are generally accompanied by physical distortion of the digestive cavity, including partial or complete expulsion of the contained fluid. Extensive movement is therefore not conducive to leisurely, thorough digestion. Finally, gonadal development often takes place within the digestive cavity, and the gametes or embryos must be released into this cavity before their release to the exterior through the mouth.

CLASS SCYPHOZOA

The Scyphozoa consist of only a few hundred species, all of which are marine and many of which are quite large (up to about 0.5 meter across). The mesoglea layer of scyphozoans is very thick and has the consistency of firm gelatin. For this reason, scyphozoans are known collectively as "jellyfish."

Jellyfish morphology is described as **"medusoid."** The body is in the form of an inverted cup, with nematocyst-studded tentacles extending downward from the cup, or **bell** (Fig. 5.3). The mouth is borne at the end of a muscular cylinder known as the **manubrium**.

Most scyphozoan species are capable of active swimming, through the exploitation of muscles and the mechanical properties of the mesoglea. When the circular and radial muscle fibers of the swimming bell contract, the volume of fluid enclosed under the bell decreases. Water is forcefully expelled from under the bell as a consequence, and the animal is propelled forward (Fig. 5.4). The muscular contraction deforms the elastic mesoglea, so that when the musculature is relaxed, the mesoglea "pops" back to its normal shape. This, of course, pulls the jellyfish downward as the volume enclosed by the swimming bell increases. Net forward movement of the animal occurs primarily because

Class Scypho·zoa

(G: CUP ANIMALS)

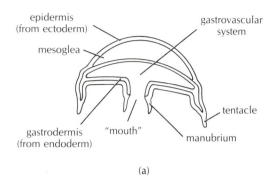

(a)

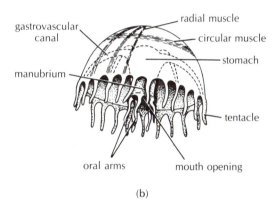

(b)

FIGURE 5.3. *(a) A medusa, seen in longitudinal section. Note the very thick layer of mesoglea and the single opening to the gastrovascular cavity; this opening serves as both mouth and anus. (After Russell-Hunter.)*
(b) Lateral view of a medusa, showing the gastrovascular canal system and the arrangement of tentacles, oral arms, and the musculature of the swimming bell.

the speed with which the bell contracts exceeds the speed with which the bell recoils to its resting state.[2]

The Scyphozoa are characterized by a

2 See Topics for Discussion, No. 10.

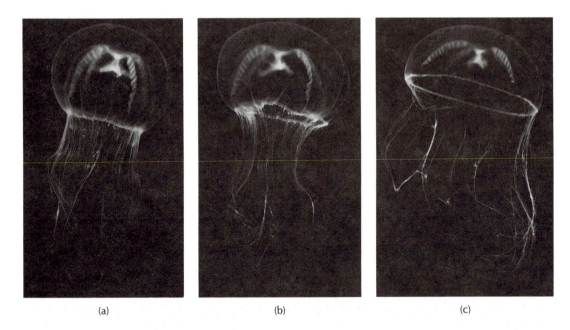

(a)   (b)   (c)

FIGURE 5.4. *Locomotion of the medusa stage of* Mitrocoma
cellularia. *As muscle contractions force water out from under the
swimming bell, the animal is propelled in the opposite direction.
Note that efficient jet propulsion relies on the incompressibility of
water; fluid must leave the bell, as it cannot "hide" under the bell
by compression. The animal shown is actually a hydrozoan rather
than a scyphozoan; the velum is seen clearly as a sheet protruding
from the bell in (a). (See page 97 for discussion of hydrozoan
medusae.) (a) Power stroke nearly completed; note that a bolus of
expelled water can be seen pushing the tentacles outwards about
midway along their length. (b) Beginning of recovery period;
swimming bell is beginning to expand. Note that the bolus of water
has moved farther down the tentacles. (c) Bell fully relaxed, ready
for the next contraction. The diameter of the swimming bell is
approximately 70 mm. (Courtesy of Claudia E. Mills.)*

well-developed system of fluid-filled **gastro-vascular canals**, ultimately connecting to the mouth through the manubrium (Fig. 5.5). Food particles captured by the tentacles and/or oral arms are ingested at the mouth and conveyed to the stomach through the manubrium. Food is then distributed among four **gastric pouches**, which contain

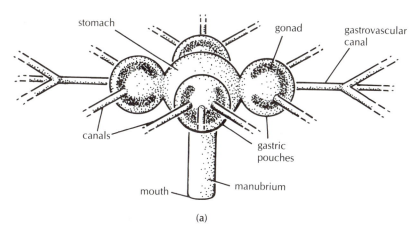

(a)

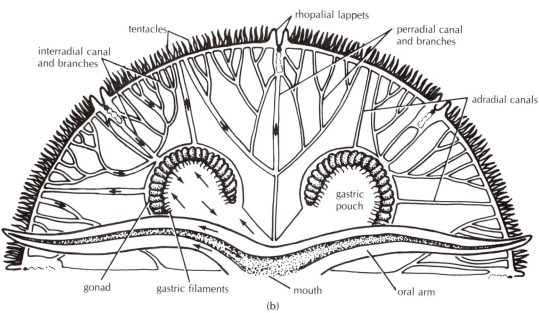

(b)

FIGURE 5.5. *Detail of the scyphozoan gastrovascular canal system:
(a) lateral view; (b) oral view. Cilia lining the canal system draw
water in through the mouth to the gastric pouches. From the
pouches, water circulates to the periphery of the bell through a
complex series of narrow canals, as shown by the arrows. Details of
the rhopalial lappets, which are sensory organs, are shown in Fig.
5.6. (After Hickman.)*

nematocyst-bearing filaments (**gastric filaments**) and secrete an array of digestive enzymes. The partially digested food particles
are then phagocytized, and digestion is completed intracellularly. Fluid within the gut is
circulated by means of cilia lining the walls of
the gastrovascular canals. The gastrovascular canals are believed to function in
the circulation of oxygen and carbon dioxide
(the "vascular" part of "gastrovascular") as
well as in the distribution of nutrients (the
"gastro" part of the term).

As befits any mobile organism, scyphozoans are equipped with fairly sophisticated sensory receptors. These include
balance organs (**statocysts**), simple light receptors (**ocelli**), and, in some species, touch
receptors (**sensory lappets**), distributed
around the periphery of the swimming bell.
The statocysts and ocelli are contained
within club-shaped structures called **rhopalia**, which are distributed along the margins of the swimming bell (Fig. 5.6). Dense
aggregations of nerve tissue are found associated with the rhopalia. These ganglia act as
pacemakers, triggering the rhythmic contraction of the swimming bell.

Statocysts operate on a beautifully simple
principle. Tubular pieces of tissue (the **rhopalia**) hang freely at several locations around
the margins of the swimming bell. Each of
these rhopalia is adjacent to (but not in contact with) a sensory cilium. Also, each tube is
weighted at the free end with a spherical calcareous mass (the **statocyst**). If the animal
tilts in a particular direction, some of the
rhopalia will press against their respective
cilia (Fig. 5.6c), causing the associated nerve
cells to generate action potentials. The
rhopalium/ statocyst system thus provides a

mechanism through which the animal can
be informed of its physical orientation—i.e.,
whether the body is horizontal or tilted—and
the jellyfish can alter its posture accordingly,
through stronger contractions of the musculature on one side of the bell or the other.

Non-image-forming light receptors (ocelli)
are also found along the bell margin. An
ocellus is simply a small area, often cup-
shaped, backed by light-sensitive pigment.
The cup-shaped ocelli of cnidarians are
sometimes covered by a lens, which serves to
concentrate incoming light (Fig 5.6b).

The life cycle of a scyphozoan is a diagnostic feature of its biology. Gonads develop
within gastrodermal tissue, and are generally
associated with the gastric pouches (see Fig.
5.5). Individual medusae are either male or
female; i.e., the sexes are separate and the
species is said to be **dioecious** (G: two
houses). This contrasts with the situation frequently encountered among other invertebrates, in which a given individual may
be both male and female, either simultaneously or in sequence. Such species are said
to be **hermaphroditic**.

A **planula** larva eventually results from the
union between sperm and egg. This larva has
the form of a heavily ciliated, very small sau-
sage. The nonfeeding planula larva soon set-
tles on a substrate and transforms into a
small polypoid individual called a **scyphistoma** (Fig. 5.7). This **polyp** form has the
same two-layered construction (plus mesoglea layer) as the medusa, but the mesoglea
layer is substantially thinner in the polyp
than in the medusa morph. The scyphistoma
is sessile, and lacks ocelli and statocysts. It is
a feeding individual, with the mouth oriented away from the substrate.

As the scyphistoma grows, it may produce additional scyphistomae by asexual budding. Eventually, a process called **strobilation** takes place in most species. During strobilation, the body column of a scyphistoma subdivides transversely, forming numerous discs that are stacked on top of each other like hotel ashtrays (Fig. 5.7). Each "ashtray" eventually breaks away from the stack as a swimming **ephyra**. As it swims, each ephyra gradually grows and changes in physical appearance, becoming an adult scyphozoan.

Reflect for a moment upon this life cycle. The scyphozoan has used the relatively inconspicuous polyp morph to achieve something quite remarkable: from a single fertilized egg, a large number of genetically identical, sexually reproducing medusae have been generated. A similar phenomenon takes place among the parasitic flatworms, as discussed in Chapter 7.

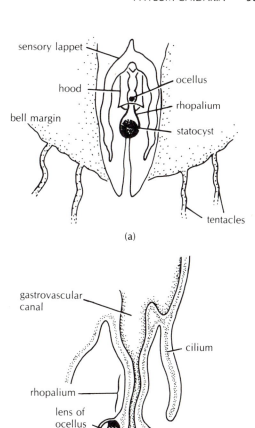

(a)

(b)

FIGURE 5.6. *(a) Detail of rhopalial system. The calcareous statocyst is shown hanging at the end of the rhopalium. The species illustrated (*Aurelia aurita*) has an ocellus and a pair of specialized sensory lappets associated with each rhopalium. (After Hyman.)*
(b) Longitudinal section through a rhopalium of Carybdea sp., *showing the structure of the ocellus. Note that the cells of the rhopalium are serviced by a branch of the gastrovascular canal system. (After Bayre and Owre; after Mayer.)*
(c) The principle of statocyst operation. When the animal tilts sideways, the statocyst swings against a cilium initiating a nerve signal to the appropriate muscles; muscle contractions restore the animal to proper orientation. (After Wells.)

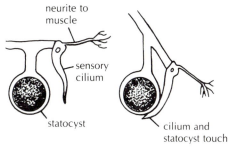

organism level organism tilted

(c)

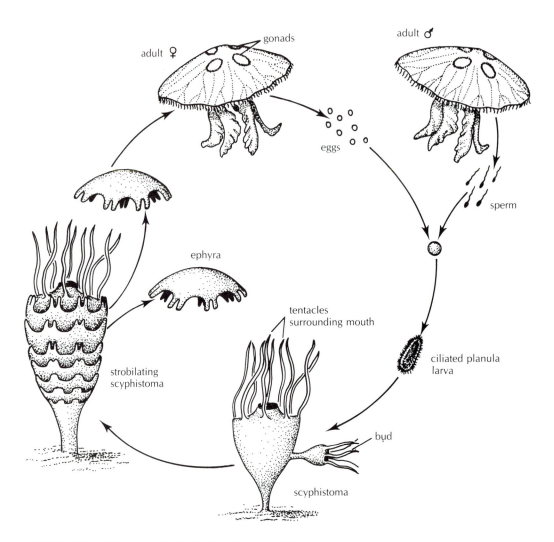

FIGURE 5.7. *Life cycle of a typical scyphozoan. The polyp stage (scyphistoma) is small—often only a few millimeters in length—and is often found hanging downward from the undersides of underwater rock ledges. (Modified from various sources.)*

CLASS HYDROZOA

Members of the Hydrozoa are characterized by generally greater representation of the polyp morph in the life cycle than is the case for scyphozoans, although the polyp and medusa morphs are about equally prominent in a number of hydrozoan species. In contrast to other cnidarians, the gastrodermal tissue of hydrozoans lacks nematocysts, and no cells are found within the mesoglea; nematocysts are restricted to the epidermis. The class Hydrozoa comprises somewhat fewer than 3000 species. Most of these species are marine.

ORDER HYDROIDA

Although most members of this order are marine, a number of freshwater species also exist—*Hydra*, for example. Hydroids are generally medusoid as adults. That is, the sexual stage of the life cycle resembles that found among the Scyphozoa. As in scyphozoans, the mesoglea layer of the hydrozoan medusa is thick, the mouth is borne at the end of a manubrium, and ocelli and statocysts are present. The sense organs may be found at the base of the tentacles, as in scyphozoans, or in between the tentacles. The medusae are dioecious, a given individual being either male or female, but never both. Hydrozoan medusae tend to be much

smaller than scyphozoan medusae (generally only a few millimeters or less across), and often possess a shelf of tissue (the **velum**) that extends inward from the edge of the swimming bell toward the manubrium (Fig. 5.8). The presence of the velum causes water to be ejected from the swimming bell through a narrower opening, and thus with greater force, when the musculature contracts and the volume of fluid under the bell decreases. The majority of scyphozoan medusae lack a velum.

As in the Scyphozoa, a planula larva (Fig. 5.9) develops from a fertilized egg, and this planula metamorphoses into a sessile polypoid individual lacking both statocysts and ocelli. The polyps of the Hydrozoa are structurally and functionally more complex than are the scyphistomae of the scyphozoan life cycle. The hydrozoan genus most familiar to

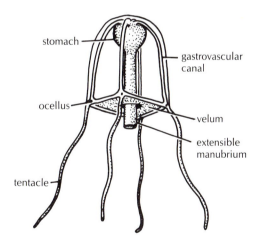

FIGURE 5.8. *Diagrammatic illustration of a typical hydrozoan medusa. Note the conspicuous velum, through which water is forcefully expelled when the musculature of the swimming bell contracts. (Modified from various sources.)*

Class Hydro•zoa

(G: WATER ANIMALS)

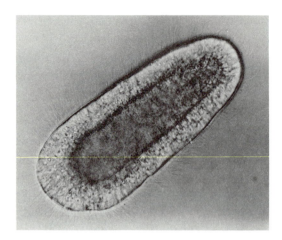

FIGURE 5.9. *The planula larva of a hydrozoan*, Mitrocomella polydiademata. *(Courtesy of V. J. Martin.)*

the reader is probably *Hydra*. In *Hydra*, each polyp is a separate, distinct being, completely responsible for its own welfare. I must impress upon you that in this, and several other respects, *Hydra* is quite the atypical hydrozoan. Most hydrozoans in the polyp stage of the life cycle are colonial; that is, a single planula often gives rise to a large number of polyps, called **zooids**, all of which are interconnected and share a continuous gastrovascular cavity (Fig. 5.10). The zooids are often connected to each other, or to a substrate, by means of a root-like **stolon**, termed the **hydrorhiza**. The oral end of a polyp, i.e., the end bearing the mouth and tentacles, is called the **hydranth**.

The stolon and stalks of the colony are commonly encased in a transparent protective tube, known as the **perisarc**, composed of polysaccharide and protein (Fig. 5.11a). The perisarc may or may not extend upward to encase the hydranth of a polyp, depending on the species. The perisarc surrounding the hydranth is known as a **hydrotheca** (Fig. 5.10b), and the hydroid is said to be **thecate** (as opposed to being **athecate**, the Greek prefix "a" meaning "not," or "without"; see Fig. 5.11.)

Several structurally and functionally distinct individuals are generally present in each hydroid colony. That is to say, the colonies are **polymorphic**, consisting of two or more types of individuals. The feeding individuals are called **gastrozooids**. Gastrozooids collect small animals using the tentacles (which are densely clothed in nematocysts) and ingest the food through the single opening into the gastrovascular cavity. Digestion is extracellular in the gastrovascular cavity, and then becomes intracellular as the partially digested food is distributed elsewhere in the colony.

Medusoids are produced asexually by budding from various regions of the polyp colony. Often, the medusoids derive from a particular type of individual, called a **gonozooid**. Some gonozooids lack tentacles, and so are incapable of feeding; they are specialized for production of medusoids and must depend upon other members of the colony for nutrition. In some hydrozoans, the medusae eventually break free of the polyp colony, swim off, and commingle gametes ("have sex") with other medusae as already discussed. More commonly, however, the gamete-producing medusoid morph remains attached to the hydrozoan colony. No free-swimming medusa stage exists in such species. In fact, the medusa morph may be reduced to little more than a mass of gonadal tissue.

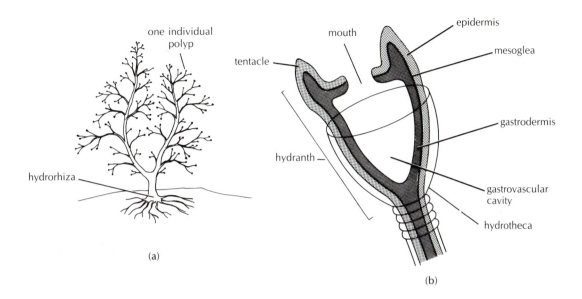

one individual polyp

hydrorhiza

(a)

mouth

epidermis

mesoglea

tentacle

gastrodermis

hydranth

gastrovascular cavity

hydrotheca

(b)

FIGURE 5.10. *(a) A typical colonial hydroid,* Campanularia *sp. The entire colony is only a few centimeters in height (After Hyman, Vol. I.) (b) Diagrammatic illustration of a single hydrozoan polyp. Note that the mesoglea layer is much thinner than in the medusoid morph illustrated in Fig. 5.3(a). (After Russell-Hunter.)*

Colonies commonly contain fingerlike members specialized for defense (**dactylozooids**; *dactylus* = G: finger), as well as members specialized for feeding and reproduction (Fig. 5.12). The dactylozooids are heavily studded with nematocysts. Dactylozooids never possess mouths, and thus, like some of the highly specialized gonozooids, are dependent upon the gastrozooids for food collection. It is well to remember that all individuals in a given colony, no matter how polymorphic the colony is, are originally derived from a single planula larva.

ORDER SIPHONOPHORA

The epitome of polymorphism is reached in this group of cnidarians. The siphonophores, including the Portuguese man-of-war, are free-floating hydrozoans in which medusoid and polypoid morphs are present simultaneously in a number of different incarnations. Modified medusae serve as gas-filled floats (**pneumatophores**); or as individuals (**nectophores**) modified to propel the colony through the water by jet propulsion; or as

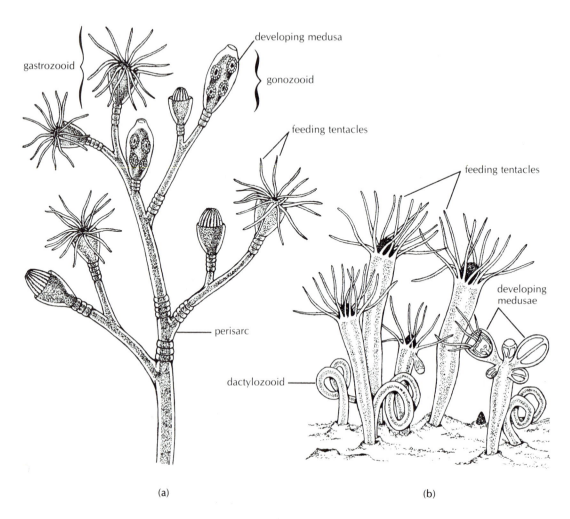

(a)

(b)

FIGURE 5.11. *(a) A thecate, marine hydroid,* Obelia commissuralis, *showing specialized reproductive and feeding polyps (gonozooids and gastrozooids, respectively). This species commonly forms a pale bushy covering on pilings and floats in protected harbors. (After McConnaughey and Zottoli; after Nutting.)*
(b) An athecate hydroid, Podocoryne carnea. *This species possesses individuals specialized for protection (dactylozooids). P. carnea is commonly encountered inside the openings of snail shells occupied by marine hermit crabs. (After McConnaughey and Zottoli; after Fraser.)*

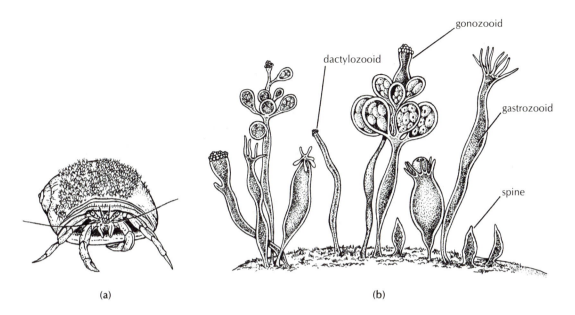

FIGURE 5.12. *(a) Colony of* Hydractinia echinata *encrusting the outside surface of a snail shell inhabited by a hermit crab. (From Smith, 1964.* Keys to Marine Invertebrates of the Woods Hole Region. Marine Biological Laboratory.*) (b) Detail of* H. echinata *colony, illustrating gastrozooids, gonozooids, dactylozooids, and individuals modified to form sharp protective spines. (After Bayer and Owre.)*

leaf-like defensive individuals (**bracts,** or **phyllozooids;** *phyllo* = G: leaf). The mesoglea layer is much reduced, or entirely absent, in pneumatophores. Nectophores lack both mouth and tentacles. Bracts are well endowed with nematocysts. The polyp morph is represented by gastrozooids, gonozooids, and dactylozooids. (Fig. 5.13).

Each gastrozooid has a single tentacle associated with it; elongate, nematocyst-bearing structures (**tentillae**) may project from these tentacles. Dactylozooids may also have associated tentacles. All siphonophore tentacles are highly retractile, and the nematocysts are often very toxic, even to humans.

Often, individuals occur in clusters, called **cormidia,** arranged upon a long stem. Each cormidium typically contains gonozooids,

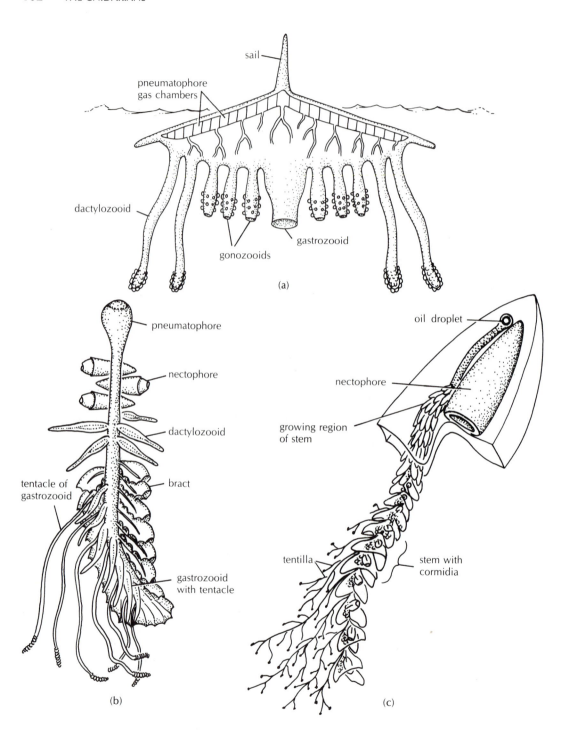

(a)

(b)

(c)

FIGURE 5.13. *Typical siphonophores. (a) In this colony* (Velella *sp.), the pneumatophore is divided into a series of chambers, and only a single gastrozooid is present. The dactylozooids are studded with nematocysts for prey capture. Locomotion is wind and current driven. (b) In this form, typical of* Nectalia *sp., the pneumatophore is relatively small. Nectophore contraction provides jet propulsion. (c)* Muggiaea *sp., a siphonophore that lacks a pneumatophore. The largest member of the colony is a nectophore. Note the long stem with cormidia (clusters of gonozooids, bracts, gastrozooids, and dactylozooids). (From Hyman, 1940. The Invertebrates, Vol I. McGraw-Hill.)*

dactylzooids, phyllozooids (bracts), and gastrozooids.

One or more morphs may be absent in some groups of siphonophores. For example, the Portuguese man-of-war lacks nectophores; movement of the animal depends entirely on wind and water currents. Other species (Fig. 5.13c) may lack a pneumatophore.

All siphonophores are voracious carnivores.

ORDER HYDROCORALLINA

This order contains a small number of species, all of which are colonial and all of which secrete an internal calcareous skeleton. The members of this order are largely restricted to warm waters. The dactylozooids are especially abundant and potent in many hydrocorals; the common name, "fire coral," is well deserved.

To avoid confusion, note that these animals are not true corals; the true corals are contained within a different class of cnidarians, the Anthozoa.

CLASS ANTHOZOA

Anthozoans (including the sea anemones and the corals) consist of about 6000 species, all of them marine. Anthozoans exploit the polyp body form and life-style exclusively; no trace of the medusa morph appears in the life cycle. Gametes are produced directly by the anthozoan polyp. A planula larva develops from the fertilized egg and metamorphoses to form another polyp. Many anthozoan species also reproduce asexually, through longitudinal or transverse **fission**, or through a process of **pedal laceration,** in which parts of the pedal disc (foot) detach from the rest of the animal and gradually differentiate to form another anemone.[3]

As with hydrozoans and scyphozoans, anthozoans are carnivorous; they capture food

3 See Topics for Discussion, No. 9.

Class Antho•zoa
(G: FLOWER ANIMALS)

by using nematocyst-studded tentacles and transfer it to a central mouth opening. However, the polyps of anthozoans differ from those of hydrozoans in several respects. The anthozoan mouth opens into a tubular pharynx, rather than directly into the gastrovascular cavity. One or two discrete, ciliated grooves, called **siphonoglyphs**, extend down the pharynx from the mouth (Fig. 5.14). The gastrovascular cavity of an anthozoan is partitioned by means of numerous sheets of tissue called **mesenteries**, whereas no mesenteries are found in the gastrovascular cavity of hydrozoan polyps. These infoldings of gastroderm and mesoglea greatly increase the surface area available for the secretion of digestive enzymes and the absorption of nutrients. Mesenteries that extend far enough from the body wall into the gastrovascular cavity to actually attach to the pharynx are called **primary mesenteries**. Those extending only part way into the gastrovascular cavity are termed secondary and tertiary mesenteries. The mesenteries are studded with nematocysts and also bear the gonads. Anthozoans are usually dioecious, but some species are sequential hermaphrodites.

Internally, near the base of an anthozoan, thin filaments called **acontia** often extend from the mesenteries. These acontia are loaded with nematocysts and secretory cells, and may be used for both defensive and offensive purposes. They can be extended outside the body through small pores in the body wall. The acontia may also function in digestion.

A number of anemone species possess rings of small spherical bulges extending around the circumference of the body column just below the tentacles. These hollow **acrorhagi** can be extended a substantial distance from the body column, possibly by the forcing of fluid into them from the gastrovascular cavity, with which they are continuous. Acrorhagi are covered with very potent nematocysts, and are used in defending a territory against invasion by other anemones (Fig. 5.15). Some anemone species that lack acrorhagi have, instead, tentacles that are specialized for fighting. These **catch tentacles** are analogous to acrorhagi, functioning in aggressive encounters among individuals.[4]

The tissues of an anthozoan contain both circular and longitudinal muscle fibers (Fig. 5.16). Provided that the animal keeps its mouth closed (by contracting appropriate sphincter muscles), the sea water in the gastrovascular cavity can serve as a hydrostatic skeleton. For example, by keeping the mouth closed, relaxing the longitudinal musculature, and contracting the circular muscles of the body wall, the animal becomes long and thin, as the longitudinal muscles are stretched by the elevated pressure within the gastrovascular cavity (Fig. 5.17). Then, by contracting the longitudinal muscles on one side of the body and not the other, the animal can bend to one side, provided that the circular muscles are not permitted to expand. In an emergency, the longitudinal muscles can be contracted while the mouth is open, causing the animal to flatten considerably as the fluid of the gastrovascular cavity is expelled. The resulting shape has been referred to as the "bubble gum on a rock" disguise (Fig. 5.17d). Reinflation is rather slow,

---

4 See Topics for Discussion, No. 8.

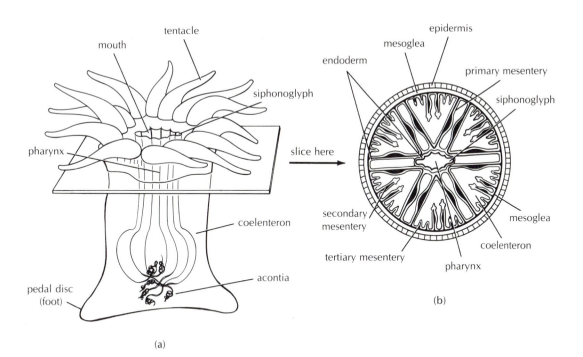

(a)

(b)

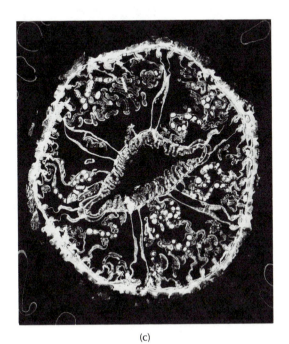

(c)

FIGURE 5.14. *(a) Schematic of a typical anthozoan. The acontia connect to the lower edge of mesenteries (not shown here). (b) Cross section made through the polyp in the area indicated in (a). (c) Photograph of a cross section through an anthozoan. Note the two siphonoglyphs and the developing embryos present in some of the mesenteries. (Courtesy of L. S. Eyster and T. Van Way.)*

FIGURE 5.15. *A sea anemone,* Anthopleura krebsi. *The individual on the left is displaying numerous stubby acrorhagi below its pointy tentacles. The acrorhagi are used by the anemone to defend territory against invasion by neighboring anemones. (Courtesy of C.H. Bigger. From Bigger. 1980. Biol. Bull. 159:117.)*

being dependent upon the activity of the cilia lining the siphonoglyphs; these cilia "pump" water back into the gastrovascular cavity.

Many species of Anthozoa are capable of moving from place to place under their own power, although usually very slowly. A fast-moving anemone might typically achieve a speed of several millimeters per minute.

Some anemones are commonly found on the backs of more mobile invertebrates and are transported adventitiously. A form of swimming has been described in several other species,[5] but most anemones are basically stay-at-homes.

---

5 See Topics for Discussion, No. 2.

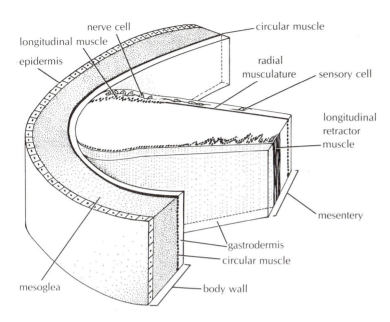

FIGURE 5.16. *Diagrammatic illustration of the musculature and tissue layers in the body wall and on the mesentery of a typical anthozoan. (After Bullock and Horridge.)*

Subclass Hexacorallia (= Zoantharia)

Hexacorallians possess numerous tentacles around the mouth opening, usually in some multiple of six, and possess a pair of siphonoglyphs associated with the pharynx. Many species in this subclass are **solitary** (i.e., they are independent individuals, rather than colonies of connected individuals) and lack any specialized protective covering. These spe-

Subclass Hexa·corallia (= Zoantharia)

(G: SIX)

cies are the sea anemones. The other species in this subclass, the corals, tend to be colonial (but individuals are not polymorphic) and secrete a calcium carbonate cup. More specifically, the corals secrete an organic matrix, which then serves as the nucleation site for the deposition of calcium carbonate. These animals comprise the true (or stony) corals. Corals may be reef-building (**hermatypic**) or not (**ahermatypic**).

Most hermatypic corals are restricted to clear, warm waters. Coral reefs are especially abundant in tropical areas of the Indo-Pacific, forming chains of islands and other structures of massive proportions. The Great Barrier Reef along Australia's northeast

FIGURE 5.17. *Shape changes in the sea anemone,* Sagartia elegans. *(a) The animal is slowly inflating, using the cilia on the siphonoglyph to drive water into the gastrovascular cavity. (b) By closing the mouth, relaxing the longitudinal muscles, and contracting the circular muscles, the anemone increases in height but decreases in body width. (c) By opening the mouth and contracting the longitudinal muscles, most of the fluid in the gastrovascular cavity is rapidly expelled. To regain its original shape, the animal must pump water back into the coelenteron by ciliary action, a much slower process. (After Batham and Pantin.)*

shoreline is more than 2000 kilometers in length and 145 kilometers in width. This reef is one of the world's highest-diversity, most complex ecosystems. It should be noted that although anthozoans may play the major role in the construction of coral reefs, other organisms (particularly calcareous red algae, foraminiferans, shelled molluscs, certain tube-dwelling polychaetes, and bryozoans) make significant contributions.

Coral reefs flourish in areas of low planktonic productivity, although, since an-thozoans are generally carnivorous, there would not seem to be sufficient zooplankton to support such luxuriant coral growth. It turns out that the growth of coral reefs is aided by endosymbiotic, unicellular photosynthesizers (modified dinoflagellates, actually) called **zooxanthellae** (Fig. 5.18).[6] The zooxanthellae are photosynthetic and provide the anthozoan host with carbohydrates.

---

6 See Topics for Discussion, Nos. 3, 4.

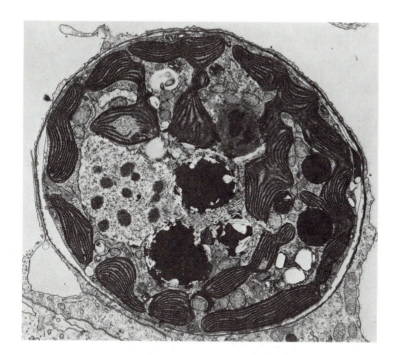

FIGURE 5.18. *A scanning electron micrograph of a zooxanthella in the tissues of the stony coral* Pocillopora damicornis. (*Courtesy of E. H. Newcomb and T. D. Pugh, University of Wisconsin/BPS.*)

In return, the zooxanthellae have intimate access to the metabolic wastes of the coral, including carbon dioxide for photosynthesis and nitrogenous wastes essential for growth of the algae. Through this symbiotic relationship, the zooxanthellae also benefit by being protected from herbivores.

The relationship between the anthozoan and the zooxanthellae has an additional effect on reef growth, quite apart from that mediated by nutritional considerations. Some 25 years ago, scientists determined that rates of calcification by hermatypic corals are considerably higher in the light than in the dark. The implication is that activities of the symbiotic algae play a role in determining the rate of calcification by the anthozoan although the mechanism through which this effect is mediated remains uncertain. The most likely possibilities are:

(1) through an effect on availability of bicarbonate ion (HCO_3^-), an essential constituent of the calcification process;

(2) through a contribution by the alga of a critical organic component of the nucleating matrix;

(3) through localized removal of soluble phosphate (PO_4^{3-}) by the alga during photosynthesis (phosphates are known inhibitors of calcification);

(4) indirectly, through the influence of elevated levels of dissolved oxygen on rates of coral metabolism.

Several ahermatypic coral species also contain zooxanthellae in their tissues.

Subclass Octocorallia (= Alcyonaria)

The members of this order possess eight tentacles (and eight primary septa) and generally have only a single siphonoglyph. The tentacles of octocorallians are **pinnate**; i.e., they bear numerous lateral outfoldings called **pinnules** (Fig. 5.19a). Pinnate tentacles are only rarely encounterd among other anthozoans.

All octocorallian species are colonial and the colonies are often polymorphic. For species with polymorphic polyps, some individuals are incapable of feeding and function solely in driving water through the gastrovascular spaces of the colony. The polyps of octocorals may be embedded in a thick matrix of mesoglea; these are the soft corals. In other species, the polyps are supported by proteinaceous or calcareous external skeletons, which are secreted by epidermal cells. This latter group (colonial species with skeletal supports) includes the sea fans and sea whips (known collectively as gorgonians, or horny corals!) and the pipe corals. Figure 5.19 shows some typical anthozoan corals.

Subclass Octocorallia (= Alcyonaria)
|
(G: EIGHT)

FIGURE 5.19. *Anthozoan diversity. (a) A hexocoral,* Heliastra heliopora, *the brain coral. (After Kingsley.) (b) Another hexocoral,* Astraea pallida, *with some of its polyps expanded and others retracted into the protective calcareous base of the colony. (After Kingsley.) (c) An octocoral,* Clavularia, *showing the pennate tentacles characteristic of this subclass. (After Gohar.) (d)* Pennatula *sp., the sea pen, an octocoral. (After Kölliker.) (e) A sea fan,* Gorgonia *sp., another octocoral. (After Bayer and Owre.) (f) A burrowing anthozoan,* Cerianthus *sp., taken from the muddy substrate in which it lives. These animals are structurally similar to sea anemones except that they lack a basal disc and have the tentacles arranged in two rings, as shown. Longitudinal muscles are especially well developed, permitting rapid withdrawal into the mucus-lined burrows. (After Bayer and Owre.)*

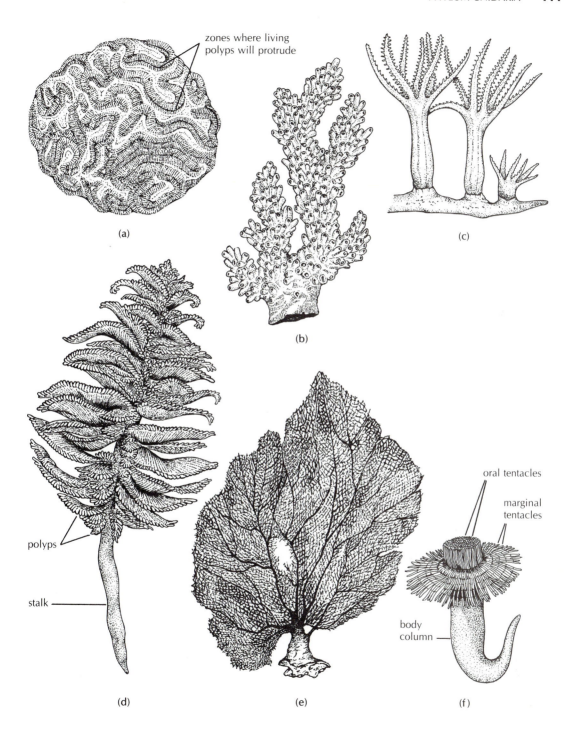

zones where living
polyps will protrude

(a)

(b)

(c)

polyps

stalk

(d)

(e)

oral tentacles

marginal
tentacles

body
column

(f)

PHYLUM CNIDARIA (= COELENTERATA)
CLASS SCYPHOZOA—THE JELLYFISH

CLASS HYDROZOA
 ORDER HYDROIDA
 ORDER SIPHONOPHORA
 ORDER HYDROCORALLINA

CLASS ANTHOZOA—THE SEA ANEM-
 ONES, SEA WHIPS, SEA
 PENS, AND CORALS
 SUBCLASS HEXACORALLIA
 (=ZOANTHARIA)
 SUBCLASS OCTOCORALLIA
 (=ALCYONARIA)

T O P I C S   F O R   F U R T H E R   D I S C U S S I O N   A N D   I N V E S T I G A T I O N

1. Investigate the evolutionary origin of the phylum Cnidaria (Coelenterata). Which class most likely represents the most primitive body plan and life history? Justify your answer.

Hyman, L.H., 1940. *The Invertebrates*, Vol. I. New York: McGraw-Hill, pp. 632-641.

Rees, W.J. (Ed.), 1966. *The Cnidaria and their Evolution*. New York: Academic Press.

2. Investigate the morphological and functional adaptations for swimming or burrowing encountered among some members of the Hexacorallia and Octocorallia, respectively.

Lawn, I.D., and D.M. Ross, 1982. The behavioural physiology of the swimming sea anemone *Boloceroides mcmurrichi*. *Proc. R. Soc. London* 216B: 315.

Mariscal, R.N., Conklin, E.J., and C.H. Bigger, 1977. The ptychocyst, a major new category of cnida used in tube construction by a cerianthid anemone. *Biol. Bull.* 152: 392.

Robson, E.A., 1961. Some observations on the swimming behavior of the anemone *Stomphia coccinea*. *J. Exp. Biol.*, 38: 343.

Ross, D.M., and L. Sutton, 1967. Swimming sea anemones of Puget Sound: swimming of *Actinostola* new species in response to *Stomphia coccinea*. *Science*, 155: 1419.

3. To what extent are the nutritional and respiratory requirements of hermatypic corals and other anthozoans met by their resident zooxanthellae?

Kevin, K.M., and R.C.L. Hudson, 1979. The role of zooxanthellae in the hermatypic coral *Plesiastrea urvellei* (Milne Edwards and Haime) from cold waters. *J. Exp. Marine Biol. Ecol.*, 36: 157.

Kinzie, R.A., III, and G.S. Chee, 1979. The effect of different zooxanthellae on the growth of experimentally reinfected hosts. *Biol. Bull.*, 156: 315.

Meyer, J.L., E.T. Schultz, and G.S. Helfman, 1983. Fish schools: an asset to corals. *Science*, 220: 1047.

Muscatine, L., and J.W. Porter, 1977. Reef corals: mutualistic symbiosis adapted to nutrient-poor environments. *Bioscience*, 27: 454.

Shick, J.M., and J.A. Dykens, 1984. Photobiology of the symbiotic sea anemone *Anthopleura elegantissima*: photosynthesis, respiration, and behavior under intertidal conditions. *Biol. Bull.*, 166: 608.

Szmant-Froelich, A., and M.E.Q. Pilson, 1980. The effects of feeding frequency and symbiosis with zooxanthellae on the biochemical composition of *Astrangia danae* Milne Edwards and Haime 1849. *J. Exp. Mar. Biol. Ecol.* 48: 85.

Taylor, D.L., 1969. On the regulation and maintenance of algal numbers in zooxanthellae-coelenterate symbiosis, with a note on the relationship in *Anemonia sulcata*. *J. Marine Biol. Ass. U.K.*, 49: 1057.

Wethey, D.C., and J.W. Porter, 1976. Sun and shade differences in productivity of reef corals. Nature, London, 262: 281.

4. Investigate the behavioral and morphological modifications of anthozoans that increase the photosynthetic capabilities of their symbiotic zooxanthellae.

Fricke, H., and E. Vareschi, 1982. A scleractinian coral (*Plerogyra sinuosa*) with "photosynthetic organs." *Marine Ecol. Progr. Series* 7: 273.

Pearse, V.B., 1974. Modification of sea anemone behavior by symbiotic zooxanthellae: phototaxis. *Biol. Bull.*, 147: 630.

Pearse, V.B., 1974. Modification of sea anemone behavior by symbiotic zooxanthellae: expansion and contraction *Biol. Bull.*, 147: 641.

Redalje, R., 1977. Light adaptation strategies of hermatypic corals. *Pacific Sci.*, 30: 212.

Sebens, K.P., and K. DeRiemer, 1977. Diel cycles of expansion and contraction in coral reef anthozoans. *Marine Biol.*, 43: 247.

5. The nematocyst is a morphologically and functionally complex structure. Through what process are nematocysts formed by the cnidoblasts?

Skaer, R.J., 1973. The secretion and development of nematocysts in a siphonophore. *J. Cell Sci.*, 13: 371.

6. What is the relative importance of chemical versus physical stimuli in triggering nematocyst discharge?

Conklin, E.J., and R.N. Mariscal, 1976. Increase in nematocyst and spirocyst discharge in a sea anemone in response to mechanical stimulation. In: G.O. Mackie (Ed.), *Coelenterate Ecology and Behavior*. New York: Plenum Publ., pp. 549–558.

Lubbock, R., 1979. Chemical recognition and nematocyte excitation in a sea anemone. *J. Exp. Biol.* 83: 283.

Pantin, C.F.A., 1942. The excitation of nematocysts. *J. Exp. Biol.* 19: 294.

7. To what extent is nematocyst discharge under direct nervous control?

Ross, D.M., and L. Sutton, 1964. Inhibition of the swimming response by food and of nematocyst discharge during swimming in the sea anemone *Stomphia coccinea*. J. Exp. Biol., 41: 751.

Sandberg, D.M., P. Kanciruk, and R.N. Mariscal, 1971. Inhibition of nematocyst discharge correlated with feeding in a sea anemone, *Calliactis tricolor* (Leseur). *Nature*, London, 232: 263.

Smith, S., J. Oshida, and H. Bode, 1974. Inhibition of nematocyst discharge in *Hydra* fed to repletion. *Biol. Bull.*, 147: 186.

8. Sea anemones must compete for space with other anemones, as well as with members of other animal and plant groups. What are the roles of acrorhagi and catch tentacles in mediating this competition for space?

Ayre, D.J., 1982. Inter-genotype aggression in the solitary sea anemone *Actinia tenebrosa. Marine Biol.*, 68: 199.

Bigger, C.H., 1982. The cellular basis of the aggressive acrorhagial response of sea anemones. *J. Morphol.* 173: 259.

Bigger, C.H., 1980. Interspecific and intraspecific acrorhagial aggressive behavior among sea anemones: a recognition of self and not-self. *Biol. Bull.*, 159: 117.

Chornesky, E.A., 1983. Induced development of sweeper tentacles on the reef coral *Agaricia agaricites*: a response to direct competition. *Biol. Bull.*, 165: 569.

Ertman, S.C., and D. Davenport, 1981. Tentacular nematocyte discharge and "self-recognition" in *Anthopleura elegantissima* Brandt. *Biol. Bull.*, 161: 366.

Francis, L., 1973. Intraspecific aggression and its effect on the distribution of *Anthopleura elegantissima* and some related anemones. *Biol. Bull.*, 144: 73.

Kaplan, S.W., 1983. Intrasexual aggression in *Metridium senile. Biol. Bull.*, 165: 416.

Purcell, J.E., 1977. Aggressive function and induced development of catch tentacles in the sea anemone *Metridium senile* (Coelenterata, Actinaria). *Biol. Bull.*, 153: 355.

Sebens, K.P., 1984. Agonistic behavior in the intertidal sea anemone *Anthopleura xanthogrammica. Biol. Bull.*, 166: 457.

9. Some form of asexual reproduction is encountered among all three classes of cnidarians. What are the adaptive benefits of asexual versus sexual reproduction?

Grassle, J.F., and Shick, J.M. (Eds.), 1979. Ecology of asexual reproduction in animals. *American Zoologist*, 19: 667.

10. How are the rhythmic swimming contractions mediated in free-living medusae?

Leonard. J.L., 1982. Transient rhythms in the swimming activity of *Sarsia tubulosa* (Hydrozoa). *J. Exp. Biol.*, 96: 181.

Lerner, J., S.A. Meleon, I. Waldron, and R.M. Factor, 1971. Neural redundancy and regularity of swimming beats in scyphozoan medusae. *J. Exp. Biol.*, 55: 177.

6

THE CTENOPHORES

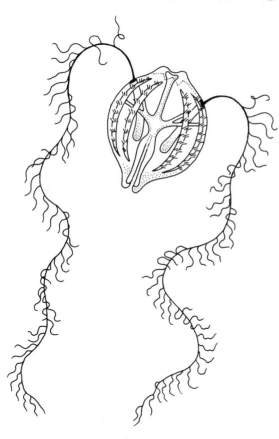

Phylum Cteno•phora
(G: COMB BEARING)

INTRODUCTION AND GENERAL CHARACTERISTICS

Ctenophores are transparent, gelatinous carnivores. All species (about 100) are marine, and most species are **planktonic**; i.e., most species are weak swimmers, carried about by ocean currents. Only one parasitic species is known. The body architecture of ctenophores is somewhat reminiscent of cnidarian medusae. Like the body of a medusa, the ctenophore body consists of an outer epidermis, an inner gastrodermis, and a thick, gelatinous middle mesoglea layer. Members of both groups are diploblastic; that is, only two distinct tissue layers form during embryogenesis. Both groups show a basic radial symmetry, with oral and aboral surfaces.

The digestive systems are similar in both groups as well. The mouth leads into a pharynx (also called a **stomodeum**), which serves as a site of extracellular digestion, and thence through a stomach into a series of **gastro-vascular canals**, where digestion is completed intracellularly. A functional excretory system has not been documented for either group, nor are any specialized respiratory organs found. The nervous system of ctenophores takes the form of a subepidermal nerve network, as in many cnidarians. Moreover, at least one ctenophore species has a planula larval stage in its life history. One species of ctenophore is known to use nematocysts in prey capture, but these nematocysts are obtained through the ingestion of cnidarian medusae; the nematocysts are not manufactured by the ctenophore. Lastly, ctenophore tentacles are solid, rather than hollow, as is generally the case among cnidarians as well.

Despite these similarities and despite a probable evolutionary relationship between the two groups, ctenophores are clearly not cnidarians. Polymorphism, an almost diagnostic characteristic of hydrozoans and scyphozoans, is never encountered among the Ctenophora. Moreover, the origin of the musculature is different in the members of the two groups, as is the swimming mechanism; the system of maintaining balance; the mechanism and mode of food capture; the means of eliminating solid wastes; the nature of sexuality; and several aspects of embryonic development. Each of these characteristics will be considered in sequence.

The muscles of ctenophores develop from amoeboid cells found within the mesoglea. Thus, the resulting muscle fibers actually reside in the mesoglea layers. In contrast, the musculature of cnidarians is found within the gastrodermis and, to a lesser extent, within the epidermis as well. The first known

giant smooth muscle fibers have recently been isolated from two ctenophore species, *Mnemiopsis leydii* and *Beroe* sp;[1] muscle preparations from these species should provide excellent material for general studies of smooth-muscle biology.

In many ctenophore species, the musculature plays little or no direct role in locomotion. Instead, swimming is commonly accomplished by the activity of many bands of partially fused cilia. Each band is called a **ctene** because of its resemblance to a comb (*ctene* = G: a comb). This explains the name of the phylum (i.e., "comb bearer"), and the common reference to its members as "comb jellies." The ctenes are organized into eight distinct rows, which are equally spaced about the body. These **comb rows**, or **costae**, extend from the oral to the aboral surface of the animal (Fig. 6.1) and are strikingly iridescent. The power stroke of the cilia comprising each ctene is toward the aboral surface, so that the typical ctenophore swims mouth-first. In summary, whereas cnidarian medusae swim by means of jet propulsion, ctenophore locomotion depends largely upon the coordinated activities of the partially fused cilia in the various comb rows.[2]

The intensity of activity in the different comb rows is under the control of a single **apical sense organ**, located at the aboral end of the ctenophore (Figs. 6.1a, 6.2). The **statolith**, a single sphere of calcium carbonate ($CaCO_3$), sits atop four tufts of fused cilia called **balancers**, or **springs**. Each balancer may consist of several hundred cilia. A cili-

---

1 Hernandez-Nicaise et al., 1984. *Biol. Bull.*, 167: 210.
2 See Topics for Discussion, No. 1.

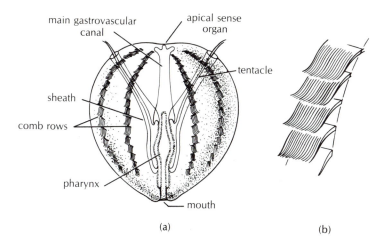

FIGURE 6.1. *(a) External anatomy of a ctenophore,* Pleurobrachia *sp., showing the comb rows and several other anatomical features. The tentacles have been withdrawn into their sheaths. (After Hyman.) (b) Detail of a comb row, showing three ctenes. (After Hardy.)*

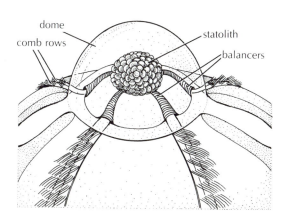

FIGURE 6.2. *Detail of an apical (aboral) sense organ and its transparent covering. Tilting of the animal causes the statolith to press against particular balancers, stimulating activity in the comb rows and thus restoring the animal's orientation. (From Hyman; after Chun.)*

ated groove radiates out from each balancer and bifurcates to service two adjacent comb rows. Experiments have shown these grooves to be agents of nerve impulse conduction from the apical sense organ to the ctenes of the comb rows. If the animal becomes tilted, the statolith presses against one of the balancers more than the others, causing the cilia of the comb rows associated with that balancer to increase their beat frequency until a satisfactory body orientation is restored. If the apical sense organ is surgically removed, the ctenophore continues to swim. However, the surgery obliterates any coordination of the beating of the cilia of different comb rows, attesting to the synchronizing role of the apical sense organ.

The epidermal nerve net also seems to play

a role in coordinating the activities of the ctenes. For example, mechanical stimulation of the oral end of the animal results in a sudden reversal of the ciliary beat in all comb rows. This response is observed even if the apical sense organ has been surgically removed.

Similarly to medusae, many ctenophore species capture their prey using tentacles. Unlike the tentacles of cnidarians, ctenophore tentacles can be completely retracted into proximal pits. Moreover, with one exception, the tentacle and/or the general epidermis of ctenophores is studded not with nematocysts but rather with quite different structures called **colloblasts** (Fig. 6.3). Each

colloblast consists of a bulbous, sticky head connected to a long, straight filament and a spiral, contractile filament. Prey organisms become stuck to the tentacles, which are then retracted. In many species, the body rotates to bring the mouth in contact with the tentacles once the food item is brought within range. The tentacles of some species may be extended more than twenty-five times the length of the body, and can be retracted in seconds.

In other species, the food-catching function of the tentacles is much reduced. Instead, the surface area of the body is increased by lateral compression, and major areas of the body are coated with a sticky

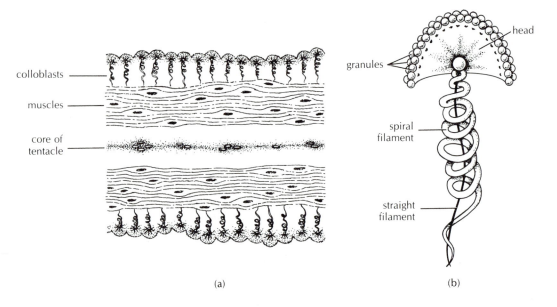

(a)

(b)

FIGURE 6.3. *(a) Longitudinal section through a tentacle. (From Hyman; after Hertwig.) (b) A single colloblast. The spiral filament is contractile. The head contains granules that when discharged, produce a sticky secretion, which traps prey. (From Hyman; after Komai.)*

mucus and with colloblast cells. The body itself thus becomes the major organ of food collection, and the small tentacles that are found in some of these species merely aid in transporting food to the mouth. The foods captured by most ctenophores are primarily small crustaceans (Arthropoda) and the larval stages of various fish and shellfish species, including some of commercial importance such as oysters. Some ctenophores are carnivorous on other ctenophores, or on gelatinous animals in other groups (especially the Cnidaria and the Urochordata).

The digestive systems of ctenophores and cnidarians differ in one interesting respect. Recall that the digestive tract of a medusa has but one opening, which serves as both mouth and anus. In ctenophores, on the other hand, four **digestive canals** lead from the roof of the stomach to the aboral surface of the animal (Fig. 6.4). Although two of the digestive canals terminate as blind sacs, the other two canals open to the outside. Undigested wastes are discharged through these **anal pores**.

All ctenophores described to date are simultaneous hermaphrodites; i.e., a single individual has both male and female gonads. In contrast, cnidarian medusae, and many anemone species, are dioecious; i.e., one sex per individual.

Ctenophore development differs in several respects from that of cnidarians. In particular, ctenophore cleavage is highly determinate; cell fates are fixed at the first cell division. Cell fates of cnidarian embryos become fixed later in development. Moreover, the mechanism of **gastrulation** in ctenophores (i.e., the formation of distinct inner and outer germ layers) is quite unlike that found in cnidarians. Among ctenophores, gastrulation is achieved either by **epiboly**, a process in which a sheet of micromeres spreads over what were the adjacent macromeres, or by **invagination**, in which groups of cells push into the blastocoelic space. In contrast, gastrulation among the Cnidaria occurs largely by **delamination** (Fig. 6.5). In delamination, the cells of the blastula divide with the cleavage plane approximately parallel to the surface of the embryo. Thus, the cells essentially divide into the blastocoel, forming an inner and outer layer, between which the mesoglea is later secreted. Gastrulation in some other cnidarians is by **ingression**, in which certain cells become detached from their neighbors and simply move into the blastocoel, creating a second layer of cells.

Ctenophore embryos rarely develop into ciliated planula larvae, in contrast to cnidarian embryos. Rather, the ctenophore embryo usually develops directly into a miniature ctenophore called a **cydippid**. The cydippid is approximately spherical in shape; is endowed with eight comb rows, a fully formed apical sense organ, and a pharynx; and usually bears a pair of branched tentacles. In many species, the cydippid closely resembles the adult. In other species, the cydippid undergoes a gradual, but substantial, alteration in morphology to attain the adult form.

A summary of certain ctenophore and cnidarian characteristics is given in Table 6.1.

All ctenophores seem to be **bioluminescent**. Unlike iridescence, in which colors are generated by the diffraction of incident light, bioluminescence is generated by a chemical reaction in which much of the

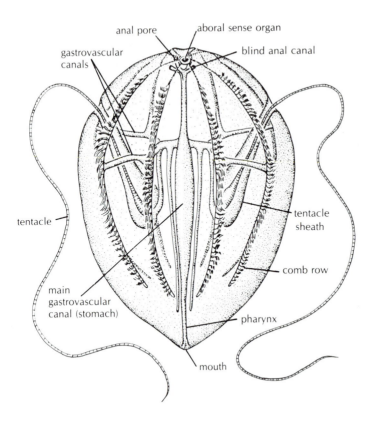

FIGURE 6.4.   *Diagrammatic illustration of the ctenophore,*
Pleurobrachia *sp., the sea-gooseberry. Note that the main gastrovascular
canal terminates as four branches near the apical sense organ. Two of
these branches end blindly, but the other two connect to the outside by
means of small anal pores. All other gastrovascular canals end blindly.
(After Bayer and Owre; after Hardy.)*

excess energy is given off as light rather than
as heat. Although the particular form of the
reaction taking place in ctenophores seems
to be peculiar to the members of this phylum, the phenomenon of bioluminescence is
not. Indeed, at least some species from most
major animal phyla display some form of bioluminescence. The functional significance
of bioluminescence is often unclear. Mate
location and species recognition, luring prey,
and startling would-be predators are possibilities that may apply to the bioluminescing
members of some phyla. Some species may
use bioluminescence to avoid detection by
visual predators, producing light equal to
ambient levels in intensity. This would break

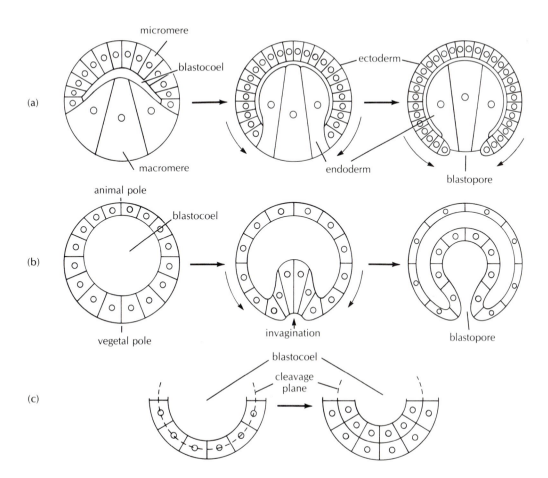

FIGURE 6.5.   *Patterns of gastrulation in ctenophores and cnidarians. (a) Gastrulation by epiboly, in which the smaller cells (micromeres) grow over the larger cells (macromeres). (b) Gastrulation by invagination, in which a group of adjacent cells indents into the blastocoel. (c) Gastrulation by delamination, in which the second cell layer is formed by mitotic division. In ctenophores, gastrulation is by epiboly or by invagination. In marked contrast, cnidarians gastrulate by delamination or ingression (in which cells on the outside of the embryo migrate inwards).*

TABLE 6.1. Similarities and Differences Between the Two Phyla of Diploblastic Gelatinous Animals.

CHARACTERISTIC	CTENOPHORES	CNIDARIANS
Cleavage	determinate	indeterminate
Gastrulation	epiboly or invagination	delamination or ingression
Common developmental stage	cydippid	planula
Digestive system	gastrovascular canals	gastrovascular canals
Nematocysts	none (unless "borrowed")	present
Colloblasts	present	none
Sexuality	hermaphroditic	dioecious
Musculature	within mesoglea	within gastrodermis

up the silhouette of the animal when observed from below by a potential predator, helping the lighted form blend into the surroundings. The adaptive significance of bioluminescence among ctenophores has not been examined, although the possibility of mate recognition can probably be ruled out by the lack of distinct photoreceptors. Regardless of its significance, the bioluminescence, together with the iridescence of the comb rows, the delicacy of the form, and the grace of movement, make ctenophores among the most glorious of living animals to observe.

CTENOPHORE DIVERSITY

Ctenophore species are divided into two classes, primarily on the basis of whether or not they possess conspicuous tentacles as adults and/or as cydippids. Representative adults are shown in Fig. 6.6.

CLASS TENTACULATA

Much of tentaculate evolution seems to be a story of modifying the mechanism of prey capture.[3] Some tentaculate species closely resemble the cydippid larvae already described, except, of course, that functional gonads are present. Long, retractable tentacles are well developed throughout life, and food is captured exclusively by these few tentacles and their side branches. These ctenophores comprise the order Cydippida (Fig. 6.7a). In other species, the body is somewhat compressed laterally; only four of the comb

3 See topics for Discussion, No. 2.

(a)

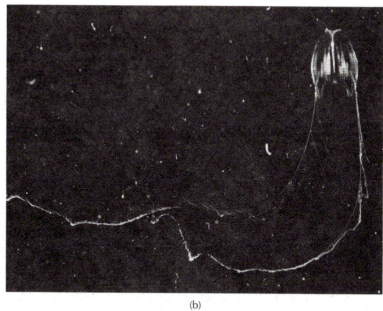

(b)

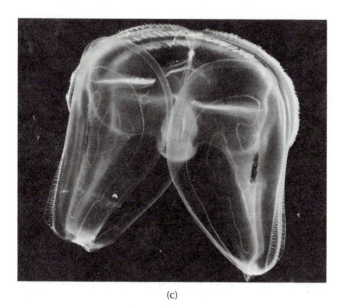

(c)

FIGURE 6.6. *Ctenophores photographed in the open ocean.*
(*a*) Eurhamphea vexilligera; (*b*) Callianira *sp.*; (*c*) Ocyropsis
maculata. (*Courtesy of G.R. Harbison and M. Jones.*)

rows are fully developed; and the tentacles are generally much reduced in length. Large **oral lobes**, covered with mucus and colloblasts, constitute the primary food collection surfaces. These are the lobate ctenophores (Order Lobata) (Fig. 6.7b). Muscular activity of the two oral lobes aids locomotion in some species.

In another group, the cestids, the body is so greatly compressed laterally that it now forms a long ribbon, with the mouth and apical sense organ on opposite sides at its midpoint (Fig. 6.7c). Swimming is accomplished primarily by sinuous, muscular movements of the body, although four of the comb rows are also well developed. Despite the great surface area of the body, prey are captured not by the ribbon itself, but rather by the numerous short tentacles extending along the extensive oral edge of the ctenophore.

Thus, selection for increased food collection capabilities seems to have favored a redistribution of colloblasts, together with an increased body surface area, in one group of tentaculate ctenophores—the Lobata—and an increased number of feeding tentacles in another group—the Cestida. Both adaptations have involved lateral compression of the body in comparison with the cydippid form, which is believed to represent the more primitive condition. In the fourth and final order of tentaculate ctenophores, the Platyctenea, the body is compressed in a different plane (Fig. 6.7d). Specifically, the oral and aboral surfaces have moved toward each other, so that the body forms a flattened plate (*platy* = G: flat). The bottom of the plate is formed largely by the pharynx, which is extensively everted. Some species simply float in the water. Other species spend their time creeping slowly over solid substrates, apparently using both the pharyngeal cilia and muscular contractions for locomotion. The only non-planktonic ctenophores are members of this order. Comb rows may be present, although these are often reduced or absent in the adults of many species. Some

FIGURE 6.7. *Ctenophore diversity. (a) The cydippid* Pleurobrachia sp., *approximately twice natural size. (After Hyman.) (b) A lobate ctenophore,* Mnemiopsis leidyi. *(After Hyman.) (c) A member of the order Cestida,* Cestum veneris, *commonly known as Venus's girdle. These animals attain lengths of about 1.5 meters. (From Hyman; after Mayer.) (d) A member of the Platyctenea,* Coeloplana mesnili, *viewed from above. The animal is about 6 cm in length, and spends its life in association with certain corals in the Indo-Pacific. (From Hyman; after Dawydoff.) (e) A member of the class Nuda,* Beroe sp. *The members of this class lack tentacles throughout life. (After Hardy.)*

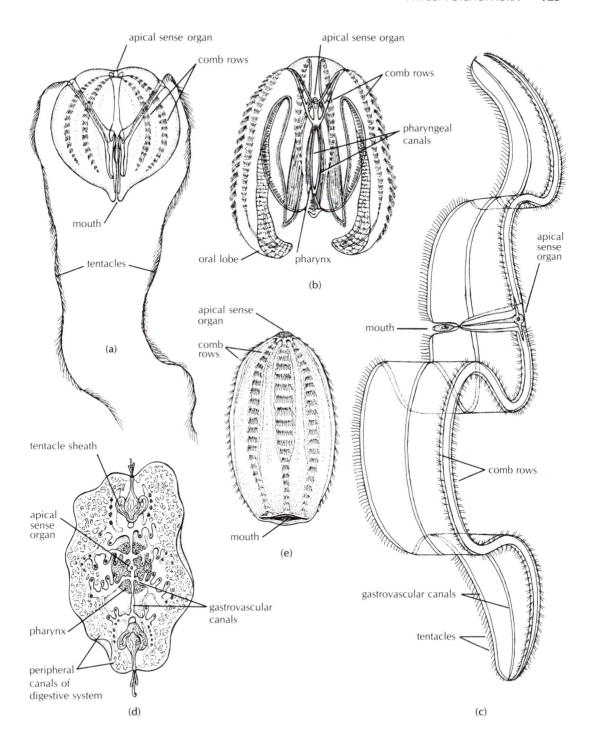

apical sense organ

comb rows

mouth

tentacles

(a)

apical sense organ

comb rows

pharyngeal canals

oral lobe

pharynx

(b)

apical sense organ

comb rows

mouth

(e)

tentacle sheath

apical sense organ

pharynx

peripheral canals of digestive system

gastrovascular canals

(d)

apical sense organ

mouth

comb rows

gastrovascular canals

tentacles

(c)

locomotion may also be accomplished by muscular flapping of the lateral lobes of the body. The adults generally bear two long tentacles.

CLASS NUDA

The members of this relatively small class are contained in a single order, the Beroida. All individuals in this group lack tentacles, even in the cydippid stage of development. Oral lobes are also lacking. The eight comb rows, however, are well developed (Fig. 6.7e). Prey, including other ctenophores, are captured and engulfed by muscular lips surrounding the mouth. **Macrocilia**, consisting of thousands of 9 + 2 axonemes enclosed by a single membrane, are located just inside the mouth; these macrocilia may be used as teeth during feeding. The mouth can be widened to accomodate prey substantially larger than the predator.   .

OTHER FEATURES OF CTENOPHORE BIOLOGY

Reproduction

Although all ctenophores appear to have substantial regenerative powers, asexual reproduction is not commonly encountered within the Ctenophora. A few species are known to reproduce asexually through a process of fragmentation and subsequent development of missing body parts by each fragment.

All ctenophores described to date are simultaneous hermaphrodites. The gonads are located on the walls of some of the gastrovascular canals, so that gametes are liberated into the digestive tract and are commonly discharged through the mouth. Fertilization of the eggs is generally external; i.e., it occurs in the surrounding sea water. All species pass through a cydippid stage of development.

◄ T A X O N O M I C   S U M M A R Y

PHYLUM CTENOPHORA
CLASS TENTACULATA
ORDER CYDIPPIDA
ORDER LOBATA
ORDER CESTIDA
ORDER PLATYCTENEA
CLASS NUDA
ORDER BEROIDA

TOPICS FOR FURTHER DISCUSSION AND INVESTIGATION

1. What role does mechanical interaction between adjacent ctenes play in coordinating the beating of cilia in a comb row?

Tamm, S.L., 1973. Mechanisms of ciliary coordination in ctenophores. *J. Exp. Biol.*, 59: 231.

Tamm, S.L., 1983. Motility and mechanosensitivity of macrocilia in the ctenophore *Beroë*. *Nature*, 305: 430.

2. Compare and contrast the feeding biology of lobate, cydippid, and cestid ctenophores.

Harbison, G.R., L.P. Madin, and N.R. Swanberg, 1978. On the natural history and distribution of oceanic ctenophores. *Deep-Sea Res.*, 25: 233.

Main, R.J., 1928. Observations of the feeding mechanism of the ctenophore, *Mnemiopsis leidyi*. *Biol Bull.*, 55: 69.

Reeve, M.R., M.A. Walter, and T. Ikeda, 1978. Laboratory studies of ingestion and food utilization in lobate and tentaculate ctenophores. *Limnol. Oceanogr.*, 23: 740

Swanberg, N., 1974. The feeding behavior of *Beroë ovata*. *Marine Biol.*, 24: 69.

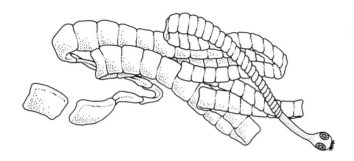

7

THE PLATYHELMINTHES

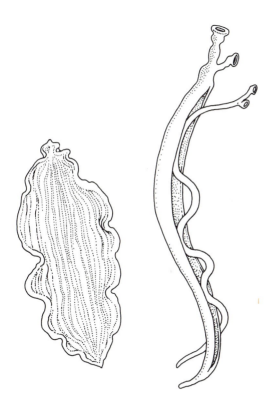

Phylum Platy•helminthes
(G: FLAT WORM)

INTRODUCTION AND GENERAL CHARACTERISTICS

This group of some 13,000 species contains one class of free-living individuals and two classes of exclusively parasitic individuals. All flatworms are triploblastic, acoelomate, and bilaterally symmetric. Their development is protostome-like, in that cleavage is spiral and determinate, and the mouth forms from the blastopore. Most species have a conspicuous anterior brain, which is connected to at least one pair of longitudinal nerve cords. In the most advanced species, only a single pair of nerve cords is present, and these are always located ventrally. The mesodermal layer of the embryo develops into a loose collection of cells known as **parenchyma** tissue. This tissue occupies the entire space between the outer body wall and the endoderm of the gut. Like the cnidarians, there is no anus found among members of this phylum; a single opening to the digestive system serves as both mouth and anus.

Perhaps the most conspicuous unifying feature of platyhelminths is that they are flat. They have no specialized respiratory organs, all gas exchange being accomplished by a simple diffusion across the body surface. The rate at which such exchange can occur (milliliters of oxygen transported from the surrounding medium into the tissues per unit of time) is dependent upon several factors: the oxygen concentration gradient across the body wall; the permeability of the body wall to gas; the thickness of the body wall; and the total exposed surface area across which diffusion can occur. By being flat, the flatworms achieve a high surface area relative to their enclosed volume, and a sufficient level of gas exchange can occur to support an active lifestyle, despite the lack of gills and an internal circulatory system.

Wastes may move out of flatworms by diffusion across the general body surface. Again, being flat is helpful. In addition, most platyhelminths contain a series of specialized organs called **protonephridia** (G: "first kidney"). The typical protonephridium consists of a group of cilia projecting into a fine-meshed cup (Fig. 7.1). The beating of the cilia within the cup has been likened to the flickering of a flame; thus, the common name for this type of cell is **flame cell**. Other protonephridia take the form of **solenocytes**, in which a single flagellum is found within the cup. In both cases, the mesh cup is attached to a long, convoluted tubule that connects to the outside of the animal through a single opening, the **nephridiopore** (G: "kidney-hole"). The inner workings of the protonephridium are not fully understood. Apparently, the beating of the flagella or cilia creates a negative pressure, drawing fluid through the mesh cup and into the nephridial tubule. The liquid entering the nephridium is thus ultra-filtered; large molecules (e.g., proteins) are excluded by their inability to pass through the openings in the wall of the cup. Measurements made on the liquid entering and leaving protonephridia indicate that the chemical composition of the fluid changes as it moves along the tubule to the nephridiopore: ions may be selectively absorbed or secreted, and the water content may be altered as well. Protonephridia thus likely play an important role in regulating ionic and water balance in flatworms, in addition to their possible role in the elimination of metabolic wastes such as ammonia.

The vast majority of flatworm species, found among all three classes, are **simultaneous hermaphrodites**. That is, each individual can, at any one time, function as both a female and a male. As a consequence, the exchange of sperm and fertilization of eggs can occur when any individual encounters another individual of the same species. A single individual generally cannot fertilize itself, although exceptions do exist, as discussed shortly.

CLASS TURBELLARIA

About 25% of all flatworms are turbellarians. The members of this class are mostly free-living (i.e., nonparasitic) individuals, and are found primarily in aquatic environments. Most species are marine; a number of species are found in fresh water; and a few species are considered to be terrestrial, although these can live only in very humid areas. The high surface-area-to-volume ratio of tur-

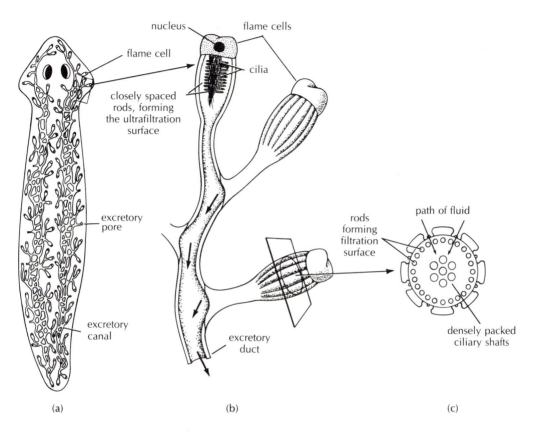

nucleus

flame cells

flame cell

cilia

closely spaced
rods, forming
the ultrafiltration
surface

excretory
pore

rods
forming
filtration
surface

path of fluid

excretory
canal

excretory
duct

densely packed
ciliary shafts

(a)

(b)

(c)

FIGURE 7.1. *(a) The elaborate branching excretory system of a freshwater free-living flatworm. Arrows indicate direction of fluid flow. (From Schmidt-Nielson, 1983. Animal Physiology: Adaptation and Environment, 3rd ed. Cambridge University Press.) (b) Detail of several flame cells emptying into a common collecting duct. (Modified from several sources.) (c) Cross section of a flame cell at the level of the ciliary bundle.*

bellarians makes them especially prone to dehydration in air. Most individuals are less than about 1 cm in length, regardless of habitat, although members of some marine species are considerably longer.

The nervous system consists of a coelenterate-style, diffuse nerve net in the most primitive turbellarian species. Increasing compactness of the system is associated with advancement in the class, culminating in the possession of a distinct brain and a single pair of longitudinal nerve cords (Fig. 7.2). Such an advanced nervous system is also characteristic of all parasitic members

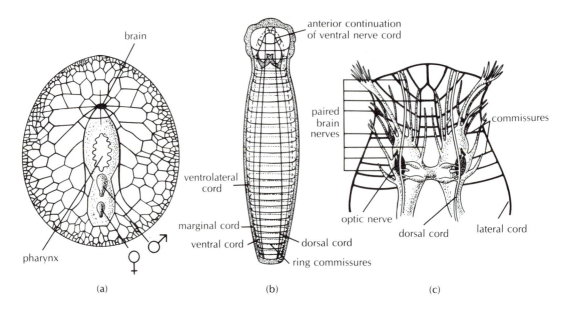

FIGURE 7.2. *The nervous system of (a)* Planocera *and
(b)* Bothrioplana. *(c) Detail of the brain of* Crenobia. *[From Bayer
and Owre, 1968.* The Free-Living Lower Invertebrates. *Macmillan;
after Lang (a); after Reisinger (b); after Micoletzky (c).]*

of the phylum (Classes Cestoda and Trematoda).

Most aquatic turbellarian species are **benthic**, i.e., they live in or on the ocean, lake, pond, or river bottom. The outer surface of the body is ciliated, often more so on the ventral surface than on the dorsal surface. Most species move at least partly by secreting mucus from the ventral surface and beating the ventral cilia within this viscous mucus. As a consequence of being flat, increased size is accompanied by a substantial increase in the amount of surface area in contact with the substrate over which the animal is moving. Thus, an increased number of cilia in contact with the substrate compensates for the increased weight of a larger animal, and

the ability to move need not suffer as the animal grows.

The locomotion of many individuals involves subtle waves of muscular contraction along the ventral surface of the animal. These **pedal waves** are unidirectional, moving from the anterior of the worm posteriorly. As a wave moves down the length of the body, small portions of the ventral surface are pulled up, away from the substrate. Appropriate muscle contractions thrust and pull the raised portion forward, in the direction of locomotion (Fig. 7.3a). The magnitude of the muscle contractions involved in generating pedal waves is quite small, and a number of waves are generally in progress down the body simultaneously. Thus the progress of the animal along the substrate is very graceful, and nearly indistinguishable from that powered entirely by the action of cilia.

The musculature of the body wall includes fibers running longitudinally, circumferentially, dorso-ventrally, and diagonally (Fig. 7.3b). All of this musculature is brought into play in the locomotory movements of some turbellarians. This movement, called **looping**, is quite pronounced. The individual attaches at the anterior end, pulls the posterior forward, attaches at the posterior end, releases the anterior end, and thrusts the body forward (Fig. 7.4).

For looping locomotion to be effective, the flatworm must be able to adhere locally to the substrate, in order to prevent sliding backward while the pulling and pushing forces are being generated. On the other hand, attachments to the substrate must be only temporary if forward progress of the animal is to occur. Flatworms typically possess a large number of paired secretory cells (**duoglands**) located on the ventral surface and opening to the exterior. One cell of each pair seems to produce a viscous glue, while the other cell of the pair presumably secretes a chemical that breaks this attachment to the substrate. (Fig. 7.5).

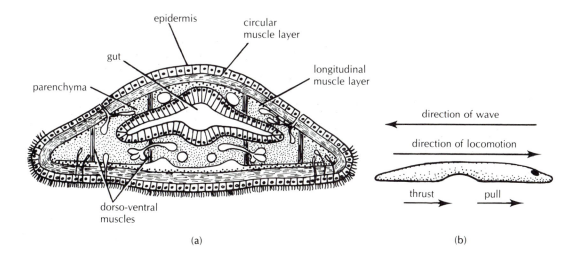

(a)

(b)

FIGURE 7.3. (a) Diagrammatic illustration of a turbellarian in cross section, showing the arrangement of muscle layers. (b) Locomotion by pedal waves in a turbellarian flatworm. A single wave is shown traversing the ventral surface of the body. A small portion of the body is both thrust and pulled forward as the wave of dorsoventral contraction lifts that region of the body away from the substrate. The thrusting and pulling forces are generated by the regions of the body that are anchored against the substrate, on either side of the contraction. In most species, many waves of contraction travel down the body simultaneously.

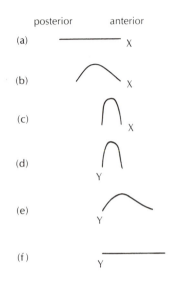

FIGURE 7.4. *Diagrammatic illustration of looping by turbellarians. Once an attachment to the substrate is made at point x (part a), contraction of the longitudinal musculature brings the rear of the animal anteriorly (b–c). The anterior attachment is released and a new attachment is made at the posterior end (d), at point y, allowing the head of the animal to be thrust forward (e–f).*

A few turbellarian species are capable of swimming, by means of vigorous but controlled waves of contraction of the body wall musculature.

The digestive system of turbellarians is fairly simple, although the details vary con-

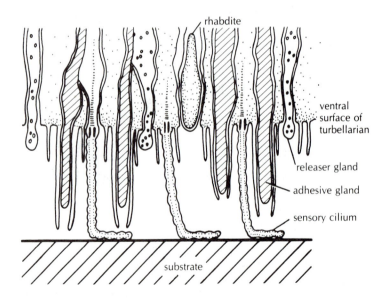

FIGURE 7.5. *Diagrammatic illustration of the ventral surface of a free-living turbellarian flatworm, showing arrangement of the duo-gland system. Adhesive glands produce a chemical that attaches part of the animal to a substrate; the releaser glands secrete a chemical that dissolves the attachment as appropriate. Rodlike rhabdites, as illustrated, are encountered in the epidermis of most turbellarians; a defensive function has been suggested. (Modified from Tyler, 1976. Zoomorphol. 84: 1)*

siderably among species. Indeed, differences in the structure of the mouth parts and the gut are used to divide the class into about half a dozen constituent orders. Some species bear a simple mouth opening on the ventral surface and have no well-formed gut cavity. In these flatworms, food is essentially thrust into a densely packed mass of specialized digestive cells. (Fig. 7.6a). Members of this order (the Acoela) lack a gut cavity

(*a* = G: without; *coel* = G: a cavity), are exclusively marine, and bear a superficial, morphological resemblance to the planula larvae of cnidarians. In fact, many zoologists suspect that both cnidarians and flatworms evolved from a planula-like ancestor, in which case the acoel turbellarians resemble the most primitive triploblastic metazoans, from which all other metazoans are presumably derived.

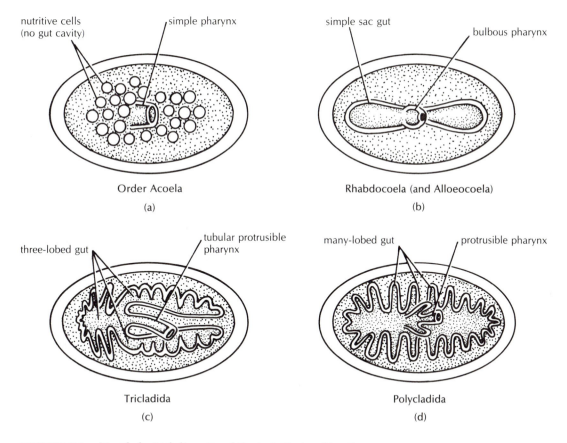

FIGURE 7.6. *Morphological diversity of the turbellarian digestive system, ranging from (a) no well-defined gut; (b) a gut that is single-branched; (c) a triple-branched gut; (d) a multi-branched gut. (Reprinted with permission of Macmillan Publishing Company from* A Life of Invertebrates *by W. D. Russell-Hunter. Copyright © 1979 by W. D. Russell-Hunter.)*

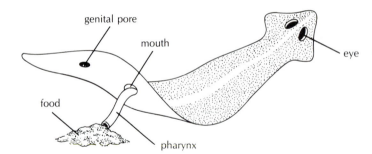

FIGURE 7.7. *Typical flatworm (Planaria) with pharynx extended for feeding. (Modified from various sources.)*

Subsequent flatworm evolution seems to have been largely a tale of increasing complexity of the digestive system and of the means of acquiring food, and increasing specialization of the reproductive system. The guts of flatworms beyond the level of acoel development may be straight, double-branched, or multibranched. (Fig. 7.6). The mouth of more advanced species is often borne at the end of a **protrusible pharynx** (Fig. 7.7); other species possess a separate proboscis that spears food items and then transfers them to an adjacent mouth opening. Most turbellarian species are active carnivores, although some species ingest detritus and algae.[1] Digestion of food is initially extracellular through secretions of enzymes. Particles are later phagocytized and digestion is completed intracellularly.

The most structurally and physiologically sophisticated system found among turbellarians is the reproductive system. Both male and female reproductive organs are found within a single individual, and both reproductive systems are anatomically complex (Fig. 7.8). During mating, generally

1 See Topics for Discussion, No. 3.

FIGURE 7.8. *(a) Triclad turbellarian reproductive system. Note the presence of both male and female reproductive organs in a single individual. The male reproductive system consists of the testes, prostate gland, and the ducts (vas deferens and seminal vesicle) that conduct sperm from the testes to the ejaculatory duct of the muscular penis. Sperm accumulates in the seminal vesicle prior to copulation. The female reproductive tract includes the ovaries, oviducts, and the gonopore, through which fertilized eggs are released. As eggs move down the oviduct toward the gonopore, they become surrounded by nutritive yolk cells manufactured by the vitellaria. The copulatory bursa serves to receive and store sperm contributed by the partner. Some species possess an additional sperm storage organ, the seminal receptacle. During copulation in triclads, the penis is inserted into the female gonopore. In most other turbellarian groups, copulation may occur by hypodermic impregnation, in which the stylets of the penis pierce the body of the partner. Cement glands coat the emerging ova with a sticky substance that glues them to a variety of solid substrates, including macroalgae and other vegetation. (b) Detail of copulatory apparatus and associated structures. (After Steinmann.)*

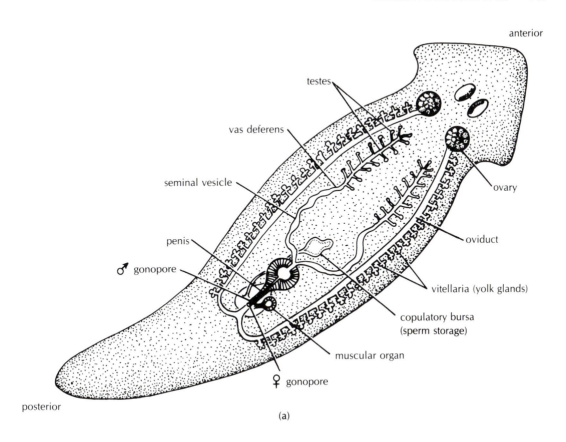

(a)

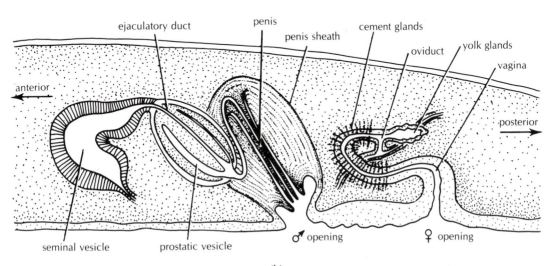

(b)

each individual inserts its penis into the female opening of the other member of the pair, so that sperm transfer is reciprocal (Fig. 7.9a). The eggs of each animal are released after fertilization and generally develop directly into miniature flatworms within a protective capsule; in the majority of turbellarian species, a free-living larval stage is absent from the life cycle. In several marine species, however, the developing embryo gives rise to a short-lived, free-swimming larval stage, most commonly a **Müller's larva** (Fig. 7.9b).

Turbellarians possess remarkable regenerative powers, as illustrated in Fig. 7.10.[2]

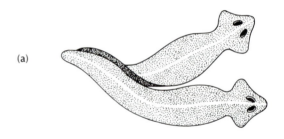

(a)

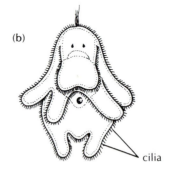

(b)

cilia

CLASS CESTODA

The cestodes, most of which are commonly known as tapeworms, are all internal parasites. They are primarily parasites of vertebrates, inhabiting various regions of the host digestive tract. Cestodes are strikingly dissimilar from turbellarians, the differences between the two reflecting an extremely high degree of cestode specialization for an endoparasitic existence. Instead of the ciliated epidermis characteristic of the Turbellaria, the cestodes are covered by a nonciliated **tegument**. The tegument contains numerous nuclei, but these are not separated by cell membranes. That is, the tegument is **syncytial**. The outer surface is outfolded into numerous cytoplasmic projections, vastly increasing the amount of exposed surface area across which nutrients can be taken up from

FIGURE 7.9. *Copulating flatworms. The penis of each individual is inserted into the female opening of the mate. (From Hyman, 1940. The Invertebrates, Vol. I. McGraw-Hill. Reproduced by permission.) (b) Müller's larva of a marine flatworm. (After several sources.)*

the gut of the host. Indeed, the cestode must receive all of its nutrients in this manner, as it has no mouth or digestive tract of its own at any point in its life cycle.

Although lacking a mouth, cestodes do have an anterior end, which in most species takes the form of a **scolex**. The scolex is studded with hooks and/or suckers and is used to maintain position within the gut of the host.

However, the essence of cestode existence lies just posterior to the scolex, in a region known as the **neck**. A seemingly endless se-

---

2 See Topics for Discussion, No. 7.

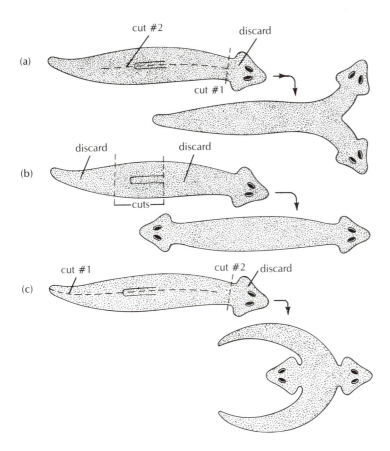

FIGURE 7.10. *Three experiments illustrating the considerable powers of regeneration possessed by free-living flatworms. Cuts are made using a clean razor blade. (Reprinted with permission of Macmillan Publishing Company from* The Invertebrates: Function and Form, *2nd ed., by Irwin W. Sherman and Vilia G. Sherman. Copyright © 1976 by Irwin W. Sherman and Vilia G. Sherman.)*

ries of sections called **proglottids** bud from the neck of most cestodes at a rate of several per day (Fig. 7.11). Each proglottid is involved primarily with the process of sexual reproduction. Not only is each cestode a simultaneous hermaphrodite, but so is each proglottid. That is, each proglottid contains both male and female reproductive systems (Fig. 7.12). The tapeworm may be thought of as a franchiser of eggs and sperm. And like the best franchising operations, a large amount of product, qualitatively similar from proglottid to proglottid, is turned out daily.

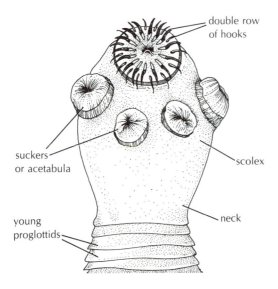

double row
of hooks

suckers
or acetabula

scolex

young
proglottids

neck

FIGURE 7.11. *Anterior end of the common tapeworm,* Taenia, *at a magnification of 12×.* (Modified from several sources.)

Each proglottid contains perhaps 50,000 eggs. The eggs are generally fertilized by sperm from a neighboring cestode, but can be fertilized by sperm from the same individual—or, in fact, from the same proglottid. One way or another, those eggs are going to be fertilized. Considering that proglottids are rarely more than 3–5 mm in length, and that the total length of a cestode may be 10–12 meters or more, this means a production of perhaps 2500–4000 proglottids per individual in many species, and an incredible number of fertilized eggs. We're talking about the production of tens of thousands, or even millions, of eggs per day per cestode.

In some species, the posterior-most pro-glottids break off periodically. In other species, these proglottids burst open, releasing the fertilized eggs into the gut of the host. Either way, the eggs leave the body of the host along with the feces. Generally, the eggs cannot take up residence in the final, or **definitive host**, immediately, but must first enter an **intermediate host**, or, in some species, a series of intermediate hosts. Different cestode species require different intermediate hosts, which include both vertebrates and invertebrates.

When a cestode egg is ingested by the appropriate intermediate host, an **oncosphere** larva commonly hatches out. Each oncosphere has muscles, flame cells, and, most significantly, three pairs of hooks with which it attaches to the wall of the host digestive tract (*oncus* = G: a barb or hook). The oncosphere then **lyses** (dissolves) its way through the intestinal wall, taking up residence as an encysted form in the coelomic space or in specific organs and tissues of the host. Further development is arrested until the intermediate host is eaten by a different host, which may be the final host or another intermediate host. Thus, the complete sequence from egg to adult is achieved only if the fertilized cestode eggs are ingested by the appropriate intermediate hosts, which in turn must be ingested by the appropriate final, or definitive, host (Fig. 7.13). Among the vertebrates, fishes, cows, pigs, dogs, and sometime birds may serve as intermediate hosts. Humans often serve as acceptable final and intermediate hosts—think twice before eating undercooked beef, pork or fish, and before letting a dog lick your face! Among the invertebrates, most intermediate hosts are arthropods.

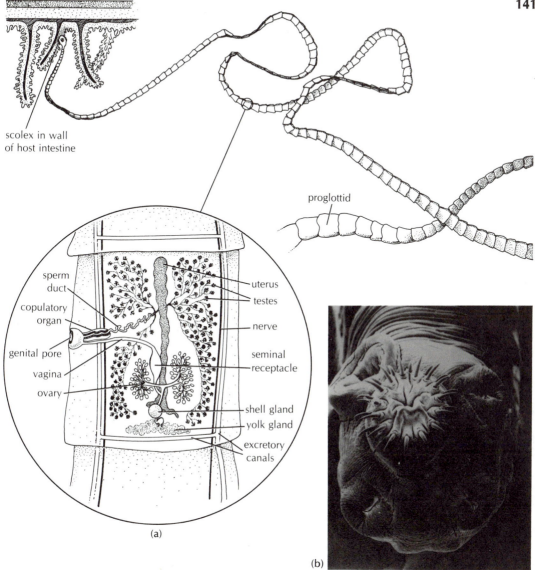

scolex in wall
of host intestine

proglottid

sperm
duct

copulatory
organ

genital pore

vagina

ovary

uterus

testes

nerve

seminal
receptacle

shell gland

yolk gland

excretory
canals

(a)

(b)

FIGURE 7.12. (a) Taenia solium, *the pork tapeworm, attached to the intestinal wall of its host. Note that the entire proglottid is dedicated to the task of reproduction. Fertilized eggs are enclosed individually in protective capsules in the shell gland. (Modified from various sources.) (b) Scanning electron micrograph of the anterior end of* Taenia hydatigena, *magnified 170×. (Courtesy of D.W. Featherston, from Featherston, 1975. Internat. J. Parastol. 5: 615, Pergamon Press.)*

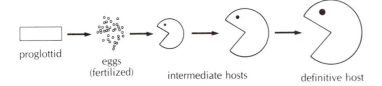

proglottid

eggs
(fertilized)

intermediate hosts

definitive host

FIGURE 7.13. *Diagrammatic illustration of the typical cestode life cycle. The parasite cannot attain sexual maturity until the definitive host is located.*

CLASS TREMATODA

A parasitic existence poses a number of problems, none of them trivial. The successful parasite must:

(a) reproduce within the host;
(b) get fertilized eggs or embryos out of the host;
(c) contact and recognize a new, appropriate host;[3]
(d) obtain entrance into the host;
(e) locate the appropriate environment within the host;
(f) maintain position within the host;
(g) withstand what is often a rather anaerobic (oxygen-poor) environment;
(h) avoid digestion or attack by the immune system of the host;[4]
(i) avoid killing the host, at least until reproduction has been achieved.

These problems are faced by all parasites, but perhaps the most remarkable adaptations to (c) and (d) above are encountered among the trematodes, or "flukes," all of which reach

adulthood as parasites in or on vertebrates.

The outer body layer of trematodes is, like that of the cestodes, a syncytial tegument. In other respects, the trematode body more closely resembles that of a turbellarian. Trematodes have a mouth opening and a blind-ended digestive tract (usually bi-lobed), and the body is never segmented (Figs. 7.14, 7.15). The animal actually ingests the tissues and blood of the host through its mouth. Schistosomiasis, an often deadly disease prominent in many regions of the world, results from an infection by trematodes (Fig. 7.15b). It has been estimated that more than 200 million persons presently suffer from schistosomiasis in more than 72 countries, making this the second most prevalent disease in the world, next to malaria (which is caused by a protozoan—see Chapter 3).

The flukes are divided into two groups; the **monogenetic trematodes** and the **digenetic trematodes**. These are also known as the **monogeneans** and **digeneans**, respectively. Members of the two groups are anatomically similar as adults, but differ greatly in the relative complexity of their life cycles and in the morphology of their larval stages. Mono-

---

3 See Topics for Discussion, No. 4.
4 See Topics for Discussion, No. 5.

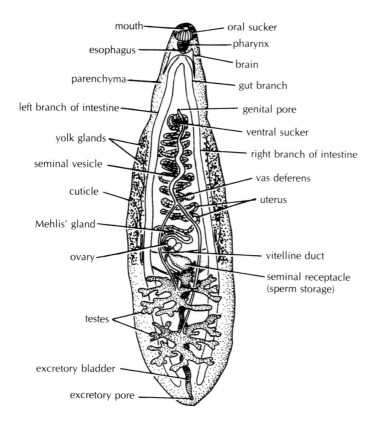

mouth
oral sucker
esophagus
pharynx
brain
parenchyma
gut branch
left branch of intestine
genital pore
ventral sucker
yolk glands
seminal vesicle
right branch of intestine
cuticle
vas deferens
uterus
Mehlis' gland
ovary
vitelline duct
seminal receptacle
(sperm storage)
testes
excretory bladder
excretory pore

FIGURE 7.14. *The Chinese liver fluke,* Opisthorchis sinensis. *Note the large percentage of the body devoted to reproduction. Mehlis' gland, also called the ootype, is a conspicuous feature of the female reproductive tract; its function is unknown (From Brown, 1950. Selected Invertebrate Types. John Wiley & Sons.)*

genetic flukes are usually parasites on the skin or gills of fishes; that is, they are **ecto-parasitic** on fishes. This is the fluke's sole host, to which it attaches by means of suckers and hooks located at the anterior and posterior ends of the fluke. There are no intermediate hosts, so that the life cycle generally involves the following stages:

sexual maturity reached in or on fish → egg production →

→ larval stage (onchomiracidium) → attachment to fish

The taxonomic status of the monogeneans is currently in question, as it has been for some time. Some workers feel that the members of this group are more closely allied with

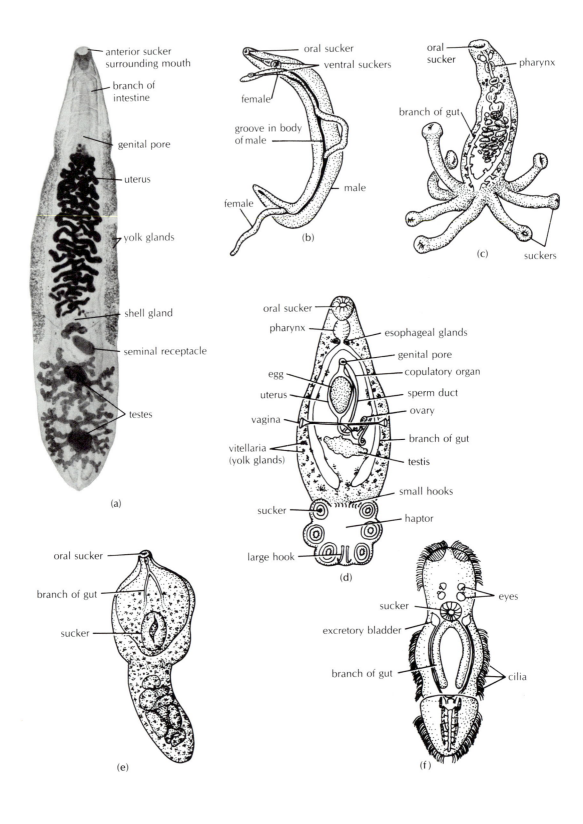

(a)

anterior sucker surrounding mouth
branch of intestine
genital pore
uterus
yolk glands
shell gland
seminal receptacle
testes

(b)

oral sucker
ventral suckers
female
groove in body of male
male
female

(c)

oral sucker
pharynx
branch of gut
suckers

(d)

oral sucker
pharynx
esophageal glands
genital pore
copulatory organ
egg
sperm duct
uterus
ovary
vagina
branch of gut
vitellaria (yolk glands)
testis
small hooks
sucker
haptor
large hook

(e)

oral sucker
branch of gut
sucker

(f)

eyes
sucker
excretory bladder
branch of gut
cilia

the cestodes than with digenetic trematodes, largely on the basis of similarity between the oncosphere of cestodes and the onchomiracidium larva of monogeneans. Others feel that the monogenes are sufficiently dissimilar from either cestodes or other trematodes to merit categorization as a separate class, the Monogenea.

Certainly the life cycles of the monogeneans and digeneans have little in common. As the name implies (*di* = G: two; *gena* = G: birth), digenetic flukes always require at least one intermediate host before reaching the final host. Unlike the passive process found among the cestodes, host location among digenetic trematodes is generally an active process, mediated by highly specialized, free-living larval stages.

Among the digenetic trematodes, each fertilized egg generally gives rise to a single, free-living, ciliated **miracidium** larva (Fig. 7.16). The miracidium is then either eaten by, or locates and bores into, an intermediate host, which is always an invertebrate—most commonly a snail. The miracidium then dedifferentiates into a **sporocyst** stage, in which most of the structures of the miracidium are lost (including the external cilia), except for the protonephridia. As no mouth is present at this stage of development, the sporocyst grows within the host through the uptake of dissolved nutrients from the surrounding body fluids. Within each sporocyst are numerous balls of cells. Each of these **germ balls** develops into another sporocyst (the daughter sporocysts), or into another larval

FIGURE 7.15. *Trematode diversity. (a) The Chinese liver fluke,* Opisthorchis sinensis. *Note the two-branched gut and the proportion of the body devoted to reproduction. (Photograph by Carolina Biological Supply Company.) (b) The blood fluke,* Schistosoma haematobium, *a dioecious species. During copulation, the female lies in a specialized groove in the body of the male. This and several other related species cause the disease known as schistosomiasis in humans. "Swimmer's itch" is caused by members of the same family. (From Hyman; after Looss.) (c)* Choricotyle louisianensis, *taken from the gills of a fish. (After Noble and Noble, 1982. Parasitology—the Biology of Animal Parisites, 5th ed. Lea & Febiger; after Hargis.) (d)* Polystomoidella oblongum, *taken from the urinary bladder of a turtle. Note the large number of suckers in* C. louisianensis *and* P. oblongum *for securing attachment to the host tissues. (From Noble and Noble; after Cable.) (e)* Neodiplostomum paraspathula *seen in ventral view. (From Noble and Noble; after Noble.) (f) Onchomiracidium larva of* Benedenia melleni. *Note the prominent eyes and the oral sucker, with which the larva attaches to a future host. (From Baer, 1951. Ecology of Animal Parasites. University of Illinois Press; after Jahn and Kuhn.) Figures (a), (b), and (e) are digenetic trematode species, while (c), (d), and (f) are monogenetic trematodes.*

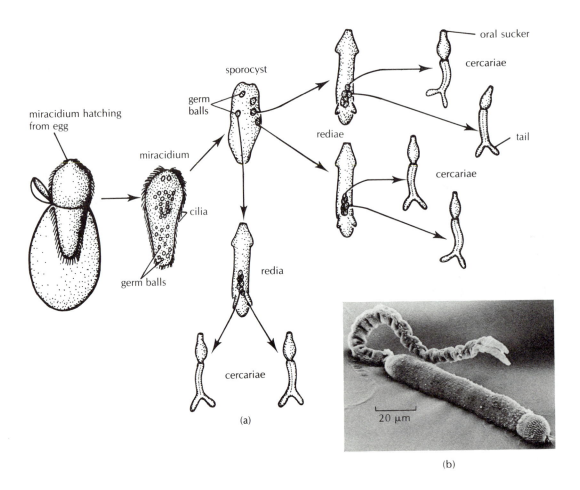

(a)

(b)

FIGURE 7.16. *(a) Replication of larval stages in the trematode life cycle. Large numbers of cercariae are contained within each redia. Many rediae are, in turn, contained within each sporocyst. All of the larvae derived from a single fertilized egg are produced asexually, and are therefore genetically identical with each other. The miracidia and cercarial stages are commonly free-swimming. The sporocysts and redia stages occur within an intermediate host, usually a snail. Note the germ balls in the miracidium and sporocyst stages; these will develop into redia larvae. Similarly the germ balls within the rediae are future cercariae. (Modified after Noble and Noble.) (b) Scanning electron micrograph of the cercaria stage of Aporocotyl simplex. Note the elongated, cylindrical head and the forked tail. (Courtesy of M. Køie, from* Ophelia. *(1982) 21: 115.)*

PHYLUM PLATYHELMINTHES **147**

stage, the **redia**. The rediae are active feeders, possessing a mouth and functional, blind-ended gut. Within each redia larva, or within each daughter sporocyst, are numerous germ balls, each of which develop into yet another anatomically distinct larval stage, the **cercaria**. The cercariae leave the "mother" by means of a birth canal, and then bore their way out of the intermediate host, or, in some species, wait within the intermediate host until that host is eaten by a predator.

Free-swimming cercaria larvae are nonciliated, but can swim actively by means of a muscular tail. Cercariae usually possess at least one sucker anteriorly, and in many species may have a ventral sucker as well. The cercaria larvae of some species transform into adults once they encounter the next host, attach to it, and enzymatically penetrate the host tissues. In most species, however, the next stop is another intermediate host. In any event, penetration by the cercaria larva into the next host is accompanied by detachment of the cercarial tail, which remains outside the host's body. Once in the second intermediate host, the cercaria may, in some trematode species, transform into an encysted waiting stage, the **metacercaria**, in which most of the specifically larval organs (e.g., adhesive suckers) degenerate. The adult trematode develops only when the second intermediate host is ingested by the appropriate definitive host. In some cases, one or more additional intermediate hosts may be involved in the life cycle before the final host is reached. Once eaten by the definitive host, the juvenile trematode migrates through the digestive tract to take up residence in the proper, species-specific location within the host, where adulthood is reached.

The adult digenetic trematode has a mouth and a blind-ended gut, usually two-branched, and often has both an anterior and a ventral sucker for maintaining position within the host. Most species are hermaphroditic, although some have separate, anatomically distinct sexes; these species are dioecious. Most of the body is taken up by the reproductive system.

The life cycle of digenetic trematodes is, to say the least, highly complex: several hosts are required for completion of the life cycle; the hosts often occupy very different habitats, complicating the movement from one host to the next; and both the intermediate larval stages and the adult stage often require specific host species (Figs. 7.17 and 7.18). The free-living larval stages are nonfeeding and can generally remain alive for only a few hours, during which time the next host animal must be located. Clearly, the probability that any given fertilized egg will reach adulthood is very small. Therein lies the adaptive significance of producing tremendous numbers of dispersal stages. Cestodes achieve a high rate of offspring production by franchising egg and sperm manufacture, and by adding on new franchisees (i.e., proglottids) daily. Among the digenetic trematodes, a high rate of offspring production is accomplished through the geometric multiplication of larval stages. Each female generally has only a single ovary, and each fertilized egg develops into only a single miracidium larva. Yet, within each miracidium are the germ balls of the redia stage (or of an additional sporocyst stage), within which are the germ balls of the

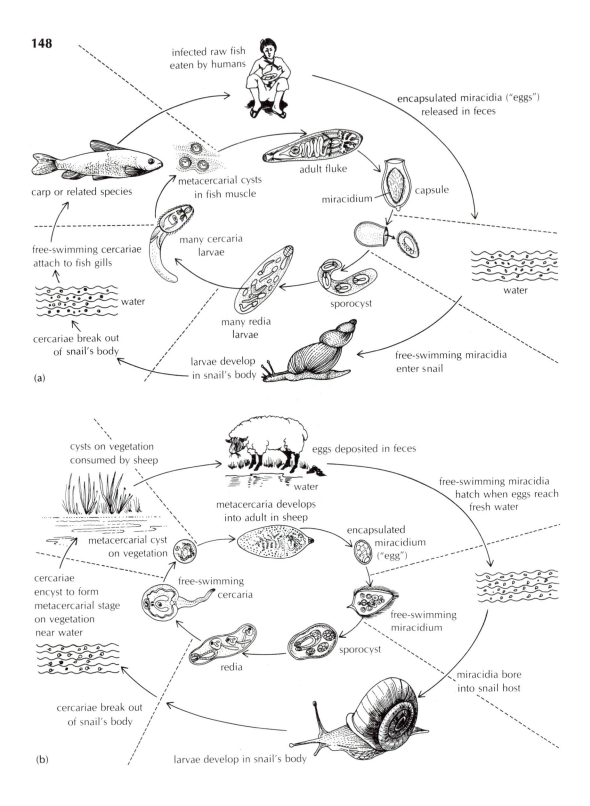

infected raw fish
eaten by humans

encapsulated miracidia ("eggs")
released in feces

metacercarial cysts
in fish muscle

adult fluke

carp or related species

miracidium

capsule

free-swimming cercariae
attach to fish gills

many cercaria
larvae

water

water

many redia
larvae

sporocyst

cercariae break out
of snail's body

larvae develop
in snail's body

free-swimming miracidia
enter snail

(a)

cysts on vegetation
consumed by sheep

eggs deposited in feces

free-swimming miracidia
hatch when eggs reach
fresh water

water

metacercaria develops
into adult in sheep

encapsulated
miracidium
("egg")

metacercarial cyst
on vegetation

cercariae
encyst to form
metacercarial stage
on vegetation
near water

free-swimming
cercaria

free-swimming
miracidium

sporocyst

redia

miracidia bore
into snail host

cercariae break out
of snail's body

larvae develop in snail's body

(b)

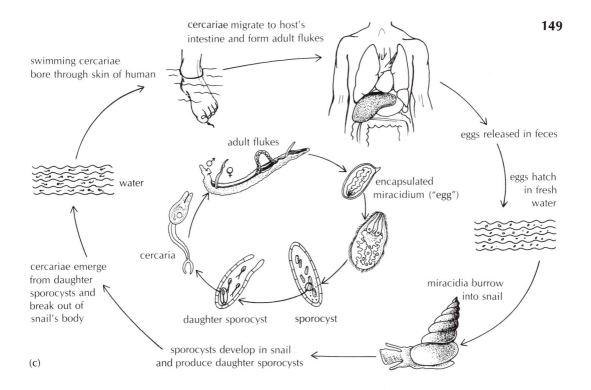

FIGURE 7.17. *Fluke life cycles. (a) Chinese human liver fluke,* Opisthorchis sinensis. *There are two intermediate hosts (a snail, usually in the genus* Bythinia, *and a fish, of the family* Cyprinidae*). This fluke can reach sexual maturity only in humans, which thus serve as the definitive hosts. Encapsulated miracidia pass out of the definitive host's body along with feces. Hatching of the miracidium occurs only following ingestion by the snail. The miracidium becomes a single sporocyst, which produces (asexually) many rediae, which, in turn, each produce (asexually) many cercariae. These leave the snail and swim freely in the surrounding water. Upon encountering an appropriate fish host, the cercariae bore in and encyst within the fish muscle. The life cycle reaches completion only when humans eat raw or undercooked fish. Adulthood is reached within the human bile duct. (b) The sheep liver fluke,* Fasciola hepatica. *This parasite has a snail as the sole intermediate host. Infected sheep deposit fluke eggs along with feces. Free-swimming miracidia hatch from the eggs if the eggs reach fresh water. The miracidia bore into an appropriate snail host; sporocyst, redia, and cercaria stages follow within the snail. The cercariae leave the snail and encyst to form a metacercarial resting stage on emergent vegetation. Adulthood is reached if the encysted stage is ingested by grazing sheep. (c)* Schistosoma mansoni, *one of the species causing schistosomiasis in humans. Members of this genus are atypical trematodes in that the adults are dioecious. Fertilized eggs leave the human host with fecal material, and miracidia hatch out if the feces contact fresh water. The miracidia must locate and bore into the tissues of the single intermediate host, a snail in the genus* Biomphalaria. *Each miracidium then develops into a single sporocyst, which then produces many daughter sporocysts. Many cercaria larvae emerge from each daughter sporocyst and exit the snail. If the free-swimming cercaria contacts a human, it bores through the skin and eventually reaches maturity within the veins of the host intestine. (All modified from several sources.)*

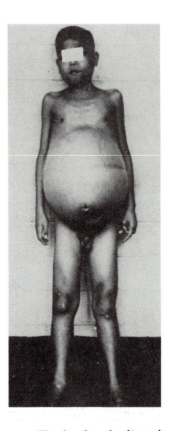

FIGURE 7.18. *This boy has the distended belly typical of infection by* Schistosoma japonicum. *(Courtesy of the Centers of Disease Control, Department of Health and Human Services, Atlanta, GA.)*

cercaria stage. A single miracidium generally gives rise to, on average, tens of thousands of cercariae, all of which are genetically identical with each other and with the original fertilized egg. A moderately infected intermediate host may release thousands of cercaria larvae per day for many years.

Several remarkable behavioral adaptations

that have arisen among trematodes increase the likelihood that a given larval stage will indeed make it to the next required host. One of the most wonderful adaptations is shown by a liver fluke found in sheep, cattle, pigs, and several other terrestrial vertebrates. The species is *Dicrocoelium dendriticum*, and I will go through its story in some detail, in order to review the events of the trematode life cycle.

The adult liver flukes are hermaphroditic, and deposit their eggs into the bile ducts of their host. The eggs, which are less than 50 μm in diameter, leave the host with the feces. The eggs will not hatch until they are eaten by one particular species, or possibly one of several species, of land snail. Miracidia emanating from the fertilized eggs lyse their way through the gut wall of the snail host and migrate to a specific region of the digestive system of the snail, where they gradually transform to sporocysts. Each mother sporocyst produces a large number of daughter sporocysts, each of which, in turn, gives birth to large numbers of cercariae; this species has no redia stage. Each cercaria is about 600 μm long. The cercariae migrate to the "lung" of the snail, where groups of about 500 larvae become engulfed in mucus. These mucus balls are expelled from the snail to the outside. The outer surface of the slime balls dries to form a water-resistant outer coat, while the cercariae persist in the watery environment within.

From here, the tale becomes even less believable. The slime balls are routinely collected and ingested by several species of ant. Within the ant, the cercariae encyst in various tissues to form the metacercarial stage, and somehow succeed in altering the behavior of the ant. Each evening, the ants are now obliged to crawl upward to the tip of a blade of grass, and to bite down firmly upon the grass blade. They apparently remain in this position, unable to open their jaws, until the air temperature rises during the following day. Meanwhile, they are especially vulnerable to grazing by local herbivores, which, after all, tend to feed primarily during the evening and early morning. Once the ants are ingested by an appropriate host, the life cycle can be completed. Adult flukes of this species can reside in the bile ducts and gall bladders of many mammals, including horses, sheep, cattle, dogs, rabbits, pigs, and humans.

Exactly how the trematode larva is able to influence the behavior of the ant is not known, but it is easy to imagine how this ability would be selected for, once it appeared. Those metacercaria possessing this capability would have a greater chance of infecting the definitive host and leaving offspring. Some proportion of these offspring would also possess the capacity to alter the behavior of the ant intermediate host, since the capability is genetically determined, and gradually the trait would become a species-specific characteristic.

The life cycle of this particular trematode is perhaps more remarkable than most. Equally remarkable is the patience and persistence of the people who have pieced this and other trematode life cycles together.[5]

---

5 See Topics for Discussion, No. 6.

TAXONOMIC SUMMARY

PHYLUM PLATYHELMINTHES—THE FLATWORMS
CLASS TURBELLARIA—THE FREE-LIVING FLATWORMS
CLASS CESTODA—THE TAPEWORMS
CLASS TREMATODA—THE FLUKES

TOPICS FOR FURTHER DISCUSSION AND INVESTIGATION

1. A new phylum the Gnathostomulida, was discovered less than 30 years ago (Fig. 7.19). The one hundred or so species that have been assigned to this phylum bear many resemblances to some of the turbellarian flatworms. Describe the morphological and developmental similarities and differences between gnathostomulids and turbellarians.

Riedl, R.J., 1979. Gnathostomulida from America. *Science*, 163: 445.

Sterrer, W., 1972. Systematics and evolution within the Gnathostomulida. *Syst. Zool.*, 21: 151.

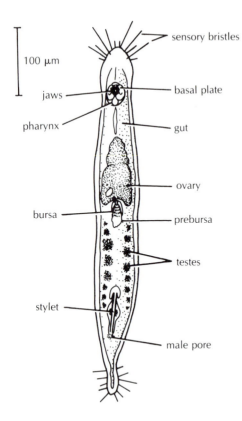

FIGURE 7.19. A gnathostomulid,
Gnathostomulida jenneri.
*Gnathostomulids live interstitially, in the
spaces between sand grains in marine
habitats. Most species are less than one
mm in length. Gnathostomulids possess a
number of flatworm-like characteristics.
(From Sterrer, 1972. Syst. Zool. 21: 151.)*

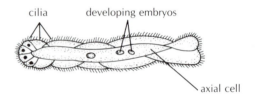

FIGURE 7.20. *A mesozoan. Mesozoans
are small (up to 7 mm) parasites of marine
invertebrates, including clams, flatworms,
brittle stars, and squid. Some authorities
believe that mesozoans have close
evolutionary ties to the Platyhelminthes,
but the degree of relationship is still
uncertain. The axial cell produces sperm
and eggs, and thus functions as a
unicellular "gonad." The embryos develop
within the axial cell following
fertilization. (After Lapan and Morowitz,
1972. Scientific American, 223: 96.)*

Lapan, E.A., and H. Morowitz, 1972. The Mesozoa. *Scientific Amer.*, 227: 94.

Stunkard, H.W., 1972. The life history and systematic relations of the Mesozoa. *Quart. Rev.
Biol.*, 29: 230.

3. Discuss the feeding biology of turbellarian
flatworms.

Calow, P., and D.A. Read, 1981. Transepidermal
uptake of the amino acid leucine by freshwater
triclads. *Comp. Biochem. Physiol.*, 69A: 443.

Jennings, J.B., and J.I. Phillips, 1978. Feeding
and digestion in three entosymbiotic graffillid
rhabdocoels from bivalve and gastropod molluscs.
Biol. Bull., 155: 542.

4. How do trematode larvae locate and penetrate their future hosts?

Blankespoor, H.D., and H. van der Schalie, 1976.
Attachment and penetration of miracidia observed by scanning electron microscopy. *Science*,
191: 291.

2. Another group of flatworm-like animals is often placed within the Phylum Mesozoa (Fig.
7.20). Adult mesozoans are all internal parasites of
marine invertebrates. In what respects do the life
cycles of mesozoans resemble those of the parasitic platyhelminthes?

MacInnes, A.J., 1965. Responses of *Schistosoma mansoni* miracidia to chemical attractants. *J. Parasitol.*, 51: 731.

Mason, P.R., 1977. Stimulation of the activity of *Schistosoma mansoni* miracidia by snail-conditioned water. *J. Parasitol.*, 75: 325.

Roberts, T.M., S. Ward, and E. Chernin, 1979. Behavioral responses of *Schistosoma mansoni* miracidia in concentration gradients of snail-conditioned water. *J. Parasitol.*, 65: 41.

Wilson, R.A.N., and J. Dennison, 1970. Short chain fatty acids as stimulants of turning activity by miracidia of *Fasciola hepatica*. *Comp. Biochem. Physiol.*, 32: 511.

5.   How do flatworm parasites evade the host immune system?

Clegg, J.A., S.R. Smithers, and R.J. Terry, 1971. Acquisition of human antigens by *Schistosoma mansoni* during cultivation "in vitro." *Nature*, 232: 653.

Damian, R.T., 1967. Common antigens between adult *Schistosoma mansoni* and the laboratory mouse. *J. Parasitol.*, 53: 60.

Dineen, J.K., 1963. Immunological aspects of parasitism. *Nature*, 197: 268.

Hopkins, C.A., and H.E. Stallard, 1974. Immunity to intestinal tapeworms: the rejection of *Hymenolepis citelli* by mice. *Parasitology*, 69: 63.

Smithers, S.R., R.J. Terry, and D.J. Hockley, 1969. Host antigens in schistosomiasis. *Proc. Roy. Soc. London.*, 171B: 483.

6.   How do researchers go about documenting the complex life history of a trematode flatworm?

Stunkard, H.W., 1941. Specificity and host-relations in the trematode genus *Zoogonus*. *Biol. Bull.*, 81: 205.

Stunkard, H.W., 1964. Studies on the trematode genus *Renicola*: observations on the life-history, specificity, and systematic position. *Biol. Bull.*, 126: 467.

Stunkard, H.W., 1980. The morphology, life-history, and taxonomic relations of *Lepocreadium aveolatum* (Linton, 1900) Stunkard, 1969 (Trematoda: Digenea). *Biol. Bull.*, 158: 154.

7.   Flatworms are capable of considerable regeneration of lost body parts, including the head. Investigate regeneration in turbellarian flatworms.

Best, J.B., A.B. Goodman, and A. Pigon, 1969. Fissioning in planarians: control by brain. *Science*, 164: 565.

Goldsmith, E.D., 1940. Regeneration and accessory growth in planarians. II. Initiation of the development of regenerative and accessory growths. *Physiol. Zool.*, 13: 43.

Nentwig, M.R. 1978. Comparative morphological studies of head development after decapitation and after fission in the planarian *Dugesia dorotocephala*. *Trans. Amer. Microsc. Soc.*, 97: 297.

8.   Discuss the similarities and differences in the life cycles of digenean flatworms and members of the cnidarian class Scyphozoa.

9.   Like the flatworms, many members of the phylum Cnidaria lack specialized respiratory structures and blood circulatory systems. How are cnidarians able to meet their gas exchange requirements without being flat?

8

INTRODUCTION TO THE HYDROSTATIC SKELETON

The word "skeleton" invariably conjures up an image of the articulated bones hanging in the corner of the high-school biology classroom, or perhaps in the corner of the general practitioner's office. However, jointed bones are only one form of skeletal system. Nearly all multicellular animals, even invertebrates, require a skeleton for movement. The only exceptions to this rule are those small, aquatic metazoans that may move exclusively by cilia. A functional definition of the word "skeleton" is:

A solid or fluid system permitting muscles to be stretched back to their original length following a contraction. Such a system may or may not have protective and supportive functions as well.

A skeletal system is essential simply because a muscle is capable of only two of the three activities required for repeated movements. A muscle can either shorten or relax; a muscle cannot actively extend itself. In order to bend your arm at the elbow, one set of muscles, the biceps, must contract. This contraction of the biceps not only causes your arm to bend at the elbow, but also serves to stretch another muscle in your arm, the triceps (Fig. 8.1). The triceps can now contract, making it possible for you to re-extend your arm. Re-extension of your arm, in turn, serves to stretch the biceps. The bones in

155

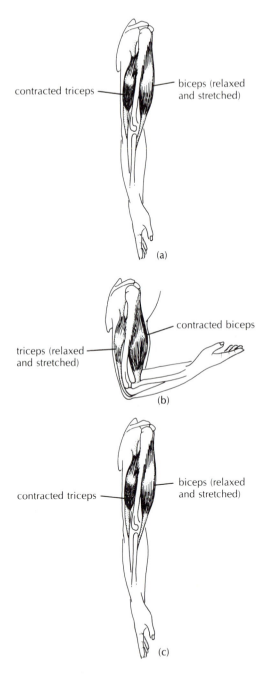

contracted triceps

biceps (relaxed and stretched)

(a)

contracted biceps

triceps (relaxed and stretched)

(b)

contracted triceps

biceps (relaxed and stretched)

(c)

FIGURE 8.1. *Antagonistic interaction between the biceps and triceps in the human arm. Contraction of the biceps (a) results not only in movement of the arm (b), but also in stretching of the opposing muscle, the triceps. Contraction of the triceps then returns the arm to its initial position (c), and stretches the biceps. In vertebrates, muscle pairs antagonize each other through a rigid skeleton, which is internal and jointed.*

your arm have functioned in these movements as the vehicle through which the triceps and biceps take turns stretching each other back to precontraction length. That is, the muscles **antagonize** each other, making controlled, repeatable movement possible. In this case, the mutual antagonism of muscles is mediated through a solid skeleton. A rigid skeletal system is essential in a terrestrial environment, in part because the skeleton must also serve to support the body in a non-supportive medium (see Chapter 1). Aquatic organisms are supported by the medium in which they live, so that a rigid skeletal system is not required.

Invertebrates commonly make use of a **hydrostatic skeleton**, a system in which fluid serves as the vehicle through which sets of muscles interact. A functional hydrostatic skeleton requires:

(a) the presence of a cavity housing an incompressible fluid that will transmit pressure changes uniformly in all directions;
(b) that this cavity be surrounded by a flexible outer body membrane, permitting deformations of the outer body wall to take place;
(c) that the volume of fluid in the cavity remain constant;

(d) that the animal be capable of forming temporary attachments to the substrate, if progressive locomotion is to occur on or within a substrate.

Assume that attributes (a)–(d) are met in the hypothetical organism shown in Fig. 8.2. This cylindrical being is equipped with **longitudinal muscles** only. Suppose that this animal attaches at point X, as shown in part (a), and then contracts its musculature. The increase in internal hydrostatic pressure will deform the outer body wall, resulting in a

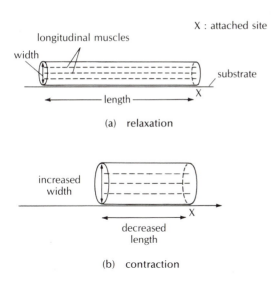

(a)   relaxation

(b)   contraction

FIGURE 8.2. *Shape changes possible in a worm-like organism equipped with only longitudinal muscles. Because the fluid volume is constant, a change in the width of the hypothetical animal must be compensated for by a change in length, brought about by the increase in internal hydrostatic pressure during muscle contraction. Similarly, a shortening of the worm is accompanied by an increase in width, as seen in (b).*

shorter, fatter animal [part (b)]. This animal can regain its initial shape only if it is surrounded by a stiff, elastic covering that will spring back to its original shape upon relaxation of the longitudinal musculature. Such a stiff covering would be difficult to deform in the first place, and is not commonly encountered among unjointed invertebrates.

Instead, we add a second set of muscles (**circular muscles**) to our hypothetical animal. Forward locomotion then results from the series of contractions illustrated in Fig. 8.3. In frame (a), the circular muscles are contracted and the longitudinal muscles are stretched. The longitudinal muscles now contract while the circular muscles relax, producing the shorter, wider animal of frame (b). In frame (b), the animal releases its anterior attachment to the subtrate and forms a new temporary attachment posteriorly. The circular muscles then contract while the longitudinal muscles relax, thrusting the animal's anterior end forward. In frame (c), we see that the animal has advanced by a distance *d* and has regained its initial shape, ready to repeat the cycle of muscular contractions.

From this discussion, we see that we must make one addition to the list of requirements for a functional hydrostatic skeleton given above:

(e) the presence of a deformable but elastic covering or the presence of at least two pairs of musculature oriented in different directions.

Clearly, the skeletal system in our hypothetical organism is a fluid. Temporary increases in internal pressure are caused by the

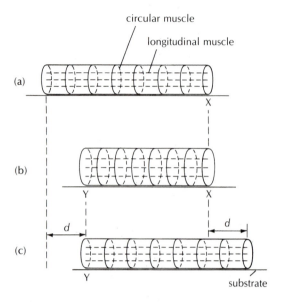

FIGURE 8.3. *Locomotion in a hypothetical worm possessing both circular and longitudinal muscles and a continuous, fluid-filled internal body cavity of constant volume. (a) The animal attaches to the substrate at point X, relaxes its circular muscles, and contracts its longitudinal muscles. The contraction increases the hydrostatic pressure within the body cavity because the volume of the cavity cannot be decreased, and the fluid within the cavity is incompressible. This pressure is relieved by permitting the circular muscles to stretch. (b) The animal releases its anterior attachment, forms an attachment at point Y, and relaxes its longitudinal muscles. (c) The circular muscles have contracted, causing another increase in pressure, which, in turn, is relieved as the longitudinal muscles extend. The animal has now advanced by distance d, and regained its initial body shape.*

contraction of one set of muscles, and this temporary pressure increase results in the elongation of another set of muscles. I emphasize that the internal pressure increase is temporary; elongation of the opposing set of muscles relieves the pressure. The fluid thus makes possible the mutual antagonism of the two sets of muscles, resulting in repeatable locomotory movements. This simple, hydrostatic skeleton plays some role in the movements made by representatives of nearly every invertebrate phylum.

TOPICS FOR FURTHER DISCUSSION AND INVESTIGATION

1. Which features of a hydrostatic skeleton do sponges possess? Which features are absent from sponges?

2. The parenchyma of turbellarians functions as a hydrostatic skeleton. What sequence of muscle contractions and relaxations is likely to be involved in mediating: (a) locomotion via pedal waves? (b) locomotion via looping?

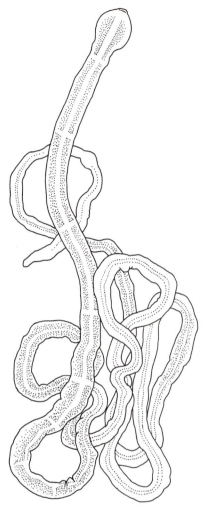

9

THE NEMERTINES

INTRODUCTION AND
GENERAL CHARACTERISTICS

The nemertines (also called nemerteans) are a small group of elongated, unsegmented, soft-bodied worms. Most of the approximately 900 described species in this phylum are marine. Some other species (mostly in a single genus) are found in fresh water, and the members of several more small genera are terrestrial. The terrestrial species are restricted to tropical, moist environments. Many nemertine species live commensally with invertebrates from other phyla, particularly the Arthropoda and Mollusca, but only

Phylum Rhyncho · coela (= Phylum Nemertea = Phylum Nemertinea)
     (G: SNOUT CAVITY)

a few parasitic nemertines are known. Nemertines are common inhabitants of marine shallow-water environments, crawling over solid substrates, burrowing into sediment, or lurking under stones, rocks, or mats of algae.

In several respects, the nemertines resemble the free-living flatworms (Turbellaria). Like the flatworms, from which they may have been derived, nemertines are ciliated externally and deposit a mucus through which they progress. Small nemertines can move over substrates exclusively by means of ciliary beating. Larger individuals often gen-

erate waves of muscle contraction for locomotion over hard substrates or through soft substrates. A few species can swim, by generating relatively violent waves of muscular contraction.

As with the turbellarians, most nemertines are flattened dorso-ventrally, and possess circular, longitudinal, and dorso-ventral muscles. Nemertines apparently lack a coelom; the area lying between the outer body wall and the gut is generally described as being solid, mesodermally derived tissue referred to as **parenchyma** (Fig. 9.1). The acoelomate

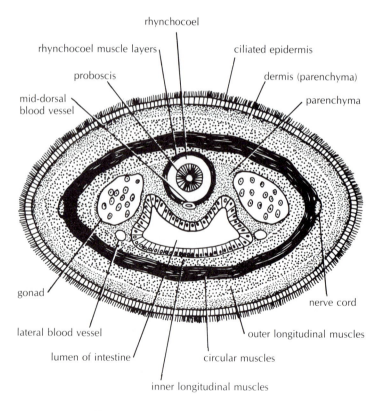

FIGURE 9.1. *Diagrammatic cross section through the body of a nemertine worm. The functions of the proboscis and rhynchocoel are discussed later in this chapter. (After Harmer and Shipley.)*

status of these animals may be questioned in the future, based upon ultrastructural studies currently in progress.[1] The nervous system of nemertines is also reminiscent of the turbellarian plan: the cerebral ganglia form a ring anteriorly, and give rise to a ladder-like arrangement of longitudinal nerves with lateral connectives. Nemertines are equipped with chemoreceptors and mechanoreceptors, located in specialized pits and grooves on the body surface, and sensory bristles. Most species also possess pigmented photoreceptors, and a few species possess balance organs (**statocysts**). As in the flatworms, nemertines generally have protonephridial excretory systems, and digestion is, as in the Platyhelminthes, largely intracellular. The body of most species is of considerable length, typically from several to twenty centimeters, and as long as several meters. "Ribbon worm," the common name for members of this phylum, is especially appropriate for such elongated nemertine species.

---

1 E. Ruppert, personal communication.

Nemertines differ from the flatworms in a number of respects, important enough to warrant the placement of these two groups in separate phyla. Specifically, the two groups differ with respect to the system of gas exchange, food capture, and digestion.

Unlike flatworms, ribbon worms have a one-way digestive tract, with a mouth anteriorly and separate anus posteriorly (Fig. 9.2). The unidirectional flow of food material through this tubular gut makes an orderly digestive process possible. Digestive enzymes may be secreted in sequence as the food is moved through the gut by ciliary activity, and the animal can continue to eat without disrupting the digestive process. As in the Turbellaria, most nemertines are meat-eaters. Their food preferences are often highly specific. Nemertines seem to have a predilection for small annelids and crustaceans, and a given nemertine species often shows a preference for a particular species of prey.

Because the bodies of nemertines are flat and permeable, a fair amount of gas exchange likely occurs by diffusion across the

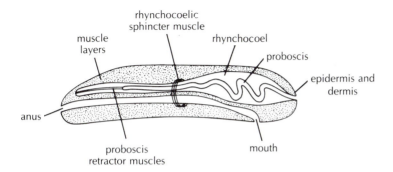

FIGURE 9.2. *Diagrammatic longitudinal section through a nemertine, showing the tubular gut. (Modified from Turbeville and Ruppert, 1983. Zoolmorphol., 103: 103.)*

general body surface. In addition, and in contrast to the platyhelminths, nemertines possess a true circulatory system. The blood circulates throughout the body through well-defined vessels, many of which are contractile (Fig. 9.3). There is no true heart and the blood vessels lack one-way valves, so that the blood does not circulate unidirectionally. Instead, it ebbs and flows erratically, propelled largely by muscular contractions associated with routine movements of the animal. A few species have hemoglobin in their blood, but most species lack any such oxygen-carrying blood pigment.

Unlike flatworms, the ribbon worms possess a hollow, muscular **proboscis**. This structure is not homologous with the turbellarian pharynx. Recall that the flatworm pharynx is essentially a protrusible extension of the gut. In contrast, the nemertine proboscis is not directly associated with the digestive tract (Fig. 9.4b, c). Rather, the proboscis floats in a separate, fluid-filled tubular cavity called the **rhynchocoel** (Fig. 9.4a). Eversion of the proboscis generally occurs through a distinct **proboscis pore**, although in some species discharge occurs through the mouth opening. The proboscis can be shot out explosively from the animal, and in one major group of species, the proboscis is armed with a piercing **stylet** (see Fig. 9.6c). A potent paralytic toxin found in the proboscis is generally discharged into wounds made by the stylet; nemertines are carnivores to be

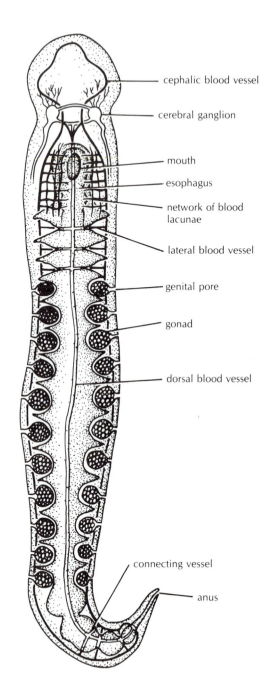

cephalic blood vessel

cerebral ganglion

mouth

esophagus

network of blood lacunae

lateral blood vessel

genital pore

gonad

dorsal blood vessel

connecting vessel

anus

FIGURE 9.3. *Schematic illustration of* Cerebratulus, *a common genus of ribbon worm. Note the well-developed circulatory system. (After Bayer and Owre; and after Joubin.)*

reckoned with, even for rapidly moving prey.

It is difficult to imagine a pharynx being shot out of the mouth of a flatworm. The turbellarian pharynx is protruded when the animal contracts the circular musculature of the tube and relaxes the longitudinal musculature, but this is not a rapid process. What, then, is the functional difference between the operation of the flatworm pharynx and the nemertine proboscis that permits the nemertine proboscis to be extended with such rapidity?

The secret of proboscis functioning lies in the fluid-filled rhynchocoel. Although the nemertine proboscis is highly muscular, it is not discharged by its own muscular activity, but rather by the musculature of the tissue surrounding the rhynchocoel. When this musculature contracts, the pressure within the rhynchocoel rises. Even a small amount of contraction creates a substantial increase in pressure, since the rhynchocoel has a constant volume and the fluid within it is essentially incompressible. That is, the fluid can neither escape nor be compressed. By default, hydrostatic pressure increases. The elevated pressure is relieved by relaxing the sphincter muscles surrounding the anterior end of the rhynchocoel, allowing the proboscis to shoot out of the cavity with great speed and power (Fig 9.5).

You can visualize the workings of a proboscis by considering a plastic glove with one of the fingers pushed inside. If the wrist of the glove is tied off so that the air cannot escape, the glove becomes a constant-volume, fluid-filled cavity—a rhynchocoel, of sorts. The inverted finger is the proboscis in this rhynchocoel. If you squeeze the glove, the proboscis will be everted, turning inside out as it goes. Poke a hole in the glove and repeat the experiment; you will then see the importance of the rhynchocoel being of constant volume.

For those nemertine species possessing a stylet on the proboscis, prey may be harpooned directly. More commonly, the everted proboscis is first wound tightly around the prey, and the struggling animal is then stabbed repeatedly with the barb (Fig. 9.6). Species lacking a stylet capture prey by simply coiling the prehensile proboscis around the victim. A sticky mucus is generally secreted by the proboscis, to help hold the prey.[2]

A **proboscis retractor muscle** runs from the tip of the everted proboscis to the inner wall of the rhynchocoel (Fig. 9.4). Contraction of this muscle draws the proboscis back inside the rhynchocoel. Captured food is usually transferred to the mouth during the process of proboscis retraction for those species lacking stylets. The species with stylets first paralyze the prey and then retract the proboscis, completely losing contact with the stricken animal. The nemertine then moves to the prey to eat. In some species, prey are ingested whole, particularly if they are **vermiform** (worm-shaped). In other species, the prey are typically too large or awkwardly shaped to be engulfed; in such cases, the nemertine inserts its foregut into the prey and, by peristaltic waves of muscular activity, pumps the juices of the victim into the gut of the victor.

---

2 See Topics for Discussion, No. 1.

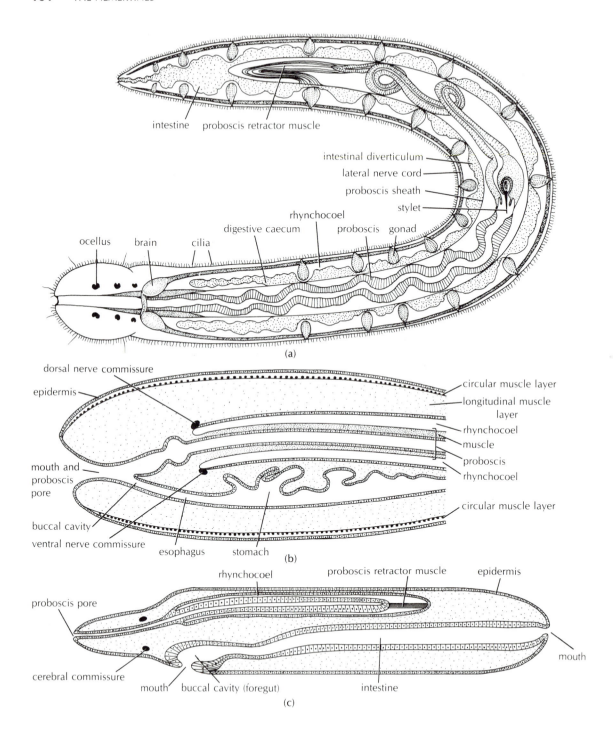

intestine   proboscis retractor muscle

intestinal diverticulum
lateral nerve cord
proboscis sheath
stylet

rhynchocoel
digestive caecum   proboscis   gonad

ocellus   brain   cilia

(a)

dorsal nerve commissure
epidermis

circular muscle layer
longitudinal muscle layer
rhynchocoel
muscle
proboscis
rhynchocoel

mouth and proboscis pore

circular muscle layer

buccal cavity
ventral nerve commissure
esophagus   stomach

(b)

rhynchocoel   proboscis retractor muscle   epidermis

proboscis pore

mouth

cerebral commissure
mouth   buccal cavity (foregut)   intestine

(c)

FIGURE 9.4. (a) Diagrammatic illustration of Prostoma graecense, a freshwater nemertine species. The intestine connects with various bulges (diverticulae) and sacs (caecae). (From Pennak, 1978. Fresh-Water Invertebrates of the United States. John Wiley & Sons.) (b) Diagrammatic illustration of Nemertopsis sp., anterior end in longitudinal section. Note that this species has a single opening for both the proboscis and the mouth. The nerve commissures connect to form a ring around the anterior portion of the rhynchocoel. (After Bayre and Owre; after Burger.) (c) In Cephalothrix bioculata, the proboscis is ejected through an opening separate from the mouth. (From Jennings and Gibson, 1969. Biol. Bull., 136: 405.)

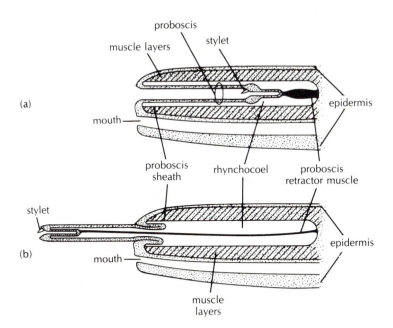

FIGURE 9.5. Diagrammatic illustration of the proboscis in (a) retracted and (b) extended positions. The digestive system has been omitted for clarity. (After Harmer and Shipley; and after Alexander.)

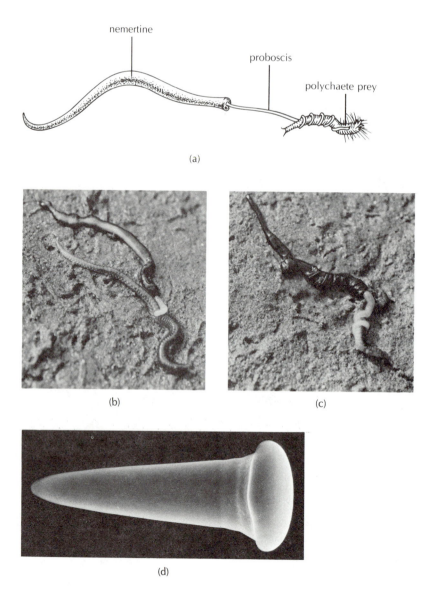

FIGURE 9.6. (a) *Diagrammatic illustration of a nemertine* (Paranemertes peregrina) *subduing its prey. (After MacGinitie and MacGinitie.)* (b,c) P. peregrina *attacking a polychaete worm* (Platynereis bicanaliculata). *(Courtesy of S.A. Stricker, from Stricker and Cloney, 1983. J. Morphol., 177: 89.* (d) *Scanning electron micrograph of the stylet of* Zygonemertes virescens; *the stylet is approximately 175 μm in length. (Courtesy of S.A. Stricker, from Stricker, 1983. J. Morphol., 175: 153.)*

The nemertine proboscis does more than make the capture of fast-moving prey possible. It also enables the animal to burrow into soft substrates, something which the flatworms cannot do. Contraction of the body wall musculature elevates the internal pressure of the rhynchocoel. The proboscis is then everted into the sediment and allowed to widen at its tip, through a relaxation of the associated circular muscles. This distal dilation anchors the end of the proboscis in the sediment. The body of the worm can then be pulled downwards, into the sediment, upon contraction of the proboscis retractor muscles and the longitudinal muscles of the body wall. Contractions of the longitudinal muscles also serve to force additional fluid into the proboscis, under considerable pressure, causing additional dilation and anchoring. We shall encounter similar burrowing mechanisms in several other animal phyla, phyla in which the coelom or blood sinuses subserve the same function as the rhynchocoel of nemertines.

Here, then, we see the principles of the hydrostatic skeleton in action. Advances in locomotion and food capture are achieved by exploiting the properties of a constant-volume, fluid-filled cavity. Such cavities have been independently evolved at least four different times in metazoans: by enterocoely; by schizocoely; by the persistence of a blastocoel; and by the formation of a rhynchocoel. That the selection pressure supporting the evolution of secondary body cavities has been strong cannot be doubted. Among the nemertines, the adaptive value would clearly seem to lie in mechanical advantages gained, and in new life-styles thereby made possible.

OTHER FEATURES OF NEMERTINE BIOLOGY

1. Classification

Nemertines are divided into two major classes: the Anopla and the Enopla. Members of the Anopla lack stylets; that is, the proboscis is unarmed. In addition, the mouth of anoplan nemertines is posterior to the brain. In contrast, the proboscis may be armed with a stylet among the Enopla, all of the armed species belonging to a single order (the Hoplonemertea), and the mouth is anterior to the brain.

2. Reproduction

Most nemertine species have separate sexes, i.e., they are **dioecious**. The few hermaphroditic species that have been described are **protandric**. That is, a single individual may be both male and female, but not simultaneously; as it ages, each male becomes a female. Following fertilization, cleavage is spiral and determinate, as in the Platyhelminthes. In some groups of nemertines, a distinctive, microscopic pilidium larva is formed (Fig. 9.7). This is a ciliated, swimming, feeding individual whose appearance has been likened to a football helmet with earflaps. Only a small portion of the pilidium larva develops into the juvenile ribbon worm, which eventually abandons the pilidium. What remains of the larva apparently swims off and eventually starves to death in some species, since the juvenile takes the mouth with it when it leaves. Alternatively, the worm may ingest the larval tissues following metamorphosis.

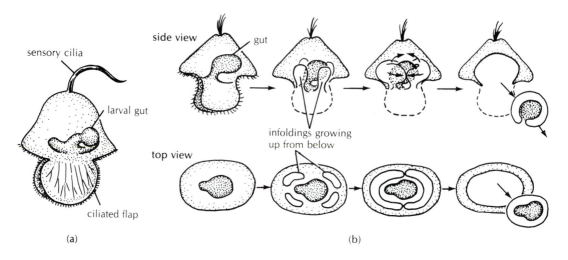

FIGURE 9.7. *(a) The pilidium larva of a nemertine worm. (After Harmer and Shipley.) (b) Stages in the metamorphosis of the pilidium larva. Infoldings form around the mouth, growing upward and toward each other. When these infoldings completely surround the gut and fuse together, the central mass detaches from the rest of the larva, to take up residence on the sea bottom as a small nemertine. The other portion of the larva continues swimming for a time, but, in the absence of a digestive system, eventually starves to death.* (From Hardy, 1965. The Open Sea: Its Natural History. Houghton Mifflin.)

TAXONOMIC SUMMARY

PHYLUM RHYNCHOCOELA (= NEMERTEA)—THE RIBBON
        WORMS
  CLASS ANOPLA—THE UNARMED NEMERTINES (PROBOSCIS
        LACKS A STYLET)
  CLASS ENOPLA—INCLUDES THE ARMED NEMERTINES (PRO-
        BOSCIS MAY BE EQUIPPED WITH A STYLET)

TOPICS FOR FURTHER DISCUSSION AND INVESTIGATION

1. Compare and contrast the different methods of feeding encountered among the Rhynchocoela.

Fisher, F.M., Jr., and J.A. Oaks, 1978. Evidence for a nonintestinal nutritional mechanism in the rhynchocoelan, *Lineus ruber*. *Biol. Bull.*, 154: 213.

Jennings, J.B., and R. Gibson, 1969. Observations on the nutrition of seven species of rhynchocoelan worms. *Biol. Bull.*, 136: 405.

McDermott, J.J., 1976. Observations on the food and feeding behavior of estuarine nemertean worms belonging to the order Hoplonemertea. *Biol. Bull.*, 150: 57.

Roe, P., 1976. Life history and predator-prey interactions of the nemertean *Paranemertes peregrina* Coe. *Biol. Bull.*, 150: 80.

Stricker, S.A., and R.A. Cloney, 1981. The stylet apparatus of the nemertean *Paranemertes peregrina*: its ultrastructure and role in prey capture. *Zoomorphol.*, 97: 205.

2. Discuss the limitations placed on shape changes in nemertines imposed by the structure of the body wall.

Clark, R.B., and J.B. Cowey, 1958. Factors controlling the change of shape of certain nemertean and turbellarian worms. *J. Exp. Biol.*, 35: 731.

Turbeville, J.M., and E.E. Ruppert, 1983. Epidermal muscles and peristaltic burrowing in *Carinoma tremaphoros* (Nemertini): correlates of effective burrowing without segmentation. *Zoomorphol.*, 103: 103.

3. How is the functioning of a nemertine proboscis like the Wild West?

10

THE NEMATODES

INTRODUCTION

The nematodes are a group of ubiquitous pseudocoelomate animals. In fact, nematodes are probably the most numerous metazoans alive today in terms of numbers of individuals. Nematode densities of one million individuals per square meter are typically encountered in shallow-water sediments in both fresh and salt water. Nematode species are also found free-living on land, and are found as major parasites of vertebrates, invertebrates, and plants. Only 10,000–20,000 species have so far been described, but some authorities estimate that there may be 10–20 times this number of nematodes awaiting description. The difficulty in describing and recognizing nematode species lies in the generally small size of these animals and in the extreme uniformity of both internal and external anatomy and morphology. Often, species determinations must be based upon biochemical attributes or morphological details that are not readily visible.

Phylum Nemato•da
(G: THREAD)

In the literature, you may see the Nematoda listed as a class within another phylum, the Aschelminthes ("bag-worms," in reference to their possession of a pseudocoel). The major groups contained within the Aschelminthes (Nematoda, Rotifera, Gastrotricha, Kinorhyncha, Nematomorpha, Acanthocephala, and Gnathostomulida) are now generally considered to be more distantly related to each other than was previously thought, so that each class has been elevated to phylum status by most workers. The relationships between the various aschelminth phyla, however, are still far from clear.[1] Here, as in most textbooks, I use the term "aschelminth" to refer to most of the pseudocoelomate invertebrates as a group, but the term has uncertain taxonomic validity.

GENERAL CHARACTERISTICS

Body Coverings

The typical nematode is about 1–2 mm long; shows no external segmentation; is pointed at both ends; and is covered by a thick, multi-layered **cuticle** (a noncellular covering) of collagen secreted by the underlying epidermis. As pseudocoelomates, nematodes have an internal body cavity lying between the outer body wall musculature and the gut, but the cavity is not lined with mesodermally derived tissue. The body organs therefore hang freely in the pseudocoel, rather than being suspended by **peritoneum** (the mesodermal lining of the body cavity of coelomates).

The epidermis of nematodes is often **syncytial**, i.e., nuclei are not separated from each other by complete cell membranes. The cuticle is composed of a highly complex meshwork of fibers that are virtually inelastic (Fig. 10.1a). The trellis-like arrangement of these fibers permits bending, stretching, and shortening of the cuticle (Fig. 10.1b). The cuticle is permeable to water and to gases. In the plus column, gas exchange can therefore take place across the entire body surface. On the other hand, the cuticle offers little protection against dehydration, so that all nematodes live in water, or at least in a film of water.

The cuticle is shed (molted) and resecreted four times during development from the juvenile to the reproductively mature adult. Unlike the situation in arthropods (Chapter 16), nematodes continue to increase in size between molts, and even after the final molt. In some parasitic nematode species, the first two juvenile stages are free-living. Prior to completion of the second molt, the animal becomes enclosed within two envelopes. The outer envelope is termed the **sheath.** Escape from the sheath (**exsheathment**), the subsequent two molts, and development to adulthood does not occur until the encapsulated form is eaten by a suitable host.[2]

1 Kristensen, 1983. Z. Zool. Syst. Evolut.-forsch., 21:163.

2 See Topics for Discussion, No.4.

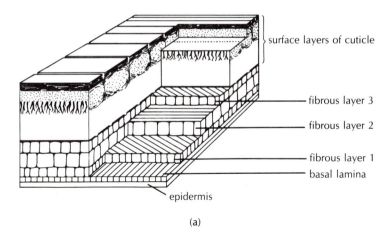

(a)

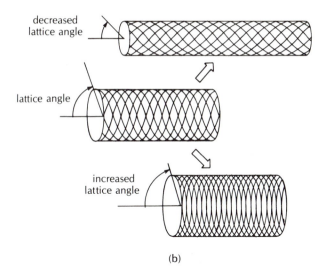

(b)

FIGURE 10.1. *Diagrammatic illustration of the multilayered nematode cuticle. Note the three layers of crossed fibers. (From Clark, 1964. Dynamics in Metazoan Evolution. Oxford University Press; after Bird and Deutsch, 1957. Parasitol., 47: 319.) (b) Although the collagenous fibers of the cuticle are individually inelastic, changes in the angle at which the fibers cross each other permit the animals to change shape. (From Wells, 1968. Lower Animals. McGraw-Hill. Reproduced by permission.)*

Musculature, Internal Pressure, and Locomotion

The body wall of nematodes contains no circular muscles. This is most unusual for animals of **vermiform** (i.e., worm-like) shape, and places great limitations on their locomotory potential. For example, peristaltic waves of contraction cannot be generated.

Another obstacle to graceful locomotion is that the body is quite turgid in most species, owing to a substantial hydrostatic pressure within the pseudocoel. Internal hydrostatic pressures as high as 225 mm Hg have been measured inside some nematode species. (For comparison, atmospheric pressure is approximately 760 mm Hg.) On average, the internal pressure within nematodes averages 70–100 mm Hg, perhaps ten times higher than pressures reported for most other invertebrates. At least two factors contribute to generating and maintaining such high internal pressures. First, the cuticle cannot swell to relieve pressure. Second, the musculature is always in a partially contracted state, trying to compress an incompressible fluid. The high internal pressure gives the nematode a very circular cross section. Thus, the common name for a member of this phylum is "roundworm."

The rigid cuticle, high internal pressure, and lack of circular muscles preclude generation of pedal waves or peristaltic waves for locomotion. Movement by means of cilia is also an impossibility, since nematodes completely lack locomotory cilia. Instead, nematodes generally must move by thrashing the body into sinusoidal waves, through alternating contractions of the longitudinal muscles

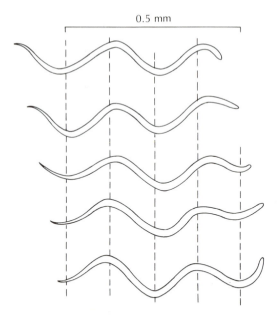

FIGURE 10.2. *Nematode locomotion on a solid surface. By contracting the muscles on each side of its body alternately, the animal forms a series of sinusoidal waves that thrust against the substrate surface, in turn propelling the animal forward. In this figure, successive silhouettes of the nematode have been displaced to the right so as not to obscure each other. The time elapsed between stages is approximately 1/3 second. (From Gray and Lissman, 1964. J. Exp. Biol., 41: 135.)*

on each side of the body (Fig. 10.2).[3] Contraction of one set of muscles causes bending and stretches out muscles elsewhere in the body. Thus, muscles antagonize each other by means of pressure changes transmitted through the fluid skeleton of the pseudocoel, according to the basic principles of the hydrostatic skeleton. Re-extension of con-

---

3 See Topics for Discussion, No. 3.

tracted muscles is attributed largely to the stiff cuticle surrounding the animal and the high pressure within the animal, both acting to spring the body back to a linear configuration when the muscles relax. Clearly, nematode design is not well-suited to a free-swimming existence. Instead, most free-living nematodes are found, often in incredibly high numbers, living in soil, aquatic sediments, fruits, on surface films, and in other similar situations where either the substrate or the surface tension of a fluid at the air-water interface can provide resistance against which a thrust can be generated.[4]

Muscle contraction is under control of a simple nervous system, consisting of an anterior brain (nerve ring plus associated ganglia) and three pairs of longitudinal nerve cords (ventral, dorsal, and lateral). Strangely, the nerve cords seem not to send out processes innervating the muscles. Rather, extensions of the muscle fibers hook up to the nerve cords. Moreover, the structure of the muscles themselves is most peculiar, and uniquely nematode. A cytoplasmic region containing the nucleus is physically separated from the contractile region, which forms a gutter (Fig. 10.3).

Organ Systems and Behavior

The high-pressure pseudocoel poses potential digestive as well as locomotory difficulties. Nematodes have a linear digestive system, with a mouth (**stoma**) at the anterior end leading, in sequence, through a

---

4 See Topics for Discussion, No. 5.

pharynx, intestine, and rectum, and thence out of the body through an anus located near the posterior end of the body. The difficulty in processing food lies in preventing the tubular, nonreinforced digestive tract from collapsing in response to the surrounding high pressure of the pseudocoel. Generally, the inner surface of the nematode gut is not lined by cilia. Indeed, it seems doubtful that ciliary beating would be very effective in countering the high positive pressure exerted on the gut by the fluid in the pseudocoel. Instead, the gut is kept open by the activity of a highly muscular pharynx, which may pump in fluid at rates of about four pulses per second in some species (Fig. 10.4). At the posterior end of the animal, the high pressure in the pseudocoel keeps the terminal end of the digestive tract tightly closed. A dilator muscle at the anus must open to allow discharge of wastes.

In addition to its role in keeping the gut lumen open against the high pressure of the pseudocoel, the pharynx (and associated glands) also adds lubricants and digestive enzymes to the food. Digestion is completely extracellular, and nutrients are absorbed by the very thin wall (only one cell thick) of the intestine. Wastes are ejected from the anus at intervals, about once every one or two minutes.

The respiratory and excretory systems are easily described. There are no specialized organs for gas exchange, no specialized circulatory system, and no nephridia. Metabolic waste products are apparently discharged along with other materials leaving the gut, and across the body wall. A glandular system, the **renette,** or a modification of this system,

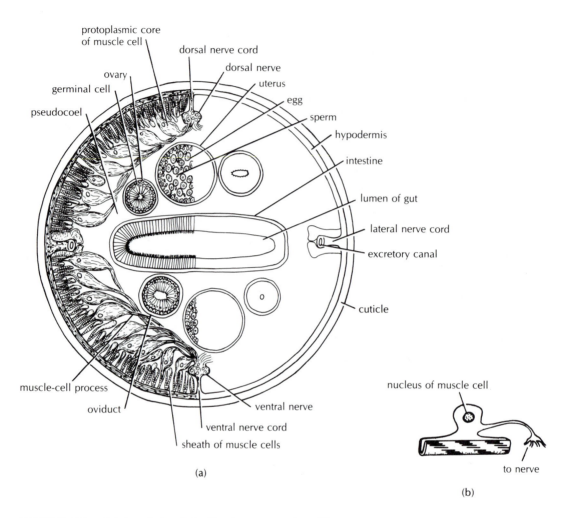

(a)

(b)

FIGURE 10.3.   (a) Ascaris lumbricoides *in cross section. The processes sent out from the muscle fibers to the nerve cords are clearly illustrated. (From Brown,1950. Selected Invertebrate Types. John Wiley & Sons.) (b) Detail of muscle fiber of a nematode. Note the long extension of the muscle cell, leading from its position in the body wall to one of the major nerve cords. (From Alexander, 1979.* The Invertebrates. *Cambridge University Press.)*

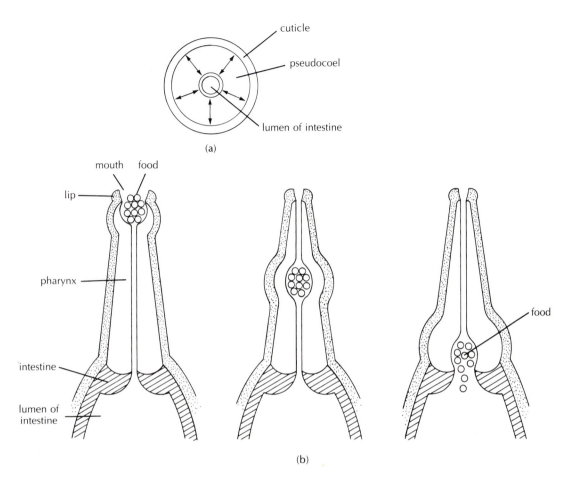

FIGURE 10.4. *(a) Diagrammatic representation of a nematode in cross section. Arrows represent the high pressure within the pseudocoel, acting to maintain the rounded body shape and to collapse the gut. (b) Sequence of movements of the pharynx musculature during swallowing. (After Sherman and Sherman; and after Clark.)*

is present in all nematodes. The renette system varies considerably in complexity among species. It is often referred to as an excretory system, but its actual function has never been convincingly documented.

Despite their small size and limited locomotory abilities, nematodes are capable of fairly sophisticated behaviors. Various species are known to respond to temperature, light, mechanical stimulation, and a variety of chemical cues, including those produced by other individuals of the same species (**pheromones**). The general body surface is often light-sensitive, possibly reflecting direct sensitivity of underlying nerve fibers. Many species have simple, pigmented light receptors (**ocelli**) as well. The major chemosensory organs, called **amphids**, are anteriorly located pits lined with highly modified, nonmotile cilia (**sensillae**) (Fig. 10.5). Similar structures, called **phasmids**, are located at the posterior ends of some nematodes, and are also thought to be chemosensory. The anterior and posterior ends of the body often have **cephalic** and **caudal papillae**, respectively, which also contain modified cilia. These structures are arranged around the mouth or anus, and are believed to be sensitive to mechanical stimulation. Many species possess external setae at various locations on the body; these are also thought to be **mechanoreceptors**.

Reproduction and Development

Most nematode species are **dioecious** (G: two houses); that is, they have separate sexes. Individuals copulate, so that fertilization is internal. Certain other aspects of reproduction and development, while not unique to nematodes, are highly unusual. For example, the sperm of nematodes are amoeboid, rather than flagellated. A second unusual characteristic is the phenomenon of **chromosome diminution**. Following fertilization of the egg, the zygote undergoes a normal first cleavage to the two-cell stage. Prior to the second cleavage, however, the chromosomes of one of the cells fragment,

and much of the chromatin from the ends of the original chromosomes is destroyed. The pieces of the chromosomes remaining after diminution replicate and are distributed in normal fashion to the two daughter cells of the subsequent division (Fig. 10.6). The other two daughter cells (**germ cells,** or **stem cells**) of the four-cell embryo retain the full chromosome complement. Chromosome diminution occurs several more times during the next few cleavages. By the 64-cell stage, only two cells retain the complete genetic information present at fertilization. These two cells will give rise to the gonad and produce the gametes for the next generation. The remaining cells produce all of the somatic (i.e., non-gamete-producing) tissues. Loss of genetic information by chromosome diminution has also been reported to occur in some insects.

Yet a third peculiar feature of nematode development is constancy of cell numbers. Once organogenesis is completed, mitosis ceases in all of the somatic cells. Thus, except for the cells of the germ line, further growth is due not to increases in cell numbers, but rather to increases in cell size. This characteristic is shared by a number of other aschelminth species.

Lastly, nematodes are among the few groups of invertebrates in which a free-swimming larval stage is never found. The young animals emerge from their sturdy egg coverings as miniatures of the adult.

Parasitic Nematodes

Many nematodes show modifications of the above description as adaptations to a para-

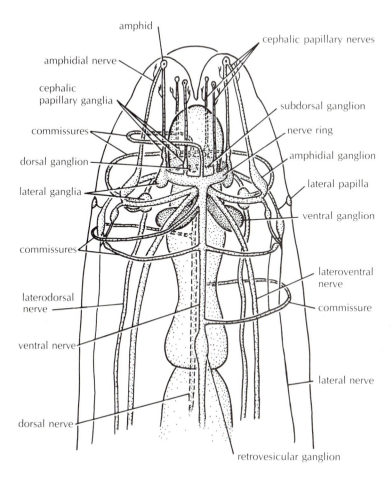

FIGURE 10.5. *Nervous system of a nematode. Note in particular the brain ganglia; lateral, dorsal, and ventral nerve cords; and innervation of the amphids. (Modified from Brown, 1950. Selected Invertebrate Types.* John Wiley & Sons; *after Hyman; after Chitwood and Chitwood.)*

sitic existence. In fact, much of the research on nematode biology has been driven by the need to control the potentially devastating impact of these parasitic species (Fig. 10.7). Nematodes are parasitic in humans, cats, dogs, and domestic animals of economic importance such as cows and sheep. Nema-

todes are also parasitic on the roots, stems, leaves, and flowers of plants, including species of great economic importance such as potatoes, oats, tobacco, onions, and sugar beets. Some of these parasitic nematodes attain great length, although they may be extremely thin. The largest nematode so far

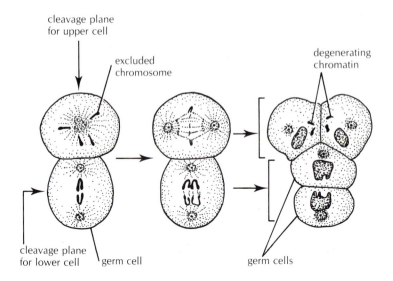

cleavage plane
for upper cell

excluded
chromosome

degenerating
chromatin

cleavage plane
for lower cell   germ cell

germ cells

FIGURE 10.6.   *Illustration of chromosome diminution in
nematodes. In the second cleavage, only the chromosomes of the
germ cell (= stem cell) undergo normal mitosis. Some of the
chromosomal material in the other cell is excluded from
participation in mitosis (a), and at the four-cell stage (c) is seen to
lie outside the nucleus. This material will later degenerate. (From*
Biology of Developing Systems *by Philip Grant. Copyright © 1978
by Holt, Rinehart and Winston. Reprinted by permission of CBS
College Publishing.)*

described is nine meters long, and resides in
the placenta of female sperm whales.

Hookworms and pinworms are two groups
of nematodes that are well known to many
humans (Fig. 10.8). The damage done by
parasitic nematodes is generally indirect, often through competition with the host for
nutrients. Hookworms, for example, may
imbibe more than 0.6 ml of blood per day per
individual. A student with a respectable infection of 100 hookworms would then be
losing perhaps 60 ml of blood daily. Infec-
tions of 1000 hookworms are not uncommon. Other nematode species do most of

their damage by becoming so densely packed
in their preferred tissues that flow of nutrients or fluids is blocked. Some ascarids
(members of the genus *Ascaris*), for example,
may completely block the intestine of the
host. Other nematode species (e.g., *Filaria
bancrofti*) plug up the lymphatic system,
sometimes resulting in a substantial build-up
of fluid in various regions of the body (Fig.
10.7). I will spare you the details.

Most parasitic nematode species make extensive migrations during their development
within the host—from intestine, to liver, to
heart, to lungs, to esophagus, and back to the

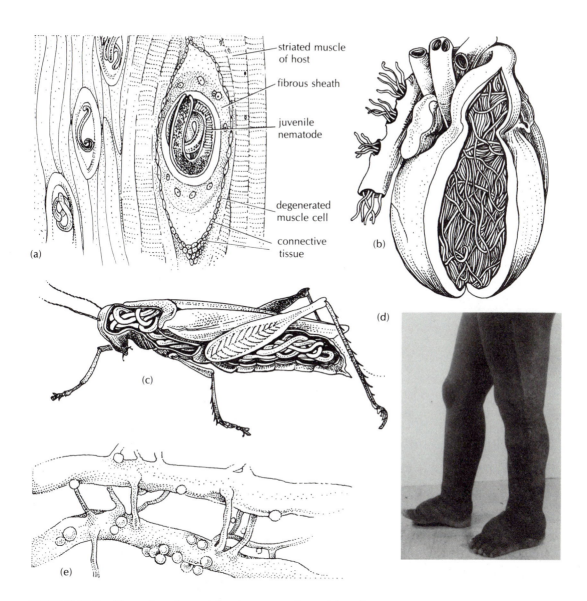

striated muscle
of host

fibrous sheath

juvenile
nematode

degenerated
muscle cell

connective
tissue

(a)

(b)

(c)

(d)

(e)

FIGURE 10.7. *Examples of parasitism by nematodes and its effects on the hosts. (a)* Trichinella spiralis, *shown encysted in vertebrate striated muscle. (Modified from Brown, 1950.* Selected Invertebrate Types. *John Wiley & Sons; and from Harmer and Shipley,* The Cambridge Natural History. *Macmillan.) (b) Heart of a dog infested with* Dirofilaria immitis *(heartworm). (From Harmer and Shipley.) (c) Cutaway view of a grasshopper infected with* Agamermis decaudata. *(From Hyman; after Christie.) (d) Human male with elephantiasis of the legs. (Courtesy of the Center for Disease Control, Department of Health and Human Services, Atlanta, GA.) (e) Gall formation by nematodes (*Heterodera rostochiensis*) parasitic in plant roots. (From Cheng, 1973.* General Parasitology. *Academic Press.)*

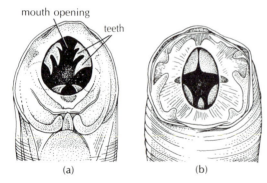

FIGURE 10.8. *Hookworm mouth regions: (a)* Ancylostoma duodenale; *(b)* Necator americanus. *(From Chandler. 1949. Introduction to Parasitology, 8th ed., John Wiley and Sons; after Looss.)*

(as when bathing), the nematode protrudes its posterior end through the sore on the host's skin, and ejects a considerable number of young from its uterus—up to about 1.5×10^7 per day! These juveniles cannot re-infect humans directly, but must first be ingested by a microscopic species of aquatic arthropod crustacean. Humans become infected by drinking water containing these crustaceans. The parasite, liberated from the intermediate host during digestion, migrates through the intestinal wall of the primary host and into the host connective tissue. Males, which are relatively small, die soon after inseminating the females.

Once females reach adulthood beneath

intestine, for instance. Damage to various organs, including the intestinal wall and the eyes, can result from these migrations. Plant parasites also tend to do their damage indirectly, by: (a) causing wounds, which are then susceptible to bacterial or fungal infection; (b) injecting plant viruses; or (c) damaging root, leaf, or stem transport systems.

The life cycles of parasitic species are generally more complex than those of their free-living relatives, and the nematodes are no exception. Often one or more hosts are obligate in the life cycle. One such example is *Dracunculus medinensis.* You can tell from the generic name that this nematode is up to no good. The female, only about one mm in width, but generally one meter or more in length, lives just beneath the skin of a human, causing a pronounced skin irritation. When the skin comes into contact with water

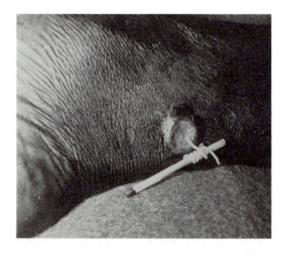

FIGURE 10.9. *Using a matchstick to wind* Dracunculus medinensis *out of an infected human leg. (Courtesy of the Center for Disease Control, Department of Health and Human Services, Atlanta, GA.)*

the skin, the cure is charmingly simple. Commonly, an incision is made and the worm is rolled out on a stick, very slowly (only a few centimeters daily), to prevent breakage (Fig. 10.9). Unless you travel widely, you need not worry about encountering this problem. *Dracunculus medinensis* is found only in Africa, South America, and Western Asia.

It is unfortunate that parasitic species have given nematodes such a bad name. Most species are free-living, innocuous detritivores. In fact, they likely play an extremely important ecological role in the cycling of nutrients and energy in a variety of environments. In addition, much of what we currently know about aging, inheritance, and the factors that control gene expression during development comes from the study of nematodes.

TOPICS FOR FURTHER DISCUSSION AND INVESTIGATION

1. A group of several hundred parasitic pseudocoelomates are placed in the phylum Nematomorpha. These animals, called horsehair worms (Fig. 10.10), are extremely long (reaching a length

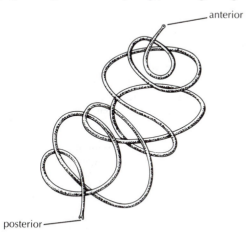

FIGURE 10.10. *A male horsehair worm, from the phylum Nematomorpha. (After Noble and Noble.)*

of one meter in some species) and extremely thin (rarely more than one mm in width). Nematomorphs are unusual parasites, in that the adult stage is nonfeeding and free-living in wet soil or in fresh water; the larval stage is parasitic in a variety of arthropod hosts, including crabs and insects. What characteristics do nematomorphs share with the nematodes? What characteristics set the Nematomorpha apart from the Nematoda? (Consult any general parasitology text.)

2. Another group of parasitic pseudocoelomates are placed within the phylum Acanthocephala (*acantho* = G: spine; *cephala* = G: head). The members of this group derive their name from their proboscis, which bears a number of spines (Fig. 10.11). In what respects does the life cycle of acanthocephalans resemble that of parasitic flatworms (Phylum Platyhelminthes)?

Nicholas, W.L., 1973. The biology of the Acanthocephala. *Adv. Parasitol.*, 11: 671.

Van Cleve, J.H., 1941. Relationships of the Acanthocephala. *Amer. Nat.* 75: 31.

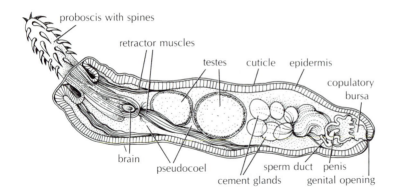

proboscis with spines

retractor muscles

testes   cuticle   epidermis

copulatory
bursa

brain

pseudocoel

cement glands

sperm duct   penis

genital opening

FIGURE 10.11. *An acanthocephalan (*Acanthocephalus sp.*). The copulatory bursa is eversible and holds the female during copulation, using secretions from the cement glands. (From Hyman; after Yamaguti.)*

3. Discuss the locomotion of nematodes.

Clark, R.B., 1964. Nematode locomotion. In: *Dynamics in Metazoan Evolution*, Oxford University Press, pp. 78–83.

Gray, J., and H.W. Lissman, 1964. The locomotion of nematodes. *J. Exp. Biol.*, 41: 135.

Harris, J.E., and H.D. Crofton, 1957. Structure and function in the nematodes: internal pressure and cuticular structure in *Ascaris*. *J. Exp. Biol.*, 34: 116.

Wallace, H.R., 1959. The movement of eelworms in water films. *Ann. Appl. Biol.*, 47: 350.

Wallace, H.R., and C.C. Doncaster, 1964. A comparative study of the movement of some microphagous, plant-parasitic and animal parasitic nematodes. *Parasitology*, 54: 313.

4. Investigate the factors that induce the hatching or exsheathment of parasitic nematodes.

Barrett, J., 1982. Metabolic responses to anabiosis in the fourth stage juveniles of *Ditylenchus dipsaci* (Nematoda). *Proc. R. Soc. London Ser. B*, 216: 159.

Clarke, A. J., and A.H. Sheperd, 1966. Picrolonic acid as a hatching agent for the potato cyst nematode *Heterodera rostochiensis* Woll. *Nature, London*, 211: 546.

Lackie, A.M., 1975. The activation of infective stages of endoparasites of vertebrates. *Biol. Rev.*, 50: 285.

Ozerol, N.H., and P.H. Silverman, 1972. Enzymatic studies on the exsheathment of *Haemonclius contortus* infective larvae: the role of leucine aminopeptidase. *Comp. Biochem. Physiol.*, 42B: 109.

Rogers, W.P., and R.I. Sommerville, 1957. Physiology of exsheathment in nematodes and its relation to parasitism. *Nature, London*, 179: 619.

Wilson, P.A.G., 1958. The effect of weak electrolyte solutions on hatching rate of *Trichostrongylus retortaeformis* (Zeder) and its interpretation in terms of a proposed hatching mechanism of Strongylid eggs. *J. Exp. Biol.*, 35: 584.

5. Investigate the distribution and abundance of the free-living nematodes found in shallow-water marine environments.

Bell, S.S., M.C. Watzin, and B.C. Coull, 1978. Biogenic structure and its effect on the spatial heterogeneity of the meiofauna in a salt marsh. *J. Exp. Marine Biol. Ecol.*, 35: 99.

Hopper, B.E., J.W. Fell, and R.C. Cefalu, 1973. Effect of temperature on life cycles of nematodes associated with the mangrove (*Rhizophora mangle*) detrital system. *Marine Biol.*, 23: 293.

Tietjen, J.H., 1977. Population distribution and structure of the free-living nematodes of Long Island Sound. *Marine Biol.*, 43: 123.

Warwick, R.M., and R. Price, 1979. Ecological and metabolic studies on free-living nematodes from an estuarine mud flat. *Estuarine Coast. Marine Sci.*, 9: 257.

11

THE ROTIFERS

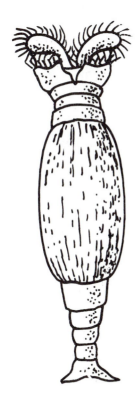

INTRODUCTION

I call it a Water Animal, because its Appearance as a living Creature is only in that Element. I give it also for Distinction Sake the Name of Wheeler, Wheel Insect, or Animal; from its being furnished with a Pair of Instruments, which in Figure and Motion appear much to resemble Wheels. It can, however, continue many Months out of Water, and dry as Dust; in which Condition its Shape is globular, its Bigness exceeds not a Grain of Sand, and no Signs of Life appear. Notwithstanding, being put into Water, in the Space of Half an Hour a languid Motion begins, the Globule turns itself about, lengthens by slow Degrees, becomes in the Form of a lively Maggot, and most commonly in a few Minutes afterwards puts out its Wheels, and swims vigorously through the Water in Search of Food.

So wrote a Mr. Baker, in a letter addressed to the President of the Royal Society in London, in 1744, on the subject of rotifers. Rotifers were so-named because of the two ciliated anterior lobes that are present in many species. This ciliated surface is called the **corona** (Fig. 11.1). The cilia of the corona do

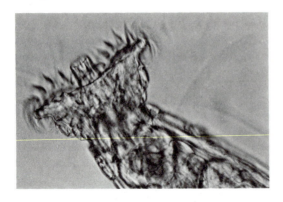

FIGURE 11.1. *Light micrograph of rotifer corona, showing pattern of ciliation. (Photo Copyright © 1975 by T. E. Adams.)*

not beat in synchrony. Instead, each cilium is at a slightly earlier stage in the beat cycle than the preceding cilium in the sequence. That is, the cilia beat **metachronally.** A wave of ciliary beating therefore appears to pass around the periphery of the ciliated lobes, giving the impression of rotation. The degree and pattern of ciliation of the corona varies considerably among species.

Approximately 2000 species of rotifer have been described, mostly (about 95%) from freshwater environments, including lakes, ponds, and the surface films of mosses and other semi-terrestrial vegetation. Rotifers also occur **interstitially,** in the spaces between the sand grains of freshwater beaches. In fact, rotifers are generally considered to be one of the most characteristic groups of freshwater animals. Typically, 40–500 individuals are found per liter of lake or pond water, with densities as high as 5000 individuals per liter having been recorded on several

occasions. About 5% of rotifer species are found in shallow-water marine environments.

GENERAL CHARACTERISTICS

Like all aschelminths, rotifers are pseudocoelomate. As in several other aschelminth phyla, mitotic divisions cease early in development, so that further increases in body size are due to an increase in cell size rather than in cell number. That is, like nematodes rotifers are **eutelic,** and the different organs and tissues are characterized by species-specific numbers of nuclei. Probably as a consequence of eutely, rotifers have poor regenerative powers. The epidermis of rotifers, like that of nematodes, is syncytial; i.e., cell membranes between nuclei are incomplete. The syncytial epidermis secretes a nonchitinous cuticle that is never molted. Rotifers possess both smooth and striated muscle fibers (Fig. 11.2), the latter being used for the rapid movement of spines and other appendages. Specialized respiratory and blood circulatory systems are lacking.

Most species are free-living and short-lived. Typically, the life span of a rotifer is between one and two weeks, although individuals of a few species can survive for up to about five weeks. Some free-living species spend their adult lives permanently attached to a substrate, while others are capable of moving from place to place. Some rotifers are parasitic, but their hosts are always invertebrates, especially arthropods and annelids (Fig. 11.3). Thus, the life histories of parasitic rotifers have never commanded the

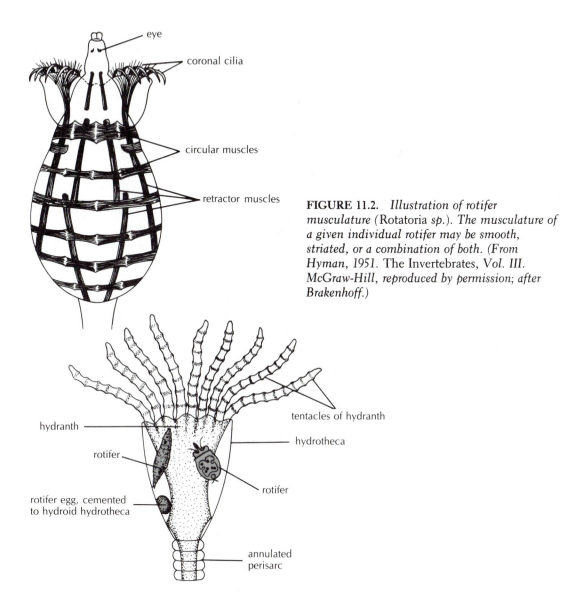

eye

coronal cilia

circular muscles

retractor muscles

FIGURE 11.2. *Illustration of rotifer musculature* (Rotatoria *sp.). The musculature of a given individual rotifer may be smooth, striated, or a combination of both. (From Hyman, 1951. The Invertebrates, Vol. III. McGraw-Hill, reproduced by permission; after Brakenhoff.)*

tentacles of hydranth

hydranth

rotifer

hydrotheca

rotifer

rotifer egg, cemented
to hydroid hydrotheca

annulated
perisarc

FIGURE 11.3. *Cnidarian hydranth with rotifer parasites* (Proales gonothyraea) *in hydrotheca. The rotifer feeds on the cnidarian tissues, and lays its eggs in the hydranth. Other species within this genus are parasites in or on protozoans, oligochaetes (Annelida), gastropod embryos (Mollusca), crustaceans (Arthropoda), and filamentous algae. Several other genera are common internal parasites of oligochaetes, leeches (Annelida), and slugs (Mollusca). (From Hyman; after Remane.)*

attention accorded those of most other parasitic animals, since parasitic rotifers have minimal impact on humans. Free-living species have never been much in the public eye either. Most are only 50–500 μm long, which is remarkably small for an adult metazoan. The largest individuals never exceed a length of 3 mm. Interest in rotifer biology seems to be increasing somewhat, as several free-living rotifer species have been found to be good food sources for the rearing of some commercially important fish and crustaceans. Rotifers may also play important roles in determining community structure and in mediating the flow of energy through freshwater ecosystems. Quite recently, rotifers have been utilized as models for research on the process of **senescence** (i.e., aging).

Nonparasitic (i.e., free-living) rotifers, feed on a variety of items. Most species are omnivores, selectively ingesting: algae of appropriate size and chemical composition; small free-living animals (**zooplankton**); and detritus of appropriate size. Other species feed on the intracellular juices of algae, and a number of species are carnivores, preying on a variety of animals smaller than themselves, including other rotifers.

Certain elements of the feeding and digestive system are unique to rotifers, notably the mastax and the trophi. The **mastax** is a prominent, muscular swelling located between the pharynx and the esophagus (Fig. 11.4). In some parasitic species, the mastax is modified for attachment to the host. Within the mastax of all species are found a number of rigid structures, the **trophi,** which are often used to grind food following ingestion (Fig 11.5). In some species, the trophi are used to suck food in through the mouth; in

other species, the trophi can be protruded from the mouth to grab and/or pierce prey. The structure and sculpturing of trophi vary considerably in different species, in accordance with how they are used. They thus serve as major tools in species identification.

Free-living rotifers are propelled primarily by the cilia of the corona. In some species, particularly well-developed appendages contribute to locomotion (Fig. 11.6). Some species are entirely planktonic; that is, the rotifers swim throughout their lives. Members of many other species stop swimming periodically and form temporary attachments to a solid substrate by secreting a cementing substance from a pair of pedal glands. The pedal glands open to the outside through pores in the **toes** of the foot. Rotifer feet possess zero to four toes, depending upon the species (Fig. 11.7). Once the rotifer is attached to the substrate, the coronal cilia serve to generate water currents for respiration and food collection. In a number of planktonic and sedentary rotifers, the cilia of the corona are arranged to form two parallel bands, with a ciliated food groove lying between. The cilia of the two bands beat toward each other, sweeping particles into the food groove (Fig. 11.8). Cilia in the food groove then conduct these captured particles to the mouth for ingestion.

Many free-living species are capable of moving upon solid substrates between bouts of swimming. The foot and toes play a role in such locomotion by forming temporary attachments to the substrate. Once the attachment is made, the body can be elongated through the contraction of circular muscles, which are distributed about the body in discrete bands. Contraction of the circular

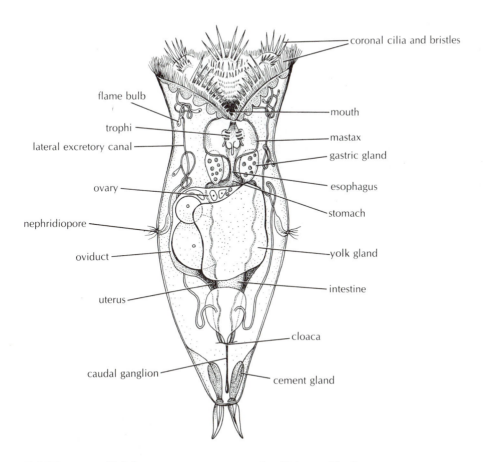

flame bulb

trophi

lateral excretory canal

ovary

nephridiopore

oviduct

uterus

caudal ganglion

coronal cilia and bristles

mouth

mastax

gastric gland

esophagus

stomach

yolk gland

intestine

cloaca

cement gland

FIGURE 11.4.   Epiphanes senta, *a common free-living rotifer from fresh water. Note the placement of the mouth and feeding apparatus. The trophi/mastax complex is used to grind ingested food into smaller particles. In some species, the trophi can be protruded from the mouth for prey capture. (After Plate, de Beauchamp, and Remane; from Brown, 1950. Selected Invertebrate Types.* John Wiley & Sons.)

musculature elongates the body and stretches out the relaxed longitudinal musculature. The pseudocoel of these rotifers thus functions as a hydrostatic skeleton, permitting the mutual antagonism of the circular and longitudinal musculature through the generation of temporary increases in hydrostatic pressure.

As the rotifer's body elongates, the corona is withdrawn turning the anterior end of the animal into a suction-generating **proboscis.** The proboscis can then be applied to the substrate, forming a new attachment site. The posterior attachment is then broken, and the body becomes shorter and fatter through contraction of the longitudinal muscles and

(a)   (b)

FIGURE 11.5. *Scanning electron micrographs of rotifer trophi.* (*Courtesy of Dr. George Salt. From Salt et al., 1978.* Trans. Amer. Microsc. Sci., 97: 469.)

stretching of the relaxed circular musculature (Fig. 11.9). Finally, the anterior attachment is released while the foot regains its grip on the substrate, and the body is again elongated. The animal can thus progress by **looping,** somewhat as described previously for free-living flatworms (Phylum Platyhelminthes, Class Turbellaria), and as yet to be described for leeches (page 219). As illustrated, some rotifers make extraordinary changes of shape during locomotion. Such changes are often facilitated by the foot and trunk of the body being divided into a number of sections that can be telescoped into

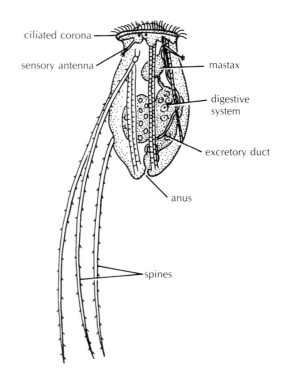

FIGURE 11.6. Filinia longiseta, *a pelagic rotifer commonly found in lakes. Note the lack of a foot, lack of toes, and the presence of conspicuous spines for swimming. (From Hyman; after Weber.)*

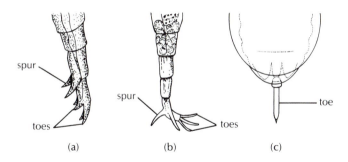

(a) (b) (c)

FIGURE 11.7. *Rotifers may bear 0, 1, 2, 3, or 4 toes on the foot, depending on the species, and may also bear a pair of nonsecretory protuberances called spurs. (a)* Philodina roseola, *with four toes and two spurs. (From Hyman, 1951.* The Invertebrates, Vol. III. *McGraw-Hill, reproduced by permission; after Hickernell.) (b)* Rotaria *sp., with three toes and two spurs. (From Hyman, Vol. III.) (c)* Monostyla *sp., with a single toe and no spurs. (From Hyman, Vol. III; after Myers.)*

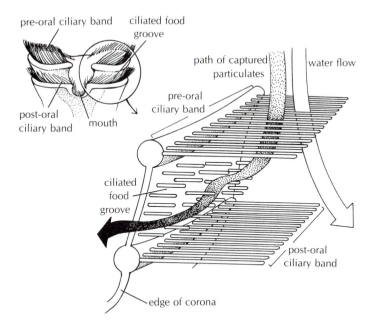

FIGURE 11.8. *Double-banded ciliary system of filter-feeding rotifers. As water is swept between the cilia of the pre-oral ciliary band, suspended food particles are captured and conducted to the mouth by the cilia of the food groove. (After Strathmann.)*

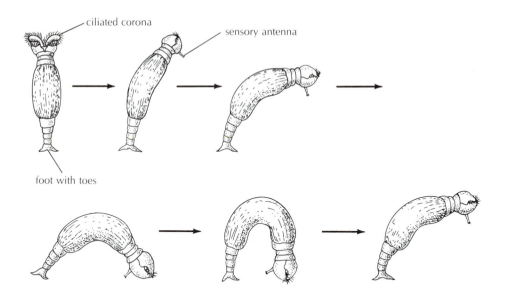

ciliated corona

sensory antenna

foot with toes

FIGURE 11.9. *Looping locomotion of* Philodina roseola. *Contraction of circular muscles elongates the body, which is then bent toward the substrate by the differential contraction of longitudinal muscles. Once the anterior end of the animal grips the substrate, the foot releases its attachment and the rear of the body may be pulled forward. The body may then reattach posteriorly and extend forward again. Note that the corona is withdrawn during this sequence of movements. (After Harmer and Shipley.)*

each other. This principle has subsequently resurfaced with the human invention of the collapsible drinking cup.

Other nonparasitic rotifers are **sessile** as adults; i.e., they are incapable of locomotion. Pedal gland secretions attach these animals to a substrate permanently. Members of many species secrete protective tubes, often incorporating debris, sand grains, or even fecal pellets into their walls (Fig. 11.10). All species that are sessile as adults have free-swimming young resembling the young of planktonic species. Thus, even sedentary species are capable of locomotion—at least for a short time between adult generations.

The reproduction of rotifers is unique in a number of respects. Like nematodes and other aschelminths, reproduction by fission or fragmentation is unknown, again a likely consequence of eutely. Where asexual reproduction does occur among rotifers, it is by **parthenogenesis,** the development of unfertilized eggs. The details of sexual and asexual reproduction among rotifers are best discussed on a class-by-class basis.

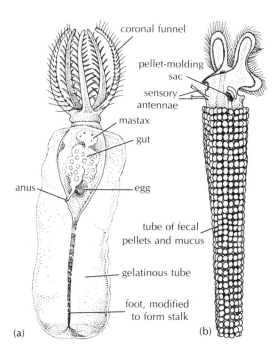

FIGURE 11.10. *Two sedentary rotifer species: (a)* Stephanoceros fimbriatus. *(After Jurczyk; and after Jägersten.) (b)* Floscularia ringens, *showing region of pellet formation. (After Pennak.)*

CLASS SEISONIDEA

This small class is composed exclusively of ectoparasites of marine crustaceans. As might be expected, the corona of these species is often greatly reduced in size (Fig. 11.11a). Reproduction seems to be exclusively sexual. Individuals have separate sexes; i.e., the members of this class are always **dioecious.** Fertilization is internal, either through true copulation or by means of hypodermic impregnation of the female by the male. In the latter case, sperm are injected by the male into the pseudocoel of the female, from which they must find their way into the ovary.

CLASS BDELLOIDEA

The members of this major class of rotifers are all free-living and mobile. That is, this class contains no tube-dwelling, sessile species and no parasitic species. Most members are omnivorous suspension feeders. The corona is well developed and bilobed (Fig. 11.11b,c). Reproduction among bdelloid rotifers appears to be exclusively by parthenogenesis, since males have never been discovered. Thus, all known rotifers in this class are female.

Bdelloid rotifers commonly inhabit environments that periodically expose the animals to physiologically stressful conditions, such as freezing, dehydration, or high temperature. Species living in polar lakes and ponds, temporary ponds, and among emergent mosses and lichens are typically capable of entering a state of extremely low metabolism, or **cryptobiosis.** Some species secrete a gelatinous covering during the early stages of drying. This covering then hardens to form a **cyst** (Fig. 11.12). Like the nematodes already discussed and the tardigrades to be discussed in Chapter 17, rotifers in the cryptobiotic state can withstand environmental extremes, including extensive desiccation, for prolonged time periods. Indeed, some bdelloid rotifers have been successfully rehydrated following 50 years or more of desiccation! How cryptobiosis permits rotifers to withstand what would otherwise be lethal environmental conditions is not yet known.

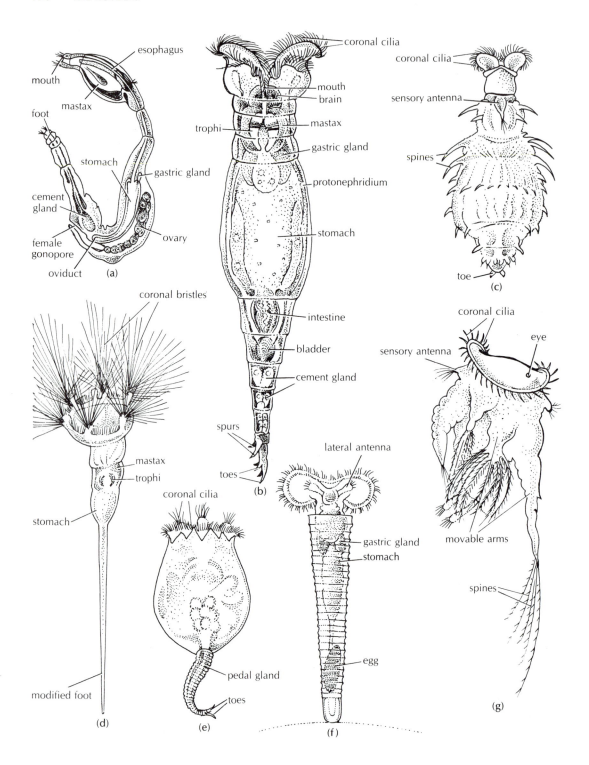

esophagus

mouth

mastax

foot

stomach

gastric gland

cement gland

female gonopore

ovary

oviduct

(a)

coronal cilia

mouth

brain

trophi

mastax

gastric gland

protonephridium

stomach

intestine

bladder

cement gland

spurs

toes

(b)

coronal cilia

coronal cilia

sensory antenna

spines

toe

(c)

coronal bristles

mastax

trophi

stomach

modified foot

(d)

coronal cilia

pedal gland

toes

(e)

lateral antenna

gastric gland

stomach

egg

(f)

coronal cilia

sensory antenna

eye

movable arms

spines

(g)

FIGURE 11.11.   *Representatives of the three rotifer classes.*
(a) Seisonidea (Seison). All species are ectoparasitic on marine
crustaceans, and are dioecious. The illustrated individual is a
female. Note the reduced ciliation associated with a parasitic
life-style. (From Meglitsch; after Plate.) (b) Bdelloidea (Philodina
roseola). All species are free-living and motile suspension-feeders,
and no males have ever been described; all reproduction is
parthenogenetic. (From Hyman; after Hickernell.) (c) Bdelloidea
(Macrotrachela multispinosus). (From Hyman; after Murray.)
(d) Monogononta (Collotheca). Note the highly modified corona,
with seven distinct lobes. The animal is sessile and secretes a
gelatinous tube (not illustrated). (After Hyman.) (e) Monogononta
(Brachionus rubens). This species is free-swimming, and is
commonly used as a food source for rearing larval fish. (From
Pennak; after Halbach.) (f) Monogononta (Limnius sp.), a sessile
tube-dweller. (From Meglitsch; after Edmondson.) (g) Monogononta
(Pedalia mira), a free-swimming species. Members of this class show
both sexual and asexual reproduction; the switch to sexual
reproduction is triggered by alterations in the physical environment.
(From Hyman; after Hudson and Gosse.)

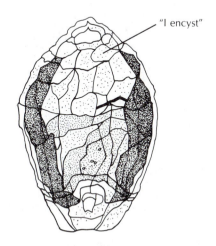

"I encyst"

FIGURE 11.12.   *Philodina roseola, encysted.*
The adult is illustrated in Fig. 11.11(b). (From
Hyman; after Hickernell.)

CLASS MONOGONONTA

Monogonont rotifers may be free-swimming
or sessile. In fact, this class contains the only
free-living sessile rotifers. The sessile rotifers
are generally found attached to macroscopic
plants, filamentous algae, or to the tubes of
other sessile rotifers of the same or of different species. Some sessile species use the corona to collect food particles as described
previously, or with a modification of that
theme. In other species, the corona is poorly
ciliated or is nonciliated. Instead, the corona
may bear long spines surrounding a funnel-
like anterior end. The spines can be moved
to entrap small metazoans that come too
close, and the prey are forced into the mouth
for ingestion. Thus, many members of this

class are carnivores. Ciliation of the **buccal field** surrounding the mouth presumably aids in ingestion.

Many sessile species live in protective tubes (Figs. 11.10, 11.11f). In some instances, these are merely gelatinous secretions from specialized glands opening exter-nally. In other species, particles are collected by the corona, coated with mucus, and cemented in place, elongating the tube as the animal grows in size. In some species, fecal pellets are incorporated into the tube. In both free-swimming and sessile monogonont species, the cuticle is commonly thick and rigid, forming a protective **lorica**. Production of a lorica is encountered only among members of this class.

The reproductive pattern of monogonont rotifers is unique as well. Typically, monogononts reproduce by means of parthenogenesis. Females generally produce diploid eggs by mitosis, usually one at a time, and each egg develops into another diploid female—all in the absence of males. Such

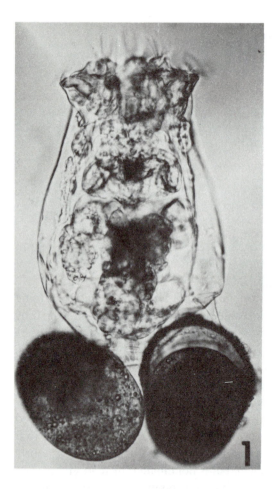

FIGURE 11.13 *Photograph of female* Brachionus calyciflorus *with mictic resting eggs. (Courtesy of John J. Gilbert. From Wurdak, Gilbert, and Jagels, 1978. Trans. Amer. Microsc. Soc. 97: 49.)*

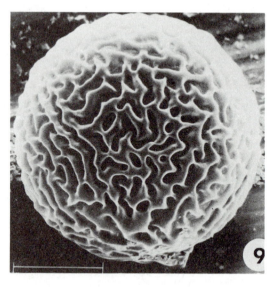

FIGURE 11.14. *Photograph of resting egg of* Asplanchna intermedia. *(Courtesy of John J. Gilbert. From Gilbert and Wurdak, 1978. Trans. Amer. Microsc. Soc., 97: 330.)*

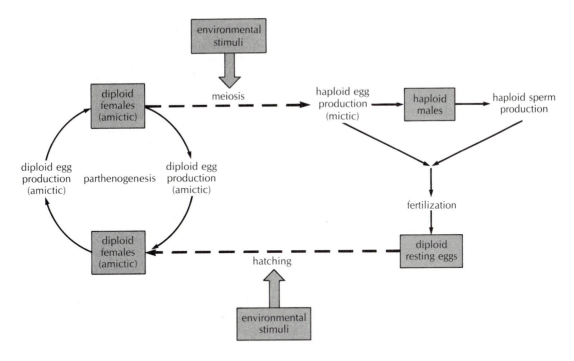

FIGURE 11.15. *The life cycle of monogonont rotifers. In the absence of specific stimuli, reproduction is exclusively asexual, by means of parthenogenesis. Environmental factors that may influence reproduction include changes in food quality and quantity, changes in photoperiod, and changes in temperature.*

females are termed **amictic,** referring to the production of eggs without the "mixing in" of genes from any other individual. Young rotifers emerge from amictic eggs soon after the eggs are released; thus, amictic eggs are also known as **subitaneous** eggs (*subit* = G: sudden). Through the production of subitaneous eggs, the population size of a given rotifer species can double in as little as 15 hours. Typically, each female produces 4–40 amictic eggs per lifetime, and 20–40 or more generations of genetically identical amictic females can occur yearly.

Under certain conditions, however, a dif-

ferent type of monogonont female is produced.[1] These **mictic** females produce their eggs through meiosis, so that the eggs are haploid (Fig. 11.13). In the absence of fertilization, mictic eggs develop into haploid males. The males are usually smaller and morphologically dissimilar to the females of the same species, but are always fast swimmers. The males are nonfeeding, lacking both mouth and anus. Their job must be done quickly, since they cannot usually survive for more than a few days. Typically, the

---

1 See Topics for Discussion, No. 4.

males are ready to fertilize eggs within an hour of hatching! The fertilized eggs form **resting eggs,** also known as **winter eggs** (Fig. 11.14). These resting eggs are highly resistant to a variety of physical and chemical stresses, permitting the developing embryos to withstand unfavorable conditions. During such periods, the embryos are in a state of developmental arrest, or **diapause.** Hatching occurs after conditions improve. Resting eggs always give rise to amictic females, which go on to reproduce by means of parthenogenesis (Fig. 11.15). Generally, only one or two mictic generations occur per year.[2]

It has recently been found that a small percentage ($< 0.5\%$) of females in the population may be characterized as neither mictic or amictic. Instead, these females are **amphoteric;** a single female produces some haploid eggs by meiosis (which become males if unfertilized) and also produces some diploid eggs by mitosis.

It should be clear by now that most free-living rotifers encountered in the field, whether mobile or sessile, are females.

OTHER FEATURES OF ROTIFER BIOLOGY

1. Digestive System

Most rotifer species have a tubular digestive system with an anterior mouth and a posterior anus (see Figs. 11.4 and 11.11). Cilia lin-

ing the inner surface of the gut move food through the digestive system. Digestion is largely extracellular and takes place in the stomach. **Gastric glands,** or **gastric caeca,** associated with the stomach, contribute digestive enzymes. Undigested wastes pass through a short intestine and discharge into a **cloaca,** which, by definition, also receives the terminal ducts of the excretory and reproductive systems. The anus opens dorsally, near the junction of the trunk and the foot. Some bdelloid rotifers have no pronounced stomach cavity and no anus. Instead, the "stomach" is a continuous syncytial mass through which food is circulated. In such species, digestion is primarily intracellular. Except for the Seisonidea, male rotifers lack a functional digestive system.

2. Nervous and Sensory Systems

The rotifer brain consists of a bilobed mass of ganglia lying dorsal to the mastax (Fig. 11.16). Nerves extend throughout the body, connecting the brain to the musculature and organ systems and to a variety of sensory receptors. Sensory bristles, and, usually, three antennae (two lateral and one median-dorsal) serve as chemo- and mechanoreceptors. A pigmented, cup-like photoreceptor often lies directly on the brain. Additional photoreceptors may occur on the corona.

3. Excretion and Water Balance

Excretion is at least partly accomplished by diffusion across the general body surface. In

---

2 See Topics for Discussion, No. 5.

addition, all rotifers contain a pair of protonephridia, resembling those of flatworms. The activity of flagella are presumed to create a negative pressure within each nephridium, drawing fluid in from the pseudocoel. As many as 50 flame bulbs may be associated with each protonephridium. The two collecting tubules, one from each nephridium, lead to a common **bladder** (Fig. 11.11b). Bladder contractions, up to six times per minute, expel the fluid into the **cloaca**, which also receives the products of the digestive and reproductive systems. The tissues and body fluids of freshwater rotifers contain a significantly higher concentration of dissolved materials than is found in the surrounding medium, so that water continually flows into the animal across the permeable body surface. The major role of the protonephridia would thus seem to be in maintaining water balance and body volume, rather than in the removal of excretion products, at least in the majority of species. It should be pointed out that the osmotic gradient that exists across the body wall of freshwater rotifers, and the accompanying diffusional influx of water, may aid the rotifer in maintaining body turgor.

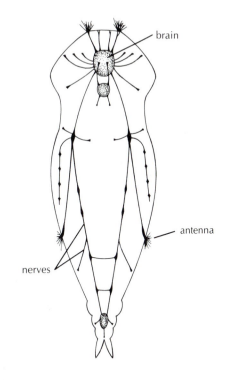

FIGURE 11.16. *Rotifer nervous system. (From Pennak, 1978. Fresh-Water Invertebrates of the United States, 2nd ed. John Wiley & Sons; after Remane.)*

◀ **T A X O N O M I C   S U M M A R Y**

PHYLUM ROTIFERA
CLASS SEISONIDEA
CLASS BDELLOIDEA
CLASS MONOGONONTA

TOPICS FOR FURTHER DISCUSSION AND INVESTIGATION

1. Another group of small, pseudocoelomate animals comprise the phylum Kinorhyncha (*kino* = G: moveable; *rhynch* = G: snout). The approximately 100 species of this phylum are exclusively marine and free-living (Fig. 11.17). In what respects are kinorhynchs like rotifers? In what respects are kinorhynchs like nematodes? In what respects are kinorhynchs dissimilar to both rotifers and nematodes?

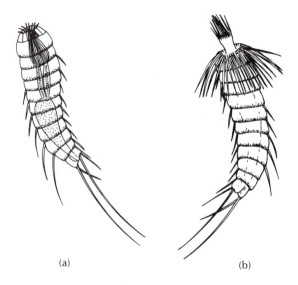

(a)   (b)

FIGURE 11.17. *Typical kinorhynch* (Echinoderella), *with head withdrawn (a) and protruded (b). (After Hyman.)*

Higgins, R. P., 1964. Redescription of the kinorhynch *Echinoderes remanei* (Blake, 1930; Karling, 1954). *Trans. Amer. Microsc. Soc.*, 83: 243.

Higgins, R. P., 1965. The homalorhagid Kinorhyncha of Northeastern U.S. coastal waters. *Trans. Amer. Microsc. Soc.*, 84: 65.

Horn, T. D., 1978. The distribution of *Echinoderes coulli* (Kinorhyncha) along an interstitial salinity gradient. *Trans. Amer. Microsc. Soc.*, 97: 586.

Hyman, L., 1951. *The Invertebrates*, Vol. 3. New York: McGraw-Hill, pp. 170–183.

Kozloff, E. N., 1972. Some aspects of development in *Echinoderes* (Kinorhyncha). *Trans. Amer. Microsc. Soc.*, 91: 119.

2. The small phylum Gastrotricha (Fig. 11.18), containing about 400 species, has been associated with the Aschelminthes for some time. However, this association has become uneasy of late, as it is no longer clear whether gastrotrichs are pseudocoelomate. In other respects, however, gastrotrichs have a number of features in common with both rotifers and kinorhynchs. All species are small and free-living. Gastrotrichs are encountered in both marine and freshwater habitats, and are especially common interstitially; that is, in the spaces between sand grains. Compare the external morphology and the digestive, excretory, and reproductive systems of gastrotrichs and rotifers.

Brunson, R. B., 1949. The life history and ecology of two North American gastrotrichs. *Trans. Amer. Microsc. Soc.*, 68: 1.

Brunson, R. B., 1950. An introduction to the taxonomy of the Gastrotricha with a study of eighteen species from Michigan. *Trans. Amer. Microsc. Soc.*, 69: 325.

D'Hondt, J. L., 1971. Gastrotricha. *Ann. Rev. Oceanogr. Marine Biol.*, 9: 141.

Hyman, L. H., 1951. *The Invertebrates*, Vol. 3. New York: McGraw-Hill, pp. 151–170.

3. In 1983, a new phylum (the Loricifera) was established to hold a single, newly discovered species of animal, *Nanaloricus mysticus* (Fig. 11.19).

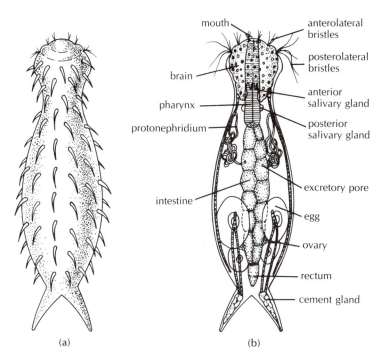

(a)   (b)

FIGURE 11.18.   *Typical gastrotrich* (Chaetonotus). *(a) External appearance; (After Brunson; and after Zelinka.) (b) Internal anatomy. (From Brown; after Remane.)*

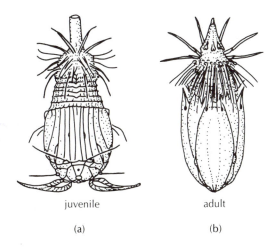

juvenile   adult

(a)   (b)

FIGURE 11.19.   Nanaloricus mysticus, *the only known member of the newly erected Phylum Loricifera. (a) Juvenile; (b) Adult. (From Kristensen, 1983. Zeit. Zool. Syst. Evol.-forsch., 21: 163.)*

These animals are only about 200 μm long and live interstitially; i.e., in the spaces between sand grains. They bear a number of morphological features in common with aschelminths, especially with members of the Kinorhyncha (see Topic 1 above) and the Nematomorpha (p. 183). In addition, they share certain features with priapulids (p. 244), a group whose affiliation with other invertebrates is presently uncertain.

Compare and contrast *Nanaloricus mysticus* with the aschelminths and priapulids.

Kristensen, R. M., 1983. Loricifera, a new phylum with Aschelminthes characters from the meiobenthos. *Zeit. Zool. Syst. Evol.-forsch.*, 21: 163.

4. Investigate the environmental factors that appear to trigger the production of mictic females in monogonont rotifers.

Birky, C. W., Jr., 1964. Studies on the physiology and genetics of the rotifer, *Asplanchna*. I. Methods and physiology. *J. Exp. Zool.*, 155: 273.

Gilbert, J. J., 1963. Mictic-female production in the rotifer *Brachionus calyciflorus*. *J. Exp. Zool.*, 153: 113.

Gilbert, J. J., and J. R. Litton, Jr., 1978. Sexual reproduction in the rotifer *Asplanchna girodi*: effects of tocopherol and population density. *J. Exp. Zool.*, 204: 113.

Pourriot, R., and P. Clément, 1975. Influence de la durée de l'eclairement quotidien sur le taux de femelles mictiques chez *Notommata copeus* Ehr. (rotifère). *Oecologia* (*Berl.*), 22: 67. [Influence of photoperiod on the mictic-female production rate in *Notommata capeus* Ehr. (rotifer).]

5. Discuss the physical and biological factors influencing the dynamics of natural rotifer populations.

Edmondson, W. T., 1945. Ecological studies of sessile Rotatoria. II. Dynamics of populations and social structures. *Ecolog. Monogr.*, 15: 141.

Gilbert, J. G., and C. E. Williams, 1978. Predator-prey behavior and its effect on rotifer survival in associations of *Mesocyclops edax*, *Asplanchna girodi*, *Polyarthra vulgaris*, and *Keratella cochlearis*. *Oecologia* (*Berl.*), 37: 13.

King, C. E., 1972. Adaptation of rotifers to seasonal variation. *Ecology*, 53: 408.

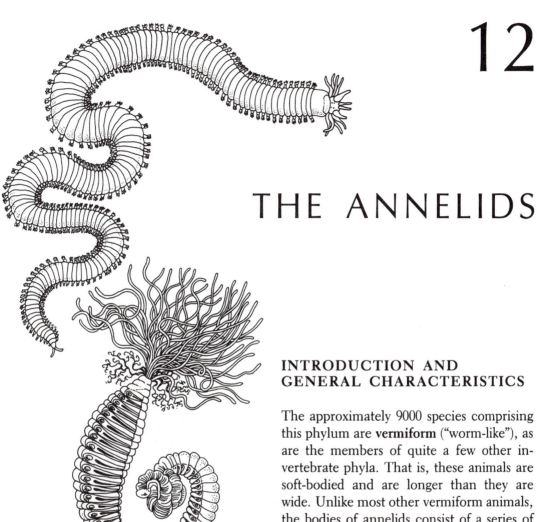

12

THE ANNELIDS

INTRODUCTION AND GENERAL CHARACTERISTICS

The approximately 9000 species comprising this phylum are **vermiform** ("worm-like"), as are the members of quite a few other invertebrate phyla. That is, these animals are soft-bodied and are longer than they are wide. Unlike most other vermiform animals, the bodies of annelids consist of a series of repeating segments. This serial repetition of segments and organ systems (skin, musculature, nervous system, circulatory system, reproductive system, and excretory system) is known as **metamerism,** or metameric segmentation (Fig. 12.1).

Most members of the phylum lack an external, rigid, protective covering. Thus, the outer body wall is generally flexible and can play an active role in locomotion. Moreover, the thin body wall can serve as a general surface for gas exchange, provided that it is kept moist. Even when the epidermis secretes a

Phylum Annelida

(ANNULUS = L: RING)

205

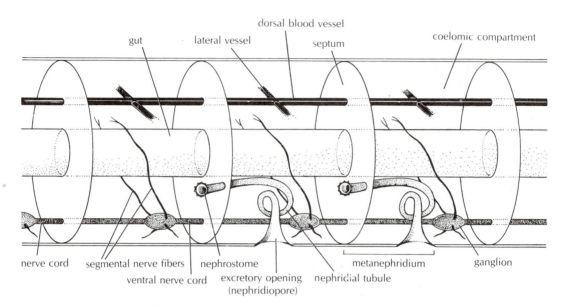

FIGURE 12.1. *Schematic illustration of metameric organization in annelids. The annelid body consists of a linear series of segments, separated from each other by transverse, mesodermally derived septa. Much of the internal anatomy, including excretory, nervous, coelomic, and muscle systems, is segmentally arranged. Body wall musculature has been omitted for clarity. (Based on various sources.)*

protective cuticle, the cuticle remains permeable to both water and gases. For this reason, annelids are restricted to moist environments.

The individual segments of an annelid are generally separated from each other to a large degree by **septa,** which are thin sheets of mesodermally derived tissue (**peritoneum**) that essentially isolate the coelomic fluid in one segment from that in adjacent segments (Fig. 12.1). This allows for localized deformation of the outer body wall, brought about by contractions of circular and longitudinal musculature within a single segment.

Although some excretion of wastes takes place across the general body surface, excretion generally occurs by means of structures called **nephridia** ("little kidneys"). Most annelid segments contain two nephridia, each of which is open at both ends. This type of nephridium is called a **metanephridium.**

Coelomic fluid is drawn into the nephridium at the **nephrostome** ("kidney-mouth"; *stoma* = G: mouth) by the action of cilia (Fig. 12.2). As the fluid passes through the convoluted tubule of the nephridium, some substances (including salts, amino acids, and water) may be selectively resorbed, and other

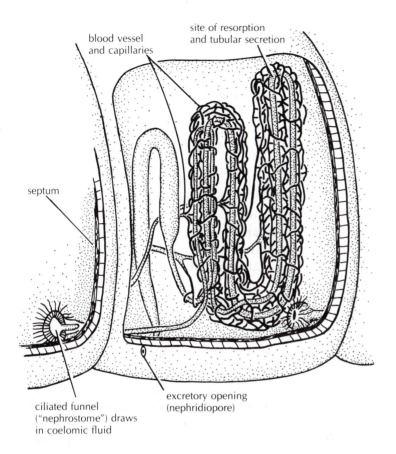

blood vessel
and capillaries

site of resorption
and tubular secretion

septum

ciliated funnel
("nephrostome") draws
in coelomic fluid

excretory opening
(nephridiopore)

FIGURE 12.2 *Diagrammatic representation of a typical metanephridium. The chemical composition of the primary urine drawn through the nephrostome is altered by selective resorption and secretion as fluid moves through the nephridial tubules. The final urine is discharged through the nephridiopore. (Modified from several sources.)*

substances (including metabolic waste products) may be actively secreted into the lumen of the tubule. The final urine emerging from the nephridiopore is thus quite different in chemical composition from the primary urine entering at the nephrostome. In addi-tion to providing an outlet for metabolic waste products, nephridia may be used to regulate the water content of the coelomic fluid.[1]

---

1 See Topics for Discussion, No. 5.

In many annelid species, ducts leading from the gonadal tissue merge with the nephridial tubule. Thus, the nephridium generally plays a role in the discharge of gametes as well as urine.

CLASS POLYCHAETA

Approximately 60% of all annelid species are placed in the class Polychaeta. Nearly all polychaetes live in salt water. Polychaetes generally possess at least one pair of eyes and at least one pair of sensory appendages (**tentacles**) on the anteriormost part of the body (the **prostomium**). Generally, the body wall is extended laterally into a series of thin, flattened outgrowths called **parapodia** (Fig. 12.3). Parapodial morphology differs significantly among species and therefore plays an important role in polychaete identification. These outfoldings increase the exposed surface area of the animal and, as parapodia are highly vascularized (Fig. 12.3b), they function in gas exchange between the worm and its environment. The parapodia also have a locomotory function in many species, being stiffened by the presence of chitinous support rods called **acicula** (Fig. 12.3a). In addition, siliceous, chitinous, or, more rarely, calcareous bristles called **setae** protrude from each parapodium. The body may be covered by a series of overlapping protective plates (**elytra**) in some species.

Class Poly•chaeta
/ /
(G: MANY SETAE)

The septa present between most segments in polychaete worms enable the hydrostatic skeletal system to function independently in each segment. Localized contractions in one part of the body result in localized deformations without interfering with the musculature in other segments of the worm. Setae serve to form temporary attachment sites and to prevent backsliding during locomotion on or within the substrate or burrow. The septa between anterior adjacent segments are often absent or incomplete (**perforate**) in active, burrowing forms, enabling a greater volume of fluid to be utilized. This, in turn, permits a greater change of shape to be associated with extending and anchoring the penetration organ (termed the **proboscis**) into the sediment.[2]

A cross section of a polychaete worm reveals a secreted, nonliving cuticle and a layer of circular muscles underlying the epidermis (Fig. 12.3a). Beneath the layer of circular muscles lies a layer of longitudinal muscle fibers. Movement is accomplished largely by the mutual antagonism of these two sets of muscles through the hydrostatic skeleton of each coelomic compartment. Oblique muscles are often found as well, serving to maintain body turgor and to operate the parapodia, which can be used as oars for locomotion through the water (i.e., swimming), over surfaces, or within burrows.

Many polychaetes form burrows in the sediment. In some burrowing polychaetes, the proboscis is everted into the sediment (Fig. 12.4a), penetrating and pushing aside

---

2 See Topics for Discussion, No. 3.

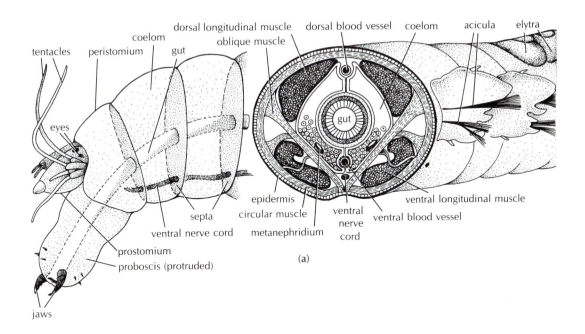

(a)

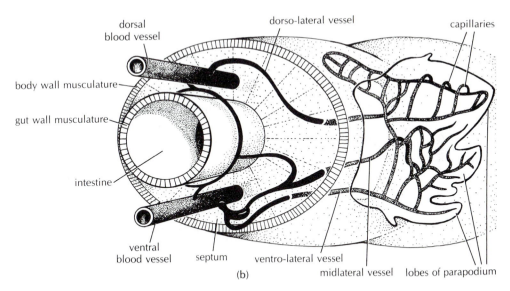

(b)

FIGURE 12.3. (a) Hypothetical polychaete worm, a composite of several species, showing typical major features. (b) Detail of parapodium, showing the high degree of vascularization. (Modified from several sources.)

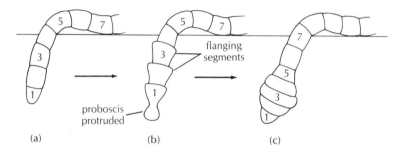

FIGURE 12.4.   *Sequence of movements involved in burrowing by the marine lugworm,* Arenicola marina. *(a) The worm has already burrowed part-way into the sediment. (b) The proboscis is protruded anteriorly. Much of the sediment is ingested; the rest is forced aside anteriorly. Appropriate muscle contractions cause adjacent segments to flange, preventing the worm from being pushed backward, out of the burrow, by the forward thrust of the proboscis. (c) Sequential contraction of the longitudinal musculature of each segment draws the worm downward. The increased width of the anterior segments, reflecting relaxation of circular muscles, anchors the worm as the more posterior segments are drawn forward. (After Trueman.)*

the sand. The segments just behind the proboscis flange markedly (Fig. 12.4b), preventing the worm from backsliding. Protrusion of setae also deters backsliding. The longitudinal muscles then contract in the anterior segments while the circular muscles in these segments relax. This results in dilation of the anterior segments of the worm (Fig. 12.4c). The anterior end is thus firmly anchored in the sediment, and the more posterior segments can then be drawn forward by contracting the longitudinal muscles. The worm is then ready to repeat the cycle. Note that burrow formation is accomplished by exploiting the properties of a compartmentalized hydrostatic skeleton.

Polychaetes may be divided into two subclasses. The subclass Errantia includes the generally active, mobile species (i.e., the **er-**rant species). In contrast to these errant polychaetes, species in the subclass Sedentaria typically spend their entire lives living in simple burrows in the sediment or in rigid, protective tubes.[3] These tubes vary in construction from simple organic secretions mixed with sand grains or mud, to tubes composed of calcium carbonate and/or complex mixtures of proteins and polysaccharides. Other species actively bore into calcareous substrates and live in the resulting burrows. The parapodia tend to be greatly reduced, highly modified, or absent among the Sedentaria (Fig. 12.5). Not surprisingly, acicula are absent as well, confirming their locomotory function in errant species. Most

_____

3 See Topics for Discussion, No. 2.

sedentary species also lack a protrusible proboscis. Movement of water through the tubes of sedentary species for respiration, feeding, and/or removal of wastes is accomplished by ciliary action; waves of muscle contractions passing from one end of the worm to the other; or rhythmic movements of modified parapodia. Many species of sedentary polychaetes display an abundance of thread-like or feathery appendages at the anterior end, some serving for food capture and others being highly vascularized and serving as gills for gas exchange (Figs. 12.5, 12.6).

Species in a number of polychaete families have protonephridia similar to those found among the Platyhelminthes and several other invertebrate groups. Thus, in these polychaete species, coelomic fluid is probably ultra-filtered as it is drawn into the nephridial tubule by ciliary beating. In other respects, protonephridia and metanephridia are believed to function similarly. All other polychaetes—indeed, all other annelids—have metanephridia.

CLASS OLIGOCHAETA

Approximately 3000 oligochaete species have been described. In contrast to the Polychaeta, only about 6.5% of oligochaete species are marine; most are found in freshwater or terrestrial habitats. The common earthworm, *Lumbricus terrestris*, is a familiar example of this class (Fig. 12.7a). Compared

with most polychaete annelids, oligochaetes have a more streamlined appearance. In particular, parapodia are lacking and the anterior-most region of the body, the prostomium, lacks conspicuous sensory structures such as eyes and tentacles. Setae are present in oligochaetes but are less densely distributed along the body. In contrast to the diversity of body plans found among polychaete worms, the oligochaete body plan is relatively invariant among species. There are usually no specialized respiratory organs; gas exchange is accomplished by diffusion across a moist body wall.

As in the Polychaeta, septa divide the body coelomic cavity into a series of semi-isolated compartments. The musculature is arranged as in the polychaetes, with an inner layer of longitudinal muscles overlain by a layer of well-developed circular muscle (Fig. 12.7b). Locomotion of oligochaetes is achieved by generating a continuous series of localized contractions and relaxations of the circular and longitudinal musculature. These cycles of contraction, known as **peristaltic waves,** are also commonly used among burrowing polychaetes, both during burrow formation and for driving water through the burrow for oxygenation and removal of wastes.

The principle of peristaltic locomotion is illustrated in Fig. 12.8. In frame (a), the musculature is partially relaxed and the worm is uniform in diameter. A portion of the body wall in each segment is in contact with the substrate. In frame (b), the circular musculature of segment 1 has contracted. Since the adjacent segments are pushing against the substrate, the anterior-most region of the body (the prostomium) moves forward, pushing off against segment 2 and other,

Class Oligo•chaeta
(G: FEW SETAE)

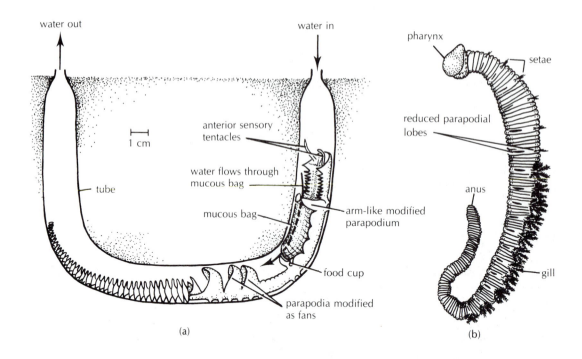

water out

water in

pharynx

setae

anterior sensory
tentacles

water flows through
mucous bag

1 cm

tube

mucous bag

arm-like modified
parapodium

reduced parapodial
lobes

anus

food cup

gill

parapodia modified
as fans

(a)

(b)

FIGURE 12.5. *Representative members of the Sedentaria, showing a variety of morphological adaptations for a nonmotile existence. Most animals are drawn removed from their tubes or burrows. (a) Chaetopterus variopedatus. Feeding currents are generated by rhythmic movements of fan-like parapodia. One pair of anterior parapodia is modified for holding open a mucous bag, which filters out food particles as water moves through the tube. The bag is held posteriorly by a ciliated food cup, which continuously gathers the food-laden mucus into a ball. Periodically, the ball is released from the rest of the net and passed forward along a ciliated tract to the mouth for ingestion. (Modified from several sources.) (b) Arenicola marina, the lugworm. This worm is a deposit-feeder and lives burrowed in sand. For* *gas exchange, water is pumped through the burrow by peristaltic waves of muscle contraction; parapodia play no role in locomotion or water flow, and are greatly reduced in size in most segments. In a number of segments, a small portion of the parapodial lobes have become modified to form gills—clusters of thin, vascularized, branching filaments. (From Fretter and Graham, 1976. A Functional Anatomy of Invertebrates. Academic Press.) (c) Sabellaria alveolata, in lateral view. Members of this species build tubes of sand grains and mucus and attach them to a variety of solid substrates, including the tubes of neighboring individuals. Segments have fused anteriorly to form a densely bristled surface that occludes the opening of the tube when the animal pulls inside. (From Fretter and*

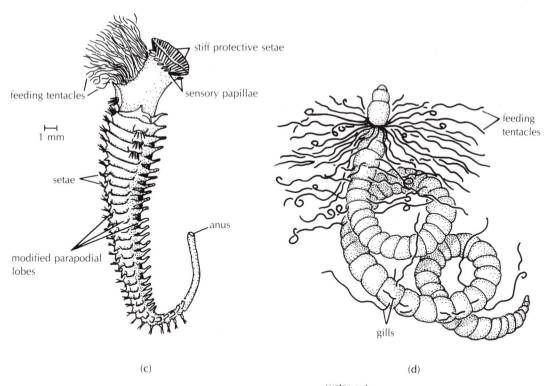

stiff protective setae

feeding tentacles

sensory papillae

1 mm

setae

anus

modified parapodial lobes

(c)

feeding tentacles

gills

(d)

*Graham.) (d) Cirratulus cirratus. These animals
live within muddy substrates. Although the
parapodia are much reduced, a portion of the
parapodium has become greatly elongated in
some segments, forming gills for gas exchange.
(From McConnaughey, Bayard H., and Zottoli,
Robert, 1983.* Introduction to Marine Biology,
*4th ed. St. Louis, MO: The C. V. Mosby Co.;
after McIntosh, W. C., 1915.* British Marine
Annelids, Vol. III. *London: Ray Society. (e)
Pectinaria belgica, in tube of mucus and sand
grains. The worm lives in a conical tube of
cemented sand grains, head down in the
sediment. Members of the genus* Pectinaria *are
selective deposit-feeders, ingesting particles high
in organic content. The undigestible fraction is
expelled at the posterior, elevated end of the
tube, as illustrated. The long tentacles, seen
anteriorly, are involved in food collection. (From
Fretter and Graham.)*

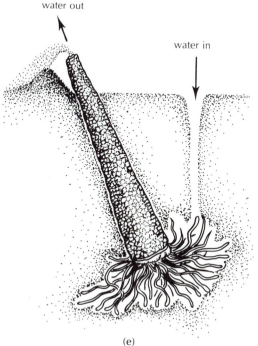

water out

water in

(e)

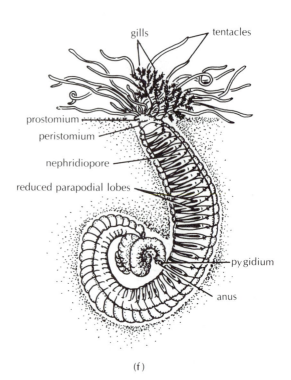

gills — tentacles
prostomium
peristomium
nephridiopore
reduced parapodial lobes
pygidium
anus

FIGURE 12.5. (cont.) *(f) Amphitrite ornata, a species living in shallow water, in burrows of sand or mud lined with mucus. The first three segments each bear a pair of highly branched gills for gas exchange. Numerous grooved tentacles project from the prostomium. When the animal is feeding, these tentacles lie against the substrate, trapping food particles. Food is moved to the mouth by cilia lining the tentacular groove, or by shortening the tentacle itself and bringing the food directly to the mouth. Note the absence of conspicuous parapodia; water movement through the burrow is accomplished by peristalsis of the body wall musculature. (After Brown; and after Barnes.)*

(f)

FIGURE 12.6. *The sedentary polychaete Sabella pavonia. The ciliated tentacles form a fan surrounding the mouth. Water currents pass between the tentacles, and captured food particles are conducted to the mouth area for sorting and subsequent ingestion or rejection. When the worm is startled, the tentacles are rapidly withdrawn within the protective tube. (Courtesy of D.P. Wilson/Eric and David Hosking.)*

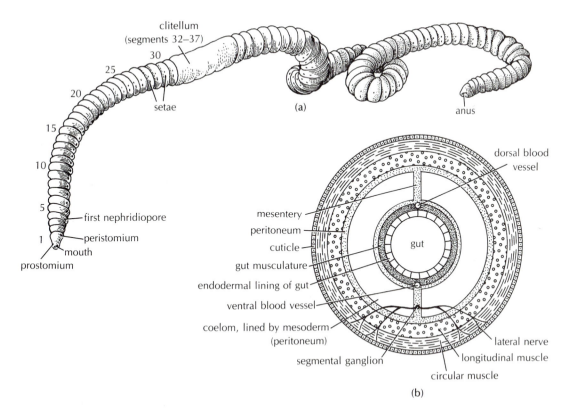

FIGURE 12.7. *(a) External morphology of the terrestrial oligochaete,* Lumbricus terrestris. *Note the absence of anterior and lateral appendages. As discussed later (p. 223), the clitellum is a glandular region characteristic of both oligochaetes and leeches. It produces a mucus that aids copulation and encloses fertilized eggs in a protective cocoon. (After Sherman and Sherman.) (b) Oligochaete in diagrammatic cross section, showing the arrangement of muscle layers of the body wall and gut. (After Russell-Hunter.)*

more posterior segments. Note that the setae are protruded in the segments behind segment 1, helping to prevent slippage. The circular muscles in segments 2, 3, and 4 then contract in succession, resulting in anterior extension of the body (and retraction of setae), as illustrated in frame (c). At about this time, a wave of contraction of the longitudinal musculature of each segment begins at the anterior end of the animal (frame d), and this wave also passes from anterior to posterior.

Note that as the peristaltic wave of circular muscle contraction passes along the body of the worm, the adjacent uncontracted segments serve as temporary anchors, so that

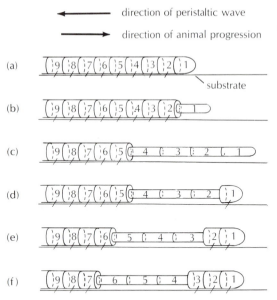

FIGURE 12.8. *Diagrammatic illustration of peristaltic locomotion in oligochaetes. A wave of contraction of the circular muscles begins passing down the body in frames a–c, resulting in forward extension of the worm. As the wave of circular muscle contraction continues to travel down the body of the worm, from segment to segment, a consolidation phase is initiated anteriorly; i.e., a wave of longitudinal muscle contractions moves posteriorly (d–f), shortening and thickening each segment in preparation for the next extension phase.*

the contraction of circular muscles generates a forward thrust. As the longitudinal musculature subsequently contracts, the anterior segments become fat and anchor the animal. The adjacent thin segments can then be pulled forward. Note also that the muscular waves pass in one direction, but the animal moves in the opposite direction

(Fig. 12.9). Such waves are known as **retrograde,** as opposed to **direct.**[4]

It is easy to see that peristaltic waves would be especially effective in burrow formation and in movement within a burrow or tube. When the longitudinal musculature is contracted, thickening the body of the worm, the entire surface area of the segments becomes tightly pressed against the walls of the burrow, greatly increasing the magnitude of the pushing and pulling forces that can be generated without causing slippage of the worm within the burrow. Protrusion of setae by the worm, at appropriate times in the cycles of muscular contraction and relaxation, aids in anchoring these portions of the worm and in preventing backward slippage when thrusting and pulling forces are generated.

CLASS HIRUDINEA

This class of approximately only 500 known species is believed to have evolved from oligochaete stock. Reflecting their presumed oligochaete ancestry, most leeches occupy freshwater or terrestrial habitats; only a small proportion of species is marine. Like the oligochaetes, the basic body plan of the Hirudinea varies little among species. Conspicuously absent are parapodia and head appendages, as in oligochaetes (Fig. 12.10).

---

4 See Topics for Discussion, No. 4.

Class Hirudinea

(G: LEECH)

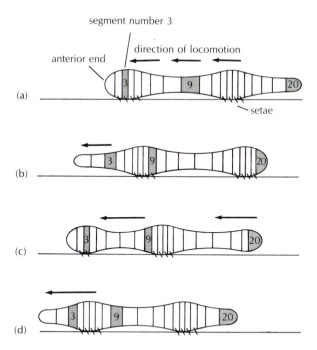

FIGURE 12.9. *Locomotion of an earthworm, based upon movie footage. (a) The circular muscles of the first 5–6 anterior-most segments are relaxed and the longitudinal muscles of these segments are fully contracted. In (b), a wave of circular muscle contraction has been initiated anteriorly; segment 3 and its immediate neighbors have become longer and thinner. Segment 3 has also moved forward, in part due to the contraction of its own circular musculature (and relaxation of its longitudinal musculature), and in part due to the forward thrust generated by the muscle contractions of segments 4 and 5. At the same time, the wave of longitudinal muscle contraction has passed posteriorly, encompassing segment 9. Note that segments 6–9 are prevented from backsliding by the protrusion of setae as the anterior segments are thrust forward. A second wave of longitudinal muscle contraction has passed from segments 12–15 to the terminal segments of the worm; consequently, segment 20 has progressed forward. In frames (c) and (d), waves of longitudinal and then circular muscle contraction continue to be generated anteriorly, thrusting and pulling the segments of the worm forward. (From Purves and Orians, 1983.* Life: The Science of Biology. *Sinauer Associates/Willard Grant Press; after Gray and Lissman.)*

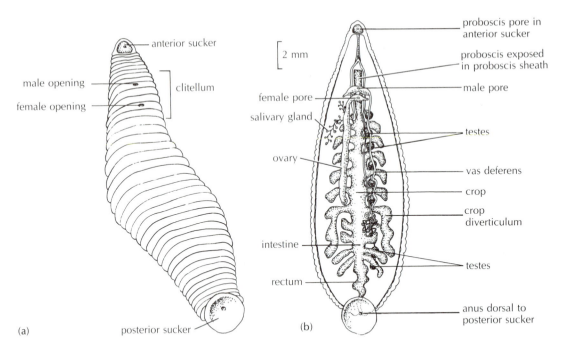

FIGURE 12.10. *(a) External anatomy of a leech. (After Pimentel.)*
(b) Internal anatomy of Glossiphonia complanata. *(From Mann,*
1962. Leeches. Pergamon Press; after Harding.)

Both oligochaetes and leeches are **hermaphroditic** (both male and female reproductive apparatus contained within a single individual), and demonstrate a specialized region of the epidermis known as the **clitellum.** In both classes, the clitellum secretes a cocoon within which embryos develop. Specialized gill structures are absent in most leech species, gas exchange being accomplished by diffusion across the general body wall surface. The morphological similarities between the oligochaetes and the leeches are sufficiently great to warrant classification of both groups as orders within

a single class, the Clitellata, according to some invertebrate zoologists.

In contrast to both polychaete and oligochaete annelids, most leeches lack setae. Moreover, the body is generally not separated into compartments by septa, and the continuous coelomic space is largely filled with connective tissue (**mesenchyme**) (Fig. 12.11). The remaining channels and sinuses within this tissue serve a blood transport function in most leech species; thus, the circulatory medium is the coelomic fluid. Not surprisingly, considering the lack of septa, the locomotion of leeches differs consid-

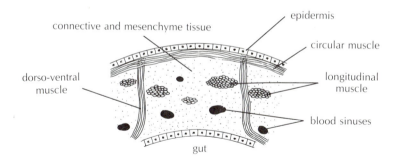

FIGURE 12.11. *Cross-section of a leech. (After Mann.)*

erably from that of other annelids; leeches do not move by the generation of peristaltic waves. Instead, groups of segments at the anterior and posterior ends of the body are modified to form ventral suckers (Fig. 12.12). The mouth is included within the anterior sucker.

Locomotion over a solid substrate is accomplished by using the anterior and posterior suckers as temporary anchors (Fig. 12.13). With the posterior sucker attached to the substrate, the worm extends itself forward by the contraction of circular muscles (Fig. 12.13a–c). The anterior sucker is then applied to the substrate; the circular muscles are relaxed; and the longitudinal muscles are contracted. The leech becomes shorter and, as the circular muscles are stretched, fatter (Fig. 12.13d–g). By further contracting the longitudinal musculature on the ventral surface, the body of the leech arches, bringing the posterior and anterior suckers in close proximity (Fig. 12.13h). The posterior sucker is then applied to the substrate, the anterior sucker is released, and the cycle can be repeated.

Leeches are generally ectoparasitic, feeding either on the blood of other invertebrates, or, more commonly, on the blood of vertebrates. Most parasitic species possess three toothed jaws within the mouth; these jaws are used by the leech to make an incision in the host. Other species have a protrusible proboscis, through which blood is removed from the host. Leeches have been

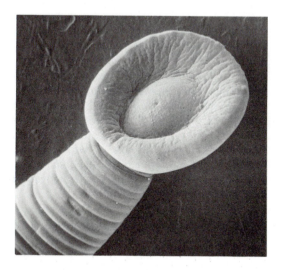

FIGURE 12.12. *Scanning electron micrograph showing anterior end of a freshwater leech,* Johanssonia arctica. *(Courtesy of Dr. R.A. Khan and Allen Press, from Khan and Emerson, 1981.* Trans. Amer. Microsc. Soc., 100: 51.)

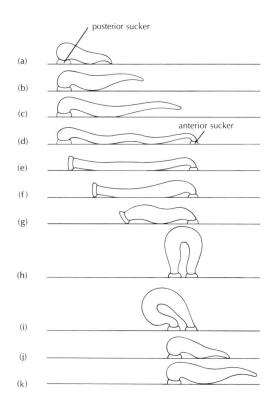

FIGURE 12.13. *Leech locomotion. In frames (a)–(d), the posterior sucker is affixed to the substrate while the circular muscles of the leech contract, extending the animal forward. The longitudinal muscles are stretched in the process. In frames (e)–(g), the posterior sucker is released while the anterior sucker is applied. Contraction of the longitudinal musculature shortens the animal and re-extends the circular muscles. In frame (h), selective contraction of the longitudinal musculature places the posterior sucker immediately behind the anterior sucker, and the entire sequence is repeated. (After Gray, Lissman, and Pumphrey.)*

widely employed in the practice of medicine, being used for bloodletting and, more recently, to alleviate fluid pressure following damage to vascular tissue (e.g., following a snake bite or the re-attachment of a severed finger). Less than 100 years ago, selling leeches to the medical establishment was a booming business. As described by the Rev. J.G. Wood, in a popular account of 1885 (*Animal Creation*), the collection of leeches was a colorful enterprise:

The Leech-gatherers take them in various ways. The simplest and most successful method is to wade into the water and pick off the leeches as fast as they settle on the bare legs. This plan, however, is by no means calculated to improve the health of the Leech-gatherer, who becomes thin, pale, and almost spectre-like, from the constant drain of blood, and seems to be a fit companion for the old worn-out horses and cattle that are occasionally driven into the Leech-ponds in order to feed these bloodthirsty annelids.

The nonparasitic leeches, about 25% of all known species, are predators upon other invertebrate animals.

OTHER FEATURES OF ANNELID BIOLOGY

1. Reproduction and Development

a. Class Polychaeta

Reproduction is exclusively sexual in most species, and most polychaetes are **dioecious** (i.e., they have separate sexes). At least six adjacent segments of a given individual are involved in gamete production, and in some species, gametes are produced within nearly all segments.

In several families of errant polychaetes, many tube-dwelling or otherwise sedentary species undergo **epitoky**, a marked morphological transformation in preparation for reproductive activity. The result of this transformation is an **epitoke**: a sexually mature being that is highly specialized for swimming and reproduction. The term "epitoke" is derived from the Greek root "tokus," meaning birth. In many species, epitoky involves asexual budding (Fig. 12.14a). One or more new, reproductive individuals (epitokes) are budded, one segment at a time, from the posterior portion of the original animal (the **atoke**).[5] These epitokes subsequently detach from the atoke and swim off to commingle with other epitokes and liberate gametes. The atoke remains safely behind in its burrow or tube. In some other epitokous polychaete species, epitoky involves remodelling of pre-existing structures rather than the budding off of new segments. The original worm thus becomes the epitoke, through adaptations for gamete production and storage, and for active swimming (Fig. 12.14c). In these species, the entire worm swims off to reproduce. Gamete maturation and epitoke formation have been shown to be under environmental and hormonal control in a number of polychaete species.

Fertilization among polychaetes occurs externally, in the surrounding water. Typically, the free-living embryo soon develops a digestive system and two rings of cilia (Fig. 12.15). These rings are the **prototroch** (G: "first wheel"), located around the equator of the animal, anterior to the mouth; and the

---

5 See Topics for Discussion, No. 10.

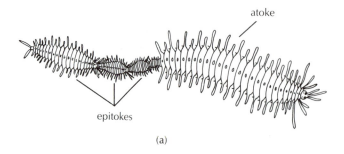

(a)

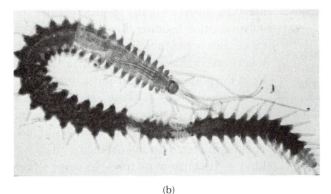

(b)

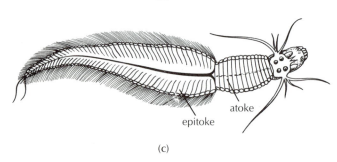

(c)

FIGURE 12.14. *(a) Epitoke formation by asexual budding in Autolytus. (From Harmer and Shipley; after Malaquin.) (b) Autolytus spp. Female atoke with several epitokes. (Courtesy of K.-L.. Schiedges.) (c) Epitoke formation via remodelling of original structures. (After Fauvel.) Note that in (a) the original worm becomes the atoke, left behind in the burrow, and does not participate directly in reproduction, while in (c) a single individual is both epitoke and atoke and participates directly in reproduction.*

telotroch ("tail wheel"), located posteriorly on what will become the terminal portion of the adult, the **pygidium.** The prototroch is the major locomotory organ of the larva. Because of the ciliated rings, the larva is called a **trochophore,** meaning "wheelbearer." A trochophore stage is not found among the other annelid classes, although it does occur in several other animal phyla. A third band of cilia, the **metatroch** ("between wheel"), later forms between the prototroch and the telotroch. As the larva swims and, generally, feeds in the water, body segments are repeatedly budded from the posterior region of the larva, just anterior to the pygidium. Each new segment bears a ring of locomotory cilia, permitting the larva to continue swimming despite the increasing body weight (Fig. 12.15b). Eventually, the larva metamorphoses to adult form and habitat.

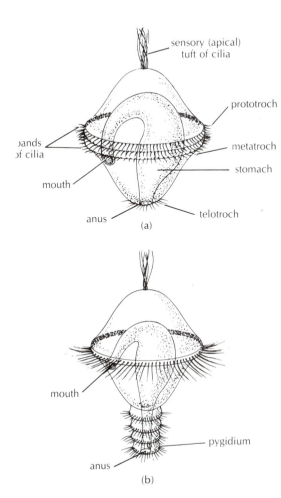

sensory (apical) tuft of cilia

prototroch

metatroch

stomach

bands of cilia

mouth

anus

telotroch

(a)

mouth

pygidium

anus

(b)

FIGURE 12.15. *(a) Polychaete trochophore larva. The larva feeds on unicellular plants collected by the ciliary bands. (From Purves and Orians, 1983.* Life: The Science of Biology. *Sinauer Associates/Willard Grant Press.)* *(b) Advanced trochophore larva, showing additional ciliated segments budded off from the initial trochophore. Parapodia will soon develop from each of these segments. New segments are added just anterior to the pygidium, which bears the anus. Thus, the youngest segments are located posteriorly, and the oldest segments of an annelid are anterior. (From Hardy, 1965.* The Open Sea: Its Natural History. *Houghton Mifflin.)*

In some species, a long-lived, free-living larval stage is lacking. Instead, the embryos may develop within a gelatinous egg mass anchored in the sediment or to the inside surface of the female's tube. In some species, embryos may be directly protected by the parent, developing to a larval or juvenile stage within specialized brooding chambers of the adult.

b. Class Oligochaeta

Unlike polychaetes, all oligochaete species are hermaphroditic and generally only a few segments of each individual produce gametes. In further contrast to the polychaetes, fertilization is usually internal and sperm are generally exchanged simultaneously between two mating individuals (Fig. 12.16a). The fertilized eggs are extruded into a complex cocoon secreted by the clitellum. The embryos develop in and feed upon a nutritive fluid found within the cocoon. When development is completed, miniature worms emerge from the cocoons; oligochaetes lack free-living larval stages, even in marine environments.

Although asexual reproduction does occur in some polychaete groups, it is more commonly encountered among oligochaetes, particularly in freshwater species. The process of asexual reproduction involves the transverse (i.e., crosswise) division of the "adult" into a number of separate sections, and the subsequent regeneration of each section into a complete individual. In addition, quite a few oligochaete species are **parthenogenetic,** especially in terrestrial environments. That is, eggs may develop normally in the absence of fertilization.

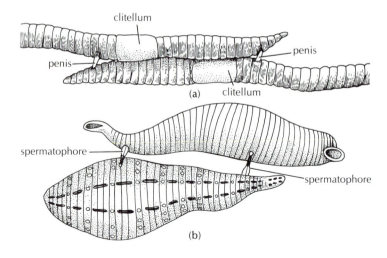

FIGURE 12.16. *(a) Copulating oligochaetes*, Pheretima communissima. *(From Fretter and Graham, 1976. A Functional Anatomy of Invertebrates. Academic Press; after Oishi.)* *(b) Copulating leeches*, Glossiphonia *sp. The members of this family generally lack a penis; spermatophores are forcefully injected into the partner and the sperm migrate to the reproductive tract for storage and subsequent fertilization of eggs. In both oligochaetes and leeches, the fertilized eggs are incubated in a gelatinous cocoon secreted by the clitellum. (After Meglitsch; and after Brumpt.)*

c. Class Hirudinea

As in the Oligochaeta, adult leeches are simultaneous hermaphrodites, and a free-living larval stage is absent from the life cycle. Only a few segments of each individual are directly involved in **gametogenesis** (production of gametes). As with oligochaetes, fertilization occurs internally, either through copulation or, in species lacking a penis, by jabbing packets of sperm (**spermatophores**) into the partner's body. Exchange of sperm is mutual between two mating individuals (Fig. 12.16b), and the fertilized eggs of each individual generally develop within external cocoons. The clitellum generally functions in producing the cocoon and a nutritive fluid, as in the Oligochaeta. A miniature leech eventually emerges from the cocoon.

In contrast to the other annelid classes, asexual reproduction is completely unknown among the Hirudinea.

2. Digestive System

The annelid gut is linear and unsegmented, with a mouth opening on the peristomum and an anus opening at the posterior end of the animal (pygidium). Food is moved through the gut by cilia and/or by muscular contractions. Digestion is primarily extracellular, although some species show an intracellular component as well.[6]

a. Class Polychaeta

The digestive tract of polychaetes is typically divided into a pharynx, esophagus, stomach, intestine, and rectum (Fig. 12.17a). In some species, evaginations of the gut form blind-ending **digestive glands** or **digestive caeca** (*caec* = L: blind), increasing the amount of surface area available for digestion and absorption.

b. Class Oligochaeta

Oligochaetes generally show more modification of the basic arrangement described above for polychaetes. The esophagus may be modified to form a **crop,** for food storage, and/or a **gizzard,** a highly muscular structure lined with hardened cuticle, used for grinding food (Fig. 12.17b). **Calciferous glands** are associated with the esophagus; these may function in regulating blood pH by controlling the concentration of carbonate ions.

The intestine of many terrestrial oligochaete species is thrown into a ridge or fold (**typhlosole**), which increases the effective surface area of the gut. (Fig. 12.18).

Associated with the intestine (and dorsal blood vessel) of oligochaetes is a characteristic yellow tissue called **chloragogen.** Chloragogen cells play major roles in the protein, carbohydrate, and lipid metabolism of oligochaetes.

c. Class Hirudinea

The mouth of a leech opens into a muscular, pumping pharynx. **Salivary glands** associated with the pharynx (Fig. 12.10b) secrete **hirudin,** an anticoagulant. A crop and digestive glands are found in some species.

3. Nervous System and Sense Organs

A mass of **ganglia** (aggregations of nerve tissue) forming a brain is present at the anterior end of an annelid (Fig. 12.19a). A solid ventral nerve cord (or a pair of nerve cords in the primitive condition) passes from the anterior to the posterior end of each individual. Swellings of the cord in each segment form segmental ganglia. A variety of sense organs are distributed along the length of the body, including touch receptors, statocysts, light receptors, vibration receptors, and chemoreceptors. These receptors are either connected to the ventral nerve cord by means of segmental nerves, or connected by nerves directly to the brain (as in the case of the eyes). The eyes of most species are not image-forming, but are sensitive to light intensity and changes in light intensity. Lateral nerves extending from the ventral nerve cord also innervate the digestive tract and the parapodial and body wall musculature of each segment.

Generally, a few of the nerves in the ventral cord are of considerably greater diameter than the others (Fig. 12.19b). These "giant" fibers may conduct nerve impulses 20 to more than 1000 times faster than other fibers,

---

6 See Topics for Discussion, No. 9.

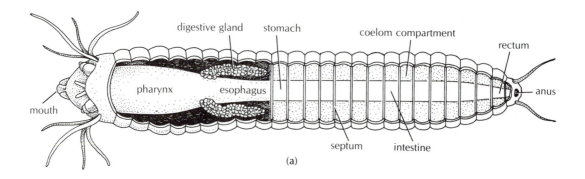

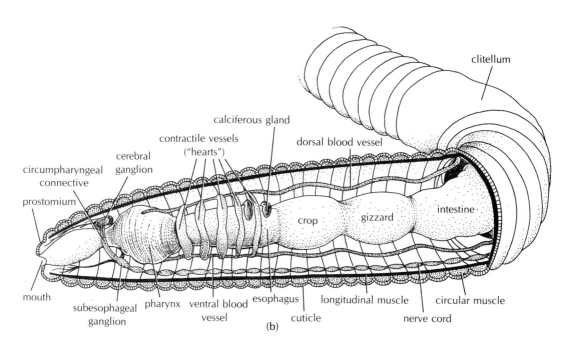

FIGURE 12.17. *(a) Digestive system of the errant polychaete,*
Nereis virens. *(Modified from Buchsbaum, 1938.* Animals Without
Backbones. *University of Chicago Press.) (b) Digestive system of an
oligochaete,* Lumbricus terrestris. *Elements of the nervous and
circulatory systems are also shown. (Modified from several sources.)*

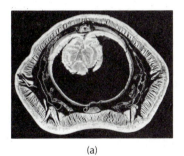

(a)

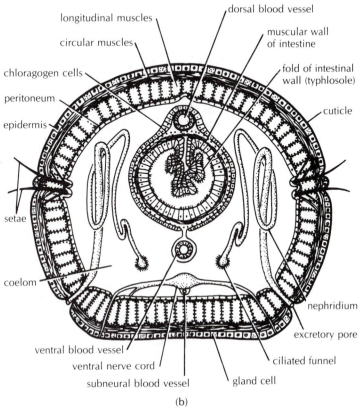

longitudinal muscles

circular muscles

chloragogen cells

peritoneum

epidermis

setae

coelom

dorsal blood vessel

muscular wall of intestine

fold of intestinal wall (typhlosole)

cuticle

nephridium

excretory pore

ciliated funnel

ventral blood vessel

ventral nerve cord

subneural blood vessel

gland cell

(b)

FIGURE 12.18.   (a) Lumbricus terrestris *in cross section. (Courtesy of L.S. Eyster.) (b) Diagrammatic illustration of* L. terrestris *in cross section. A pronounced infolding of the intestinal wall is characteristic of terrestrial oligochaetes; this typhlosole increases the gut surface area available for absorption of nutrients. Note the chloragogen tissue surrounding the intestine. This tissue, found in oligochaetes and polychaetes, plays a role in excretion, hemoglobin synthesis, and basic metabolism. (After Buchsbaum and other sources.)*

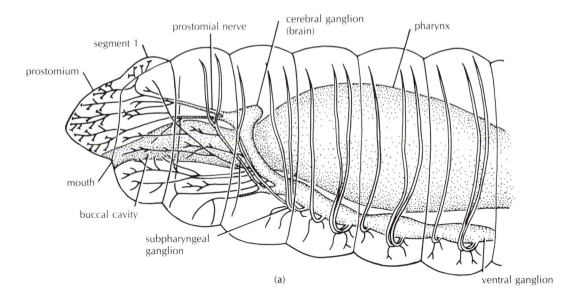

(a)

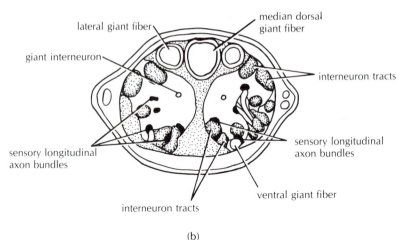

(b)

FIGURE 12.19. *(a) Nervous system of* Lumbricus terrestris, *anterior end of animal. (From Barnes; after Hess; from Avel.) (b) Diagrammatic illustration of ventral nerve cord in cross section, showing a complex array of nerve fibers, including giant fibers. The giant fibers mediate rapid motor responses. The interneuron tracts are primarily involved in coordinating the activities of the various body segments. The giant fibers conduct nerve impulses up to about 1600 times faster than the smaller nerve fibers. (From P.J. Mill, 1982.* Comp. Biochem. Physiol., *73A: 641; after Dorsett.)*

making possible the nearly simultaneous contraction of appropriate musculature throughout the body of the worm. These giant fibers permit a very rapid, coordinated response to potential predators.

4. Circulatory System

Annelids generally have a closed circulatory system, consisting of a dorsal vessel (carrying blood anteriorly), a ventral vessel (carrying blood posteriorly), and capillaries connecting the two. (See Figs. 12.10b and 12.17b.) There is no specialized heart. Instead, circulation is maintained by contractions of the blood vessels themselves, especially of the dorsal vessel. Valves assure a unidirectional flow of blood throughout the body. This form of blood circulatory system is much reduced or even absent among members of the Hirudinea. Among the leeches, coelomic fluid assumes all or part of the circulatory role, reaching the tissues via contractile coelomic sinuses and channels.

Oxygen-carrying blood pigments are found in the circulatory fluid of most (but not all) annelid species in each of the three classes. Several chemically and functionally distinct pigments are found among annelids.

The blood of most species contains **hemoglobin,** in which iron atoms serve as the binding sites for oxygen. The hemoglobins may occur in corpuscles or may be found in solution in the blood fluid, depending upon the species examined. Hemoglobin-containing blood is especially characteristic of leeches and oligochaetes, although it is found as well in the blood of some polychaete species. In addition to hemoglobin, two other blood pigments have been encountered among the Polychaeta. **Chlorocruorin** is another iron-containing pigment, found in solution in the blood of several polychaete species. This pigment is chemically quite similar to hemoglobin, but has a greenish coloration. Yet a third iron-containing pigment, **hemerythrin,** is found in at least one polychaete species. Hemerythrin is structurally quite dissimilar to the other two pigments, and is always contained within cells. More than one type of blood pigment (including several structurally and functionally different types of hemoglobin) may occur simultaneously in the blood of a single annelid.[7]

---

7 See Topics for Discussion, No. 6.

◀ T A X O N O M I C   S U M M A R Y

TOPICS FOR FURTHER DISCUSSION AND INVESTIGATION

1. A small group of annelids, the Archiannelida, is believed to be most closely related to the Polychaeta. Most archiannelids are interstitial; i.e., they are adapted for life in the spaces among sand grains. What are the morphological specializations associated with the interstitial life-style in this group of annelids?

Hermans, C.O., 1969. The systematic position of the Archiannelida. *Systematic Zool.*, 18: 85.

2. Explore the morphological, behavioral, and physiological adaptations associated with a sedentary, suspension-feeding existence among polychaetes.

Aller, R.C., and J.Y. Yingst, 1978. Biogeochemistry of tube dwellings: a study of the sedentary polychaete *Amphitrite ornata* (Leidy). *J. Marine Res.*, 36: 201.

Barnes, R.D., 1965. Tube-building and feeding in chaetopterid polychaetes. *Biol. Bull.*, 129: 217.

Brenchley, G.A., 1976. Predation detection and avoidance: ornamentation of tube-caps of *Diopatra* spp. (Polychaete: Onuphidae). *Marine Biol.*, 38: 179.

Brown, S.C., 1975. Biomechanics of water-pumping by *Chaetopterus variopedatus* Renier. Skeletomusculature and kinematics. *Biol. Bull.*, 149: 136.

Brown, S.C., and J.S. Rosen, 1978. Tube-cleaning behavior in the polychaete annelid *Chaetopterus variopedatus* (Renier). *Anim. Behav.*, 26: 160.

Busch, D.A., and R.E. Loveland, 1975. Tube-worm-sediment relationships in populations of *Pectinaria gouldii* (Polychaeta: Pectinariidae) from Barnegat Bay, New Jersey, USA. *Marine Biol.*, 33: 255.

Dales, R.P., 1955. Feeding and digestion in terebellid polychaetes. *J. Marine Biol. Assoc. U.K.*, 34: 55.

Daver, D.M., 1983. Functional morphology and feeding behavior of *Scolelepis squamata* (Polychaeta: Spionidae). *Marine Biol.*, 77: 279.

Flood, P.R. and A. Fiala-Médioni, 1982. Structure of the mucous feeding filter of *Chaetopterus variopedatus* (Polychaeta). *Marine Biol.*, 72: 27.

Hedley, R.H., 1956. Studies of serpulid tube formation. I. The secretion of the calcareous and organic components of the tube of *Pomatoceros triqueter*. *Q.J. Microsc. Sci.*, 97: 411.

Hoffmann, R.J., and C.P. Mangum, 1972. Passive ventilation in benthic animals? *Science*, 176: 1356.

MacGinitie, G.E., 1939. The method of feeding of *Chaetopterus*. *Biol. Bull.*, 77: 115.

Merz, R.A., 1984. Self-generated *versus* environmentally produced feeding currents: a comparison for the sabellid polychaete *Eudistylia vancouveri*. *Biol. Bull*, 167: 200.

Mitterer, R.M., 1971. Comparative amino acid composition of calcified and non-calcified polychaete worm tubes. *Comp. Biochem. Physiol.*, 38B: 405.

Nott, J.A., and K.R. Parker, 1975. Calcium accumulation and secretion in the serpulid polychaete *Spirorbis spirorbis* L. at settlement. *J. Marine Biol. Assoc. U.K.*, 55: 911.

Strathmann, R.R., R.A. Cameron, and M.F. Strathmann, 1984. *Spirobranchus giganteus* (Pallas) breaks a rule for suspension-feeders. *J Exp. Marine Biol. Ecol.*, 79: 245.

Whitlatch, R.B., and J.R. Weinberg, 1982. Factors influencing particle selection and feeding rate in the polychaete *Cistenides* (*Pectenaria*) *gouldii*. *Marine Biol.*, 71: 33.

Zottoli, R.A., and M.R. Carriker, 1974. Burrow morphology, tube formation, and microarchitecture of shell dissolution by the spionid polychaete *Polydora websteri. Marine Biol.*, 27: 307.

3. What mechanical advantages are gained and lost through the elimination of septa among annelids?

Chapman, G., 1958. The hydrostatic skeleton in the invertebrates. *Biol. Rev.*, 33: 338.

Chapman, G., and G.E. Newell, 1947. The role of the body fluid in relation to movement in soft-bodied invertebrates. I. The burrowing of *Arenicola. Proc. R. Soc. London, Series B*, 134: 431.

Elder, H.Y., 1973. Direct peristaltic progression and the functional significance of the dermal connective tissues during burrowing in the polychaete *Polyphysia crassa* (Oersted). *J. Exp. Biol.*, 58: 637.

Gray, J., H.W. Lissmann, and R.J. Pumphrey, 1938. The mechanism of locomotion in the leech (*Hirudo medicinalis* Ray). *J. Exp. Biol.* 15: 408.

Trueman, E.R., 1966. The mechanism of burrowing in the polychaete worm, *Arenicola marina* (L.). *Biol. Bull.* 131: 369.

4. Compare the locomotion of errant polychaetes with that of oligochaetes.

Gray, J., 1939. Studies in animal locomotion. VIII. *Nereis diversicolor. J. Exp. Biol.*, 16: 9.

Seymour, M.K., 1969. Locomotion and coelomic pressure in *Lumbricus terrestris. J. Exp. Biol.*, 51: 47.

5. What differences in nephridial structure and function might you expect to see among marine, freshwater, and terrestrial annelids?

6. A variety of blood pigments are found within the Annelida. All blood pigments must have one key feature to be functional: they must all be able to combine reversibly with oxygen. The amount of oxygen potentially carried per ml of blood, and the conditions under which a blood will become saturated with oxygen, are determined by the abundance and properties of the particular blood pigment present. In addition, blood pigments differ in how readily they give up oxygen under any given set of environmental conditions. What are the structural and functional differences among the various annelid blood pigments, and how do these differences relate to the environments in which the different species occur?

Baldwin, E., 1964. *An Introduction to Comparative Biochemistry.* Cambridge University Press, pp. 88–106.

Kayar, S.R., 1981. Oxygen uptake in *Sabella melanostigma* (Polychaeta: Sabellidae): the role of chlorocruorin. *Comp. Biochem. Physiol.*, 69A: 487.

Wood, S.C. (Ed.), 1980. Respiratory pigments. *Amer. Zool.*, 20: 3.

7. What is the influence of deposit-feeding polychaetes on the distribution and abundance of other annelids in the community?

Wilson, W.H., Jr., 1981. Sediment-mediated interactions in a densely populated infaunal assemblage: the effects of the polychaete *Abarenicola pacifica. J. Marine Res.*, 39: 735.

8. Do annelids show an immune response? What is the evidence?

Anderson, R.S., 1980. Hemolysins and hemagglutinins in the coelomic fluid of a polychaete annelid, *Glycera dibranchiata. Biol. Bull.*, 159: 259.

Chain, B.M., and R.S. Anderson, 1983. Antibacterial activity of the coelomic fluid of the polychaete, *Glycera dibranchiata.* I. The kinetics of the bactericidal reaction. *Biol. Bull.*, 164: 28.

Çotuk, A., and R.P. Dales, 1984. The effect of the coelomic fluid of the earthworm *Eisenia foetida* Sav. on certain bacteria and the role of the coelomocytes in internal defence. *Comp. Biochem. Physiol.*, 78A: 271.

9. The amount of carbon found dissolved in a given volume of sea water is approximately 100 times more than the amount of carbon found in particles (including unicellular plants, phytoplankton) in that volume of seawater. What is the potential role of this dissolved organic material in meeting the nutritional requirements of aquatic annelids?

Ahearn, G.A., and S.J. Townsley, 1975. Transport of exogenous D-glucose by the integument of a polychaete worm (*Nereis diversicolor* Müller). *J. Exp. Biol.*, 62: 243.

Stephens, G.C., 1968. Dissolved organic matter as a potential source of nutrition for marine organisms. *Amer. Zool.*, 8: 95.

Stephens, G.C., 1975. Uptake of naturally occurring primary amines by marine annelids. *Biol. Bull.*, 149: 397.

Taylor, A.G., 1969. The direct uptake of amino acids and other small molecules from sea water by *Nereis virens* Sars. *Comp. Biochem. Physiol.*, 29: 243.

10. In what respects is epitoke formation similar to the strobilation of polyps in the life cycle of scyphozoans (jellyfish)?

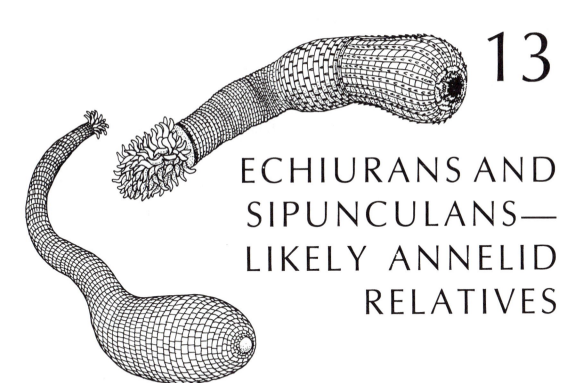

13

ECHIURANS AND SIPUNCULANS— LIKELY ANNELID RELATIVES

INTRODUCTION AND GENERAL CHARACTERISTICS

Two phyla are considered in this chapter—the Echiura and the Sipuncula. Members of both these phyla are entirely marine. All species in both groups are **vermiform** ("worm-shaped"), sedentary, and found primarily in shallow waters. Neither echiurans nor sipunculans secrete tubes, but both may form burrows. Most species of both phyla are deposit/detritus feeders. The body wall of both groups is annelid-like (i.e., composed, in sequence, of an outer cuticle, an epidermis, a layer of circular muscles, a layer of longitudinal muscles, and a peritoneum lining the coelom), as are the nervous system, excretory system, mode of gamete formation, and embryology. Echiurans and sipunculans depart from the annelid plan significantly, however, in their lack of metameric segmentation.

PHYLUM ECHIURA

About 100 species of echiurans are known, living in sandy or muddy burrows or, more rarely, in rock crevices. Most species are found in shallow water. The length of the cylindrical sausage-like body varies considerably among species, from several milli-meters to more than eight centimeters. Perhaps the most conspicuous feature of echiurans is an anterior cephalic projection commonly called the **proboscis** (Fig. 13.1). This organ contains the brain, and may well be homologous with the prostomium of annelids; i.e., the annelid prostomium and the echiuran proboscis may have had the same evolutionary origin. The proboscis is mus-

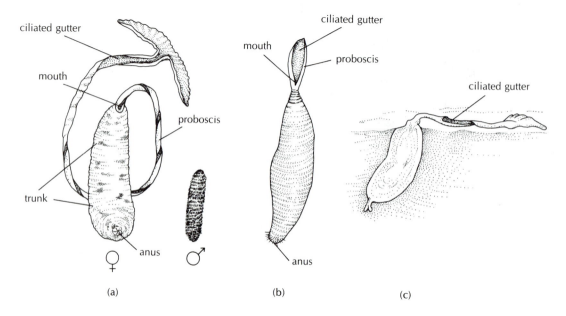

(a) (b) (c)

FIGURE 13.1. *Representative echiurans. (a)* Bonellia viridis *(female). The trunk is about 2.5 cm in length, and the proboscis may reach one meter in length when fully extended. The nondescript male has a total body length of only a few cm. Note the ciliated gutter of the proboscis, leading to the mouth. Members of the genus* Bonellia *are unique among echiurans in possessing but a single, unpaired nephridium. (Modified after several sources.) (b)* Echiurus echiurus, *in ventral view. Males and females are similar in shape. The trunk length may reach 15 cm, while the proboscis extends about 2–4 cm. Both species live in soft sediments, in burrows. (After Harmer and Shipley.) (c) Typical echiuran in feeding position. (After Zenkevitch; modified from Grassé,* Traité de Zoologie, *vol. 5.)*

cular and quite mobile—in some species, it can be extended up to twenty-five times the length of the body proper (two hundred centimeters in an eight-centimeter individual).

The proboscis serves as the organ of food collection in all echiuran species. Although the length of the proboscis can be altered greatly by its owner, the proboscis can never be retracted into the body. The edges of the proboscis curl to form a gutter, along which are cilia that move mud and detritus, trapped by mucous secretions of the proboscis, posteriorly toward the mouth. The mouth is located at the base of the proboscis, near the junction of the proboscis and the body proper (the **trunk**). The digestive tract is contained within the trunk and is very long and convoluted, with the anus opening posteriorly. Although most species are indeed deposit-feeders, a few exceptions are known. These exceptions, all members of the genus *Urechis* (Fig. 13.2), are suspension-feeders, straining out food particles as water passes through a mucous net that the animals secrete across the lumen of their burrows. Water flow through the U-shaped burrow characteristic of this genus is maintained by peristaltic waves of muscular contraction of the body (Fig. 13.3).[1]

---

1 See Topics for Discussion, No. 4.

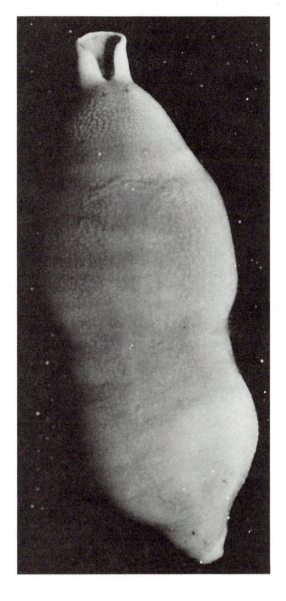

FIGURE 13.2. *Photograph of* Urechis caupo *(about 8 cm long), removed from its U-shaped burrow. The conspicuous constriction of the body wall represents a peristaltic contraction. These waves of muscle contraction drive water through the tube for food collection, gas exchange, and waste removal, and circulate the blood, which lies free in the body cavity; blood vessels are lacking in this species. To facilitate gas exchange, water is periodically taken into the thin-walled intestine through the anus, and expelled through this same opening. (Courtesy of C. Bradford Calloway and B. D. Opell.)*

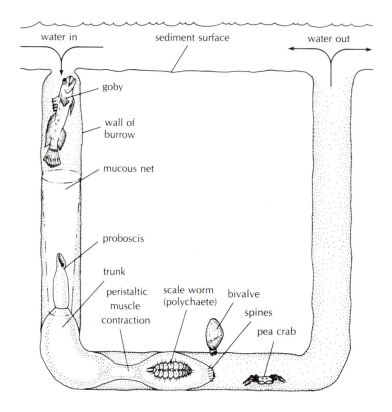

FIGURE 13.3. Urechis caupo, *the innkeeper worm, driving water through its U-shaped tube by waves of muscle contraction (peristaltic waves). The adults reach approximately 50 cm in length. The tubes of* U. caupo *are commonly shared with various symbionts, most notably a polychaete worm (*Hesperonoe adventor); *several species of pea crab (especially* Pinnixa franciscana *and* P. schmitti); *a small bivalve (*Cryptomya californica); *and the goby fish (*Clevelandia ios). *(From Pimentel; after Fisher and MacGinitie.)*

Aside from the **papillae** (small external bumps) found in some species, the only conspicuous projections from the echiuran trunk are a single pair of large, chitinous, annelid-like setae located just posterior to the proboscis. Some echiuran species bear additional setae in the form of a ring just anterior to the anus.

The trunk contains a single, uninterrupted, large coelomic space that extends into the proboscis (Fig. 13.4). There is no metameric segmentation of the coelomic space, nor of the ventral nerve cord that runs through it. In addition to the digestive system, the coelomic space houses from one nephridium to several hundred pairs of

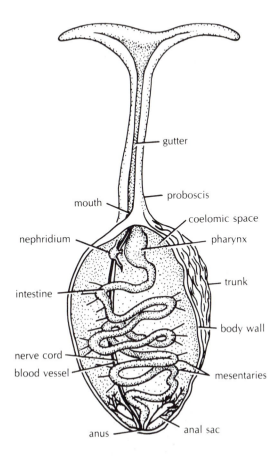

gutter

proboscis

mouth

coelomic space

nephridium

pharynx

intestine

trunk

body wall

nerve cord

blood vessel

mesentaries

anus

anal sac

FIGURE 13.4. *Internal anatomy of* Bonellia viridis. *Note the capacious coelomic cavity. (After Harmer and Shipley.)*

anal sacs outpocket from the rectum (Fig. 13.4). The sacs are muscular, and the surface of the sacs is dotted with many thousands of ciliated funnels resembling nephrostomes. Like metanephridia, these funnels collect coelomic fluid, but there is no indication that the collected fluid is then modified by either secretion or resorption. Moreover, the fluid is discharged to the outside through the anus rather than through a nephridiopore. Discharge is accomplished by periodic muscular contractions of the sacs; one-way valves at the base of the funnels prevent fluid from moving back into the coelom.

The echiuran adult generally possesses a closed circulatory system with heart and blood vessels. Although the circulatory system itself lacks any oxygen-carrying blood pigment, hemoglobin is found in the coelomic fluid of some species. No specialized respiratory organs are found; diffusion across the general body surface apparently satisfies requirements for gas exchange in most species.[2]

No distinct gonads are found in echiurans. Instead, gametes are produced by the peritoneal lining of the coelom and are released to the coelomic cavity. This is highly reminiscent of the situation encountered among many polychaetes. The gametes of echiurans leave the body by passing through the metanephridia, exiting at the nephridiopores. Gamete formation and release are by essentially identical methods among the Sipuncula.

Echiuran sex lives appear to be rather unexciting, with only a few exceptions (but, as you will see, the exceptions are exceptional

annelid-like metanephridia; most species possess at least one pair. Coelomic fluid is presumably drawn into each nephrostome by ciliary action, and a final urine is excreted through nephridiopores opening to the surrounding sea water.

One additional organ is encountered within the echiuran coelom and it, too, is believed to serve an excretory role. A pair of

2 See Topics for Discussion, No. 5.

indeed!). Gametes are always liberated through nephridiopores into the surrounding sea water, where fertilization occurs. Development is typically protostomous, culminating in the production of a **trochophore larva,** as in the polychaetes (Fig. 13.5). Of particular phylogenetic interest is the observation that segmented coelomic pouches form during embryogenesis. Although this condition does not persist into adulthood, even the transient appearance of metameric segmentation in echiurans is a tantalizing indication of probable annelid relationships. In contrast, there is no sign of metameric segmentation at any stage in the life history of sipunculans.

All echiuran species are **dioecious** (one sex per individual), and males and females are generally similar in appearance. However, here is where several echiuran species begin to stand out. *Bonellia viridis* and a few close relatives are sexually **dimorphic** (characterized by "two body forms"). Compared with the females, the males are much smaller, lack a proboscis, lack a specialized circulatory system, and possess a degenerate digestive tract. These males are not free-living; rather, they dwell within the body of the female—in the nephridia, in fact, which take on the function of uteri. A single female may commonly have twenty males living within her body at any one time, awaiting their opportunity to fertilize eggs. Perhaps the most bizarre aspect of this peculiar life history is that at fertilization the sex of most embryos is not yet determined. Instead, for the majority of individuals, maleness is only induced following contact of the larva with the proboscis of an adult female. In the absence of such contact, the larva generally becomes a female.[3]

PHYLUM SIPUNCULA

Like the echiurans, sipunculans are all marine. These vermiform creatures are commonly found in burrows in shallow-water muddy or sandy sediments, and in empty mollusc shells or polychaete tubes. One very small species is known to inhabit the calcareous shells built by foraminiferans (Protozoa). Some other species are found in rock crevices and a few species can bore into the calcareous substrates of coral reefs. There are about 3.5 times as many sipunculan species as there are echiuran species. Most of the 350 sipunculan species are only a few

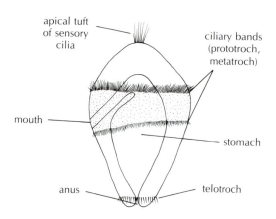

FIGURE 13.5. *A typical echiuran trochophore larva, with three conspicuous bands of cilia.* (*From Jägersten, 1972. Evolution of the Metazoan Life Cycle. Academic Press.*)

apical tuft of sensory cilia

ciliary bands (prototroch, metatroch)

mouth

stomach

anus

telotroch

---

3 See Topics for Discussion, No. 2.

millimeters in length, although some species attain lengths in excess of one meter.

The sipunculan body consists of a plump, unsegmented trunk and an anterior **introvert.** Unlike the echiuran proboscis, the sipunculan introvert is fully retractable into the body (*intro* = L: within; *verta* = L: turn; i.e., "turn within"), and the mouth opens at the end of the introvert (Fig. 13.6). Contraction of the body wall musculature (Fig. 13.6b) causes the introvert to be thrust out of the body. Contraction of well-developed introvert retractor muscles draws the anterior end of the animal back inside the trunk. The introvert commonly bears numerous mucus-covered tentacles surrounding or adjacent to the terminal mouth. These tentacles trap particles from the surrounding water, or are pressed into the substrate, trapping mud and detritus. The entire introvert may then be withdrawn into the body and the captured particulate material ingested, or the particles may be moved to the mouth by ciliary tracts on the tentacles. Thus, most sipunculans, like echiurans, are deposit-feeders. During burrow formation, mud may be ingested directly by sipunculans.

Unlike the digestive tract of echiurans, that of sipunculans is U-shaped, with the anus opening at about the mid-point of the

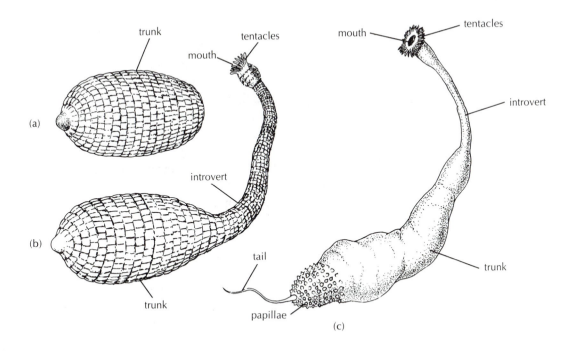

FIGURE 13.6. *A typical sipunculan, with the introvert withdrawn (a) and extended (b). (c)* Golfingia flagrifera, *a sipunculan with conspicuous papillae. (After Stephen and Edmonds; after Selenka.)*

trunk. This would seem to be an adaptation for a sedentary existence in burrows having only a single opening to the outside; solid wastes are discharged near the mouth of the burrow.

Other than the tentacles borne on the introvert, the sipunculan body lacks substantial projections; setae and appendages are both absent.

The body of sipunculans contains an enormous coelomic cavity (Fig. 13.7b). This cavity shows no sign of segmentation or septa at any stage in the life history. Surprisingly, the hollow tentacles of the introvert are not associated with the coelomic cavity. Instead, the tentacles are connected to a series of sacs, called **contractile vessels** or **compensatory sacs.** These sacs are attached to the surface of the esophagus; they may be quite extensive, ramifying throughout the trunk coelom and forming a type of circulatory system in some species. Contraction of the sac musculature drives fluid into the tentacles, bringing about their extension. Muscular retraction of the tentacles drives fluid back into the sacs. A similar system operates in the locomotion of sea urchins and sea stars and in operating the tentacles of sea cucumbers (Phylum Echinodermata), as will be discussed in Chapter 19.

In contrast to the echiurans, a circulatory system with heart and blood vessels is lacking among the members of the Sipuncula. However, like the echiurans, oxygen-binding pigment is found in the coelomic fluid. In the case of sipunculans, this pigment is not hemoglobin, but rather another iron-containing compound of quite different structure called **hemerythrin.** Hemerythrin is found among only a few other invertebrate species, including a small number of polychaetes.

Excretion by sipunculans is accomplished largely by metanephridia, which resemble those of echiurans and annelids. However, all sipunculan species bear only a single pair of these organs (Fig. 13.7b). The nephridial system is supplemented by a peculiar system of **urns.** These urns are clusters of cells that arise and detach from the peritoneal lining of the coelom (Fig. 13.8). Floating freely in the coelomic fluid, urns collect solid wastes and eventually deposit these wastes in the body wall of the animal or pass out through the nephridial system.

The reproduction and development of sipunculans is very similar to that of echiurans. Cleavage is spiral and determinate, coelom formation is by schizocoely, and the embryo generally develops into a free-swimming trochophore larva (Fig. 13.9a). In some sipunculans, the trochophore develops further to form a **pelagosphera** larva (Fig. 13.9b), which at one time was thought to be a free-swimming adult sipunculan and placed in the now-defunct genus *Pelagosphaera.*

Both sipunculans and echiurans bear a number of resemblances to annelids and to each other, as summarized in Table 13.1. The lack of metameric segmentation in adults, however, clearly distinguishes the Sipuncula and Echiura from the Annelida. The brief metameric condition encountered during the embryological development of echiurans suggests that the Echiura diverged from the presumed ancestral annelid stock later than did the Sipuncula.

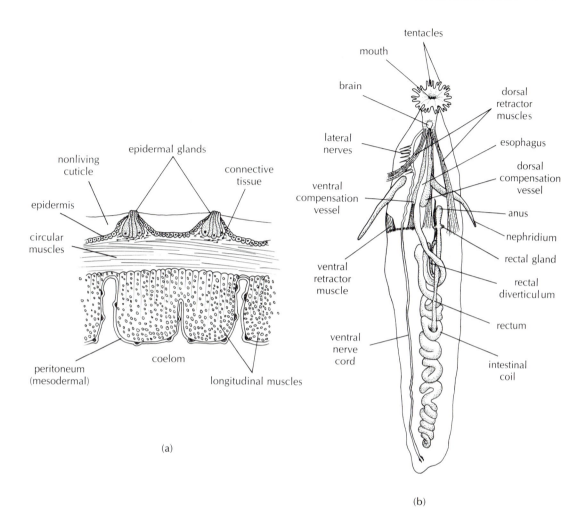

(a)

(b)

FIGURE 13.7. *(a) Cross section through the body wall of* Phascolosoma gouldi, *illustrating the typical worm-like arrangement of circular and longitudinal musculature. (From Brown, 1950. Selected Invertebrate Types. John Wiley & Sons; after Andrews.) (b) Internal anatomy of* Sipunculus nudus. *Note that the anus terminates anteriorly rather than posteriorly; i.e., the gut is "U-shaped." Also note the spacious coelomic cavity. (From Meglitsch, 1972. Invertebrate Zoology, 2nd ed. Oxford University Press.)*

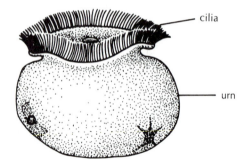

FIGURE 13.8. *An urn cell of* Sipunculus nudus *recovered from the coelomic fluid. (From Hyman; after Selensky.)*

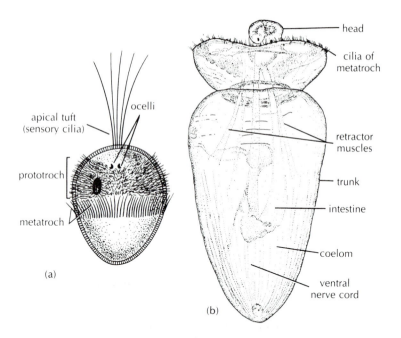

FIGURE 13.9. *(a) Trochophore larva of* Golfingia *sp. (From Hyman; after Gerould.) (b) Pelagosphera larva of* Sipunculus polymyotus. *The trochophore larva is always a nonfeeding stage in sipunculans, whereas the pelagosphera larva may feed on microscopic algae in the plankton. (Courtesy of R.S. Scheltema.)*

TABLE 13.1. Comparison of Features Encountered among
Annelids, Echiurans, and Sipunculans.

CHARACTERISTIC	ANNELIDA	ECHIURA	SIPUNCULA
Body wall musculature	outer: circular inner: longitudinal	outer: circular inner: longitudinal	outer: circular inner: longitudinal
Excretory system	metanephridia (many pairs)	metanephridia (one to many pairs) plus anal sacs	metanephridia (one pair) plus urns
Nervous system	brain with ventral nerve cord	brain with ventral nerve cord	brain with ventral nerve cord
Anterior-most feature	prostomium	proboscis (not re- tractable)	introvert (fully re- tractable)
Blood circulatory system	heart plus dorsal and ventral blood vessels	heart plus dorsal and ventral blood vessels	none
Oxygen-binding pigments	hemoglobin hemerythrin chlorocruorin	hemoglobin	hemerythrin
Setae	present	present	none
Metamerism	throughout life	only during development	none
Coelom formation	schizocoely	schizocoely	schizocoely
Cleavage pattern	spiral, determinate	spiral, determinate	spiral, determinate
Larval form	trochophore	trochophore	trochophore, pelagosphera
Gamete production	arise from peri- toneum; mature in coelom; may exit through nephridiopores	arise from peri- toneum; mature in coelom; exit through nephridiopores	arise from peritoneum; mature in coelom; exit through nephridiopores
Digestive tract	linear (i.e., mouth and anus at op- posite ends of body)	linear	U-shaped

OTHER FEATURES OF ECHIURAN AND SIPUNCULAN BIOLOGY

Nervous and Sensory Systems

There are no specialized sensory systems in the Echiura, other than sensory cells stud-ding the proboscis. Sensory cells are also found on the surface of the sipunculan introvert. In addition, some sipunculan species possess a pair of presumed chemoreceptors, called **nuchal organs,** on the introvert, and have several pairs of light-sensitive ocelli within the brain.

PHYLUM ECHIURA
PHYLUM SIPUNCULA

◄ T A X O N O M I C   S U M M A R Y

T O P I C S   F O R   F U R T H E R   D I S C U S S I O N   A N D   I N V E S T I G A T I O N

1. The Priapulida is a phylum of about one dozen worm-like animals (Fig. 13.10) found in sandy and muddy substrates in both shallow and deep water. All species are marine. The priapulids have been grouped, at various times, with both the Aschelminthes and with the Sipuncula and Echiura. Until the early part of the twentieth century, the Echiura, Sipuncula, and Priapulida were grouped together in a single phylum, the Gephyrea. In what respects are the priapulids similar to sipunculans and echiurans? In what respects are they similar to the pseudocoelomates?

Calloway, C.B., 1982. Priapulida. In S.P. Parker (Ed.), *Synopsis and Classification of Living Organisms*, Vol. I. New York: McGraw-Hill pp. 941–944.

Hyman, L., 1951. *The Invertebrates*, Vol. 5. New York: McGraw-Hill, pp. 183–196.

Shapeero, W.L., 1961. Phylogeny of Priapulida. *Science*, 133: 879.

2. Recently, a green pigment called bonellin has been isolated from the body wall of the echiuran *Bonellia viridis*. Discuss the evidence implicating bonellin as the factor that causes metamorphosing larvae to become males.

Aguis, L., 1979. Larval settlement in the echiuran worm *Bonellia viridis*: settlement on both the adult proboscis and body trunk. *Marine Biol.*, 53: 125.

Jaccarini, V., L. Aguis, P.J. Schembri, and M. Rizzo, 1983. Sex determination and larval sexual interaction in *Bonellia viridis* Rolando (Echiura: Bonellidae). *J. Exp. Marine Biol. Ecol.*, 66: 25.

3. Compare and contrast burrow formation by sipunculans and polychaetes.

Trueman, E.R., 1966. The mechanism of burrowing in the polychaete worm, *Arenicola marina* (L.). *Biol. Bull.*, 131: 369.

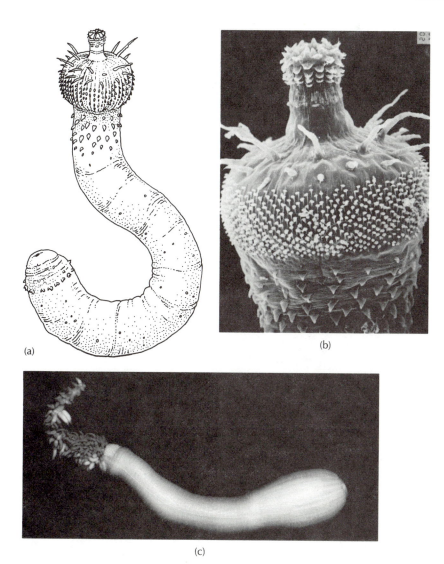

(a)

(b)

(c)

FIGURE 13.10.   *(a) Diagram of the external morphology of*
Meiopriapulus fijiensis, *an interstitial priapulid species.(Courtesy Dr. M.
P. Morse, from Morse, 1981. Trans. Amer. Microsc. Soc., 100: 239.)*
(b) Scanning electron micrograph of the anterior end of M. fijiensis.
(Courtesy of Dr. M.P. Morse.) (c) Priapulus caudatus. *The animal
is 4 cm long, excluding the tail. The longitudinal nerve cord is
visible on the ventral surface of the body. (Courtesy of C. Bradford
Calloway.)*

Trueman, E.R., and R.L. Foster-Smith, 1976. The mechanism of burrowing of *Sipunculus nudus*. *J. Zool., London*, 179: 373.

4. Compare and contrast the feeding mechanism of the echiuran *Urechis caupo* with that of the polychaete *Chaetopterus variopedatus*.

Barnes, R.D., 1965. Tube-building and feeding in chaetopterid polychaetes. *Biol. Bull.*, 129: 217.

MacGinitie, G.E., 1939. The method of feeding of *Chaetopterus*. *Biol. Bull.*, 77: 115.

MacGinitie, G.E., 1945. The size of the mesh openings in mucous feeding nets of marine animals. *Biol. Bull.*, 88: 107.

5. The innkeeper worm, *Urechis caupo* (Echiura) supplements gas exchange across its body surface by pumping water in and out of the cloaca. Discuss the mechanism by which this flow of water is accomplished.

Wolcott, T.G., 1981. Inhaling without ribs: the problem of suction in soft-bodied invertebrates. *Biol. Bull.*, 160: 189.

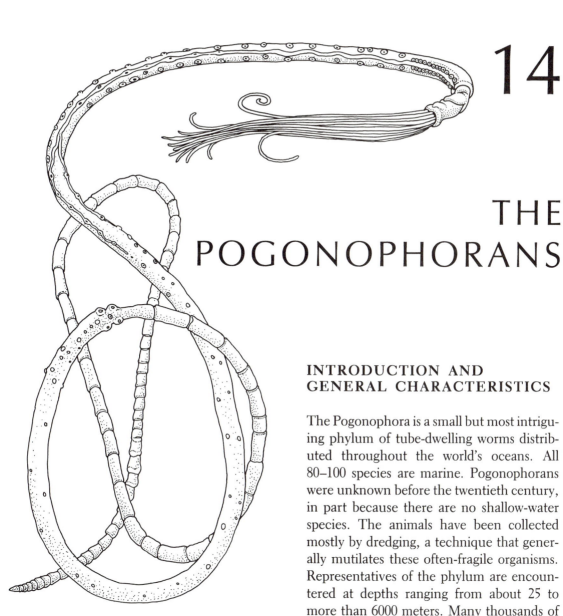

THE POGONOPHORANS

INTRODUCTION AND GENERAL CHARACTERISTICS

The Pogonophora is a small but most intriguing phylum of tube-dwelling worms distributed throughout the world's oceans. All 80–100 species are marine. Pogonophorans were unknown before the twentieth century, in part because there are no shallow-water species. The animals have been collected mostly by dredging, a technique that generally mutilates these often-fragile organisms. Representatives of the phylum are encountered at depths ranging from about 25 to more than 6000 meters. Many thousands of individuals per square meter have been reported from some areas of the ocean bottom.

Most of the pogonophoran body bears little hint of what we now suspect to be an evolutionary relationship between pogonophorans and annelids. The anterior-most region of the body bears a **cephalic lobe;** a "beard," consisting of one to many thousands of ciliated tentacles; and a glandular area re-

247

sponsible for secretion of a chitinous tube, within which the animal spends its life (Fig. 14.1). Although pogonophorans are sedentary, they are free to move up and down within their tubes, which always exceed the length of the animal's body. The tentacles are serviced by several blood vessels, and probably serve as the primary gas exchange surface. The anterior section of the animal contains a single internal body cavity. This cavity extends into each of the tentacles. Because the body cavities of pogonophorans are not completely lined with peritoneum, it is not clear that they are true coelomic spaces as strictly defined. Embryological studies may help resolve the issue.

The bulk of the pogonophoran body is composed of the **trunk** (Fig. 14.2), which contains a second, uninterrupted internal "coelomic" space. The body wall of the trunk contains both circular and longitudinal muscles. Externally, the trunk is often marked by large numbers of **papillae** (small bumps); several regions of ciliation; and two conspicuous rings of setae about half-way down the body (Fig. 14.2). The trunk is unsegmented, and the body cavity of the trunk is not septate. The major organs found within this cavity are the gonads and a multilobed, hollow structure called the **trophosome** (Fig. 14.3a). The trophosome of at least some species is composed largely of

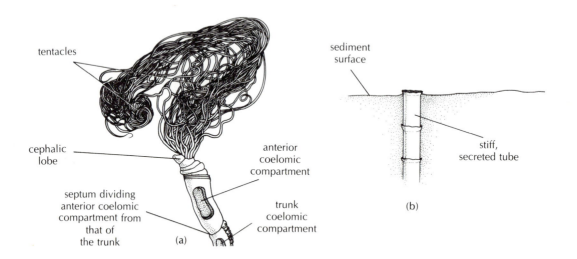

FIGURE 14.1. (a) Anterior end of a pogonophoran (Polybrachia).
(b) A portion of the tube secreted by the same animal. Much of the
tube is anchored in soft sediment. (From Hyman; after Ivanov.)

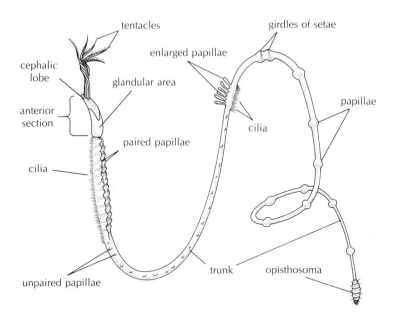

FIGURE 14.2. *Intact pogonophoran. Note the three conspicuous body divisions: cephalic lobe, with tentacles; trunk; and segmented opisthosoma. (From George and Southward, 1973. J. Marine Biol. Assoc. U.K., 53: 403.)*

closely packed bacteria (Fig. 14.3b), which may play a major role in the nutrition of the Pogonophora, as discussed below (*tropho* = G: nourish).

The probable annelid affinities of the group were made clear only about twenty years ago, when the first intact specimens were collected. At this time, the third and most posterior body region, the **opisthosoma** (*opistho* = G: hind; *soma* = G: body), was described. The opisthosoma is conspicuously segmented. Each of the approxi-

mately six to twenty-five segments contains paired coelomic compartments, which are isolated from adjacent compartments by muscular septa, as in annelids. Moreover, each segment bears annelid-like setae. The opisthosoma is believed to play a role in anchoring the animal within its blind-ended tube.

Little is presently known regarding the behavior and physiology of most pogonophorans. Perhaps the single most fascinating component of the biology studied to date

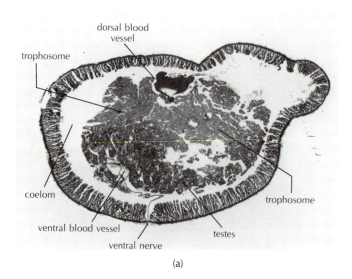

(a)

(b)

FIGURE 14.3. *(a) Transverse section through the trunk of* Riftia pachyptilia *Jones, showing location of the trophosome. In fact, most of the coelomic space is occupied by the bacteria-laden trophosome. (Courtesy of Dr. M.L. Jones, from Jones, 1981. Science, 213: 333. (b) Scanning electron micrograph of the trophosome, showing many spherical prokaryotic cells believed to be symbiotic, sulfide-reducing bacteria. (Courtesy of C.M. Cavanaugh, from Cavanaugh et al., 1981. Science, 213: 340.)*

relates to the digestive system—there is none. Pogonophorans absolutely lack a mouth, an anus, and anything resembling a digestive tract. Consequently, a number of attempts have been made to determine how pogonophorans meet their nutritional requirements. Several possibilities exist, and there is some evidence to support the likelihood of each. For example, the tentacles, at the anterior end of the animal, bear many surface microvilli and gland cells. Food particles might be trapped by the tentacles and digested externally. The solubilized nutrients could then be absorbed across the microvillous surface of the tentacles. The evidence for this mode of nutrition is strictly morphological.

On the other hand, a substantial amount of experimental evidence indicates that pogonophorans, like a considerable number of other soft-bodied invertebrates, can take up dissolved organic matter (DOM) from sea water, even at the low concentrations found naturally. Basically, the experiments consist of demonstrating that radioactively labeled amino acids, carbohydrates, and other organic molecules are accumulated by pogonophorans against remarkable concentration gradients. Some compounds can be taken up from concentrations as low as one μM (micromolar, i.e., 1×10^{-6} moles per liter). Uptake is believed to be mediated through an active, energy-requiring process. Calculations based upon rates of respiration and rates of DOM uptake suggest that for some, but not all, species, the uptake of dissolved organic molecules from the surrounding water may be sufficient to meet all of the pogonophoran's metabolic maintenance requirements.

Most pogonophorans are fairly small, perhaps 6–36 cm in length and only a millimeter or so wide. These species thus resemble long, thin threads. The ratio of surface area to volume in such an animal is very high, increasing the likelihood that the general body surface can play a major role in taking up dissolved nutrients from sea water. The tubes of many species are thin-walled, and there is some evidence that these tubes may be permeable to DOM. Moreover, most pogonophoran species are **infaunal;** i.e., they live with most of the body (and tube) implanted into soft, muddy substrates, in which dissolved nutrients accumulate to high concentrations relative to the concentrations found in sea water.

Very recently, however, much larger pogonophorans have been discovered in the deep sea, at a depth of about 2500 meters. These pogonophorans are up to 2 meters in length and 35–40 mm in width. They were first discovered by researchers aboard the deep-sea submersible *Alvin* in 1977, during exploration of a major ocean-spreading center near the Galapagos Islands, in the Pacific Ocean off the coast of Peru. The deep-water

rifts of this area are regions of the ocean bottom where new crustal material is being generated from deep within the earth. The rifts are characterized by scattered **hydrothermal vents,** small chimney-like openings in the crust from which pour hot, sulfur-rich water. A giant pogonophoran, *Riftia pachyptilia*, lives upright in tubes attached to solid rock directly in the path of this outflow (Fig. 14.4).

The surface-area-to-volume ratio of these larger animals is considerably less than in the smaller species. Moreover, the tubes of these pogonophorans are thick-walled, and could serve as barriers to uptake of DOM across much of the body surface. Although uptake of DOM has not been discounted in these species, current evidence suggests an even more intriguing mode of obtaining nutrients.

The trophosome of these "hot vent" species is composed almost entirely of bacteria (Fig. 14.3). The suspicion is that these bacteria fix carbon from carbon dioxide through the oxidation of hydrogen sulfide gas contained in the vent water. The growing bacterial population could then serve directly as a food source for the host pogonophoran, or could provide nutrition indirectly through the liberation of metabolic by-products. In either case, these deep-sea pogonophorans appear to be participants in a completely novel food chain, one that is not at all dependent upon light as an energy source. Similar bacteria have recently been described within the tissues of a non-vent pogonophoran and in a member of an entirely different phylum (the Mollusca) as well. A chemotrophic mode of nutrition may not be uncommon among animals living in waters and sediments high in hydrogen sulfide gas.

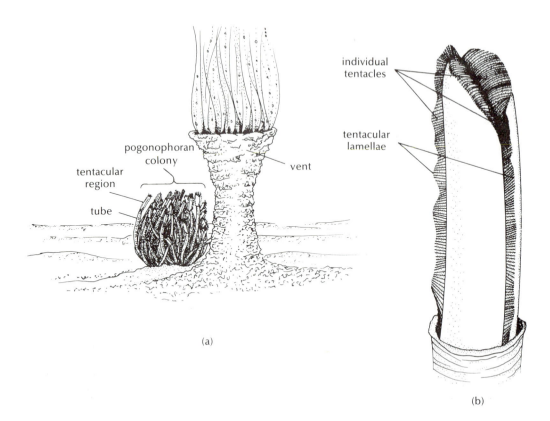

(a)

(b)

FIGURE 14.4. *(a) Animal life thrives in the warm, Sulfide-rich water next to hydrothermal vents. These pogonophorans (Riftia pachyptilia) were found at 2500 meters depth by a team of scientists aboard the research submersible Alvin in 1977. The tubes of these worms approach 3 meters in length, and are white, cylindrical, and flexible. (Modified from Science at Sea: Tales of an Old Ocean, by T. van Andel. W.H. Freeman and Company. Copyright 1981.) (b) The tentacles of Riftia pachyptilia are fused for over 50% of their length in groups of several hundred, forming flat sheets or leaflets (lamellae). More than 600 such sheets are found at the anterior end of the animal, which thus possesses a total of several hundred thousand short tentacles. (After Jones.)*

OTHER ASPECTS OF POGONOPHORAN BIOLOGY

1. Reproduction

Asexual reproduction has not been reported for this phylum; all reproduction is sexual.

Until recently, it was thought that all pogonophoran species were **dioecious** (i.e., that sexes were separate). A single hermaphroditic species was documented in 1983. In a number of dioecious species, sperm are packaged in specialized containers called **spermatophores** (i.e., "sperm-carriers"). The details of sperm transfer are not known for any species.

Embryological details are uncertain, even as to whether the body cavity forms by enterocoely or schizocoely. The cleavage patterns are neither clearly spiral nor clearly radial. The larval stage, if any, has not been described.

2. Circulatory System

A blood circulatory system is well-developed, consisting of a distinct heart and blood vessels. Both the blood and the "coelomic" fluid contain the oxygen-binding pigment hemoglobin.

3. Nervous System

A brain is located anteriorly in the cephalic lobe. A giant axon, probably involved in a rapid-withdrawal response, has been described in some species

4. Respiration and Excretion

There are no specialized respiratory structures other than the vascularized tentacles. A presumed excretory organ has been described in *Riftia pachytilia*, but the true function of this structure has yet to be demonstrated.

TOPIC FOR FURTHER DISCUSSION AND INVESTIGATION

Discuss the evidence indicating that sulfur-reducing bacteria may play a major role in meeting the nutritional needs of *Riftia pachyptilia*.

Cavanaugh, C.M., S.L. Gardiner, M.L. Jones, H.W. Jannasch, and J.B. Waterbury, 1981. Prokaryotic cells in the hydrothermal vent tube worm *Riftia pachyptilia* Jones: possible chemoautotrophic symbionts. *Science*, 213: 340.

Felbeck, H., 1981. Chemoautotrophic potential of the hydrothermal vent tube worm, *Riftia pachyptilia* Jones (Vestimentifera). *Science* 213: 336.

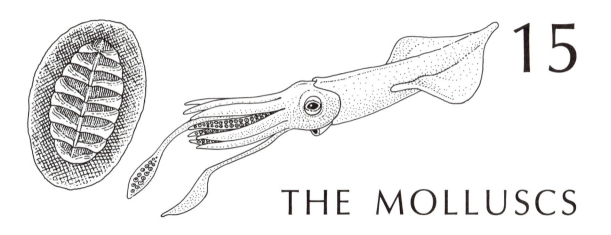

15

THE MOLLUSCS

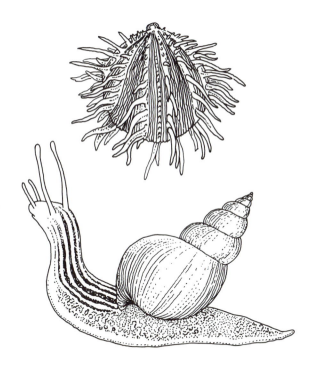

Phylum Mollusca
(L: SOFT)

INTRODUCTION AND GENERAL CHARACTERISTICS

The phylum Mollusca is one of the largest of all animal phyla, including in excess of 110,000 living species. These species are distributed among some extremely dissimilar-looking organisms, making the molluscan body plan probably the most malleable in the animal kingdom. Remarkably, clams, snails, and octopuses are all members of the Phylum Mollusca!

It is impossible to describe a "typical" mollusc. Most, but not all, molluscs have shells comprised primarily of varying amounts of calcium carbonate set in a protein matrix. Organic material may comprise as much as about 35% of the dry weight of the shell in some species of gastropod, and up to about 70% of the dry weight in bivalves. The shells of most molluscs (including all gastropods and bivalves) have a thin, outer organic layer (the **periostracum**), a thin, innermost calcareous layer (the **nacreous layer**), and a thick, calcareous middle layer (the **prismatic layer**) (Fig. 15.1). Both the organic and inor-

255

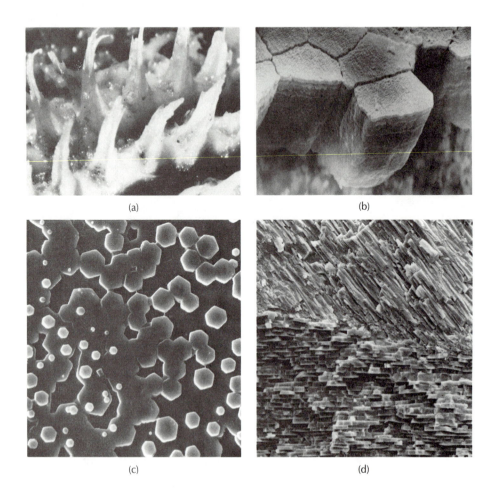

(a)

(b)

(c)

(d)

FIGURE 15.1. *Molluscan shell structure. (a) Periostracal spines from the shell of the gastropod* Trichotropis cancellata. *(Courtesy of D.J. Bottjer.) (b) Tall calcite prisms from the prismatic layer of an oyster,* Crassostrea virginica *(Gmelin). The shell has been fractured and treated briefly with Chlorox to dissolve away the proteinaceous matrix (conchiolin) that normally occurs between individual prisms. (Courtesy of M.R. Carriker, from M.R. Carriker et al., 1980. Proc. Nat. Shellf. Assoc., 70: 139.) (c) Individual calcareous tablets from the nacreous layer of a mussel shell,* Geukensia demissa *(Dillwyn). Each nacreous tablet is about 7 μm from side to side. (Courtesy of R.A. Lutz, from Lutz and Rhodes, 1977. Science, 198: 1222. Copyright © 1977 by the American Association for the Advancement of Science.) (d) Fractured shell section of a deep-sea mussel, illustrating the outer prismatic layer (top) and calcareous tablets of the underlying nacreous layer. (Courtesy of R.A. Lutz.)*

ganic components of the shell are secreted by specialized tissue known as the **mantle.** If a grain of sand or other foreign particle becomes trapped between the mantle and the inner surface of the shell, a pearl may form over a period of years. Although the mantle is a major molluscan characteristic, its role varies substantially in different molluscan groups. Similarly, most molluscs have a **foot,** but this is highly modified for a variety of functions in different groups.

In the "average" mollusc, one usually finds a characteristic cavity lying between the mantle and the viscera. This **mantle cavity** usually houses the comb-like molluscan gills, known as **ctenidia** (*ctenidi* = G: comb), and also generally serves as the outfall site for the excretory, digestive, and reproductive systems. The ctenidium, when present, may have a purely respiratory function or may function in the collection and sorting of food particles as well. A chemoreceptor/tactile receptor known as the **osphradium** (*osphra* = G: a smell) is generally located adjacent to the ctenidium.

The molluscan coelom is much reduced, being restricted largely to the area surrounding the heart and gonads. It has no locomotory role in molluscs. On the other hand, blood sinuses comprising the **hemocoel** are well-developed. This hemocoel is conspicuously involved in the locomotion of several species of mollusc.

Many molluscs possess a feeding structure known as the **radula.** The radula consists of a firm ribbon, composed of chitin and protein, along which are found rows of sharp, chitinous teeth (Fig. 15.2). The ribbon is produced from a **radular sac,** and is underlain by a supportive cartilage-like structure called the **odontophore** (literally, G: "*toothbearer*"). The odontophore–radular assembly, together with its complex musculature, is known as the **buccal mass** (*bucca* = L: cheek). For feeding, the buccal mass is protracted so that the odontophore extends just beyond the mouth. The radular ribbon is then moved forward over the tip of the supporting odontophore. As each row of teeth bends at the tip of the odontophore, the teeth automatically stand upright and rotate laterally. The teeth, when applied to a substrate by movement of the buccal mass, rasp off food particles, and bring them into the mouth when the radula is withdrawn. As old teeth are worn down or broken off at the anterior end of the radular ribbon, new teeth are continually being formed and added onto the posterior end of the ribbon in the radular sac.

As you might infer from the cautious wording in the preceding paragraphs, generalizations about molluscs are difficult to make.

CLASS GASTROPODA

This is the largest molluscan class, comprising about 75,000 living species of snails and slugs distributed among marine, freshwater, and terrestrial environments. The typical snail consists of a **visceral mass** (i.e., all of the internal organs), sitting atop a muscular **foot** (Fig. 15.3c). The visceral mass is

Class Gastro•poda
(G: STOMACH FOOT)

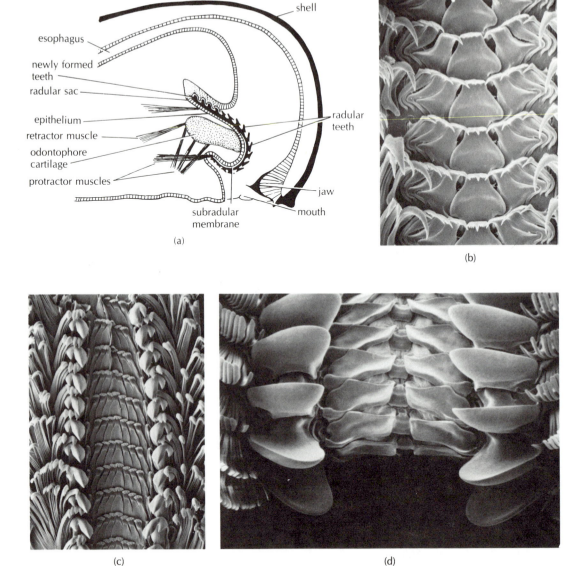

FIGURE 15.2. *(a) Longitudinal section through the anterior of a gastropod, showing the relationship between the odontophore, radular ribbon, and mouth. (Modified after Runham and various sources.) (b–d) Scanning electron micrographs of radular teeth of three different snail species: (b)* Scissurella crispata; *(c)* Montfortula rugosa; *(d)* Nerita undata. *(Courtesy of C.S. Hickman, from Hickman, 1981. Veliger, 23: 189.)*

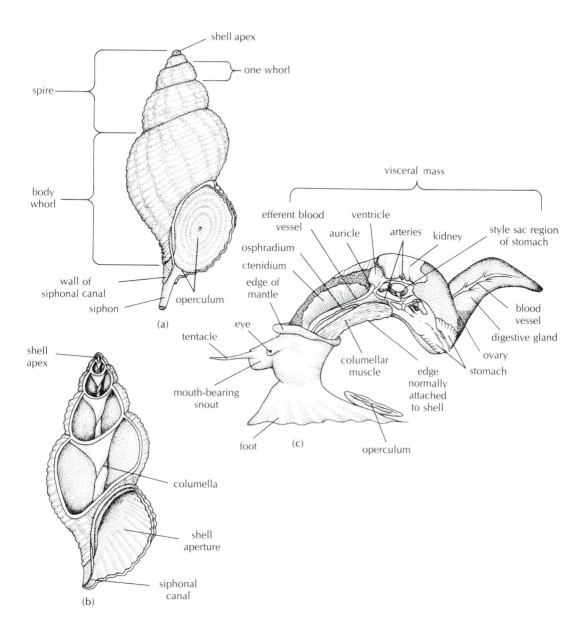

FIGURE 15.3. *(a) Major external features of a gastropod shell. The siphon is a fold of mantle tissue through which water enters the mantle cavity. The operculum is a rigid, proteinaceous shield attached to the foot; when the animal fully withdraws into its shell, the operculum seals the aperature, protecting the animal from predators and physical stresses. (b) Longitudinal section through a typical gastropod shell. As the snail grows, its body coils around the columella. (After Hyman; after Dakin.) (c) Internal anatomy and gross morphology of the female periwinkle,* Littorina littorea, *removed from its shell. (From Fretter and Graham, 1976. A* Functional Anatomy of Invertebrates. *Academic Press.)*

259

commonly protected by a univalved shell that is typically coiled, probably as an adaptation for efficient packaging of the visceral mass. Shell morphology differs considerably among species.[1] The animal is attached to the inside of the shell by a **columellar muscle,** which extends from the animal to the central axis of the shell. This central axis is known as the **columella** (Fig. 15.3a, b). The shell is typically carried so that it leans to the left side of the body. The shell axis is thus oblique to the long axis of the body, balancing the center of mass of the animal over the foot. The shells of most gastropod species coil to the right; that is, coiling is usually **dextral** (*dextro* = L: the right-hand side) rather than **sinistral** (*sinister* = L: the left-hand side). Probably as a consequence of space limitations within the shell, the ctenidium, nephridium, and heart auricle on the right side of dextrally coiled snails tend to be reduced or absent.

In addition to the shell, many gastropod species possess fairly elaborate behavioral or chemical defenses against predators. These adaptations commonly take one of the following forms: (1) the gastropod senses the presence of potential predators, either chemically or by touch, and initiates appropriate escape, avoidance, or deterrent behavior; (2) the gastropod chemically senses the presence of injured individuals of its own species (i.e., conspecific individuals) and initiates appropriate escape behavior; (3) the gastropod accumulates noxious organic compounds in its tissues, thereby becoming distasteful to potential predators.[2]

Of great importance in the evolutionary history and present-day biology of gastropods is the phenomenon of **torsion,** an anticlockwise twisting of most of the body through 180° during development. As a consequence of torsion, the nervous and digestive systems become obviously twisted, and the mantle cavity moves from the rear of the animal to become positioned over the head (Fig. 15.4). This dramatic rearrangement of the internal and external anatomy is brought about through the particular orientation and asymmetric development of the retractor muscles attaching the head-foot of the animal to its shell. Torsion may occur in a matter of hours or even minutes in some species. It must be emphasized that torsion has no direct relationship to shell coiling; the two are separate and independent processes.

The adaptive significance of torsion has been the subject of considerable speculation.[3] The controversy focuses largely on whether torsion benefits the larva or the adult, or both, and part of the difficulty in interpreting the "why" of torsion is that movement of the mantle cavity anteriorly would seem to be a mixed blessing at best. Certainly, through torsion, the ctenidia and osphradia come to be located at the front of the animal, in the direction of locomotion; but torsion also shifts the anus so that it discharges over the head, creating a potentially serious (and seemingly distasteful) sanitation problem. Moreover, the story of subsequent gastropod evolution clearly involves undoing the results of embryonic or larval torsion; that is, the more advanced gastropods contained in the subclasses Opisthobranchia and

---

1 See Topics for Discussion, No. 1.

2 See Topics for Discussion, No. 5.

3 See Topics for Discussion, No. 3.

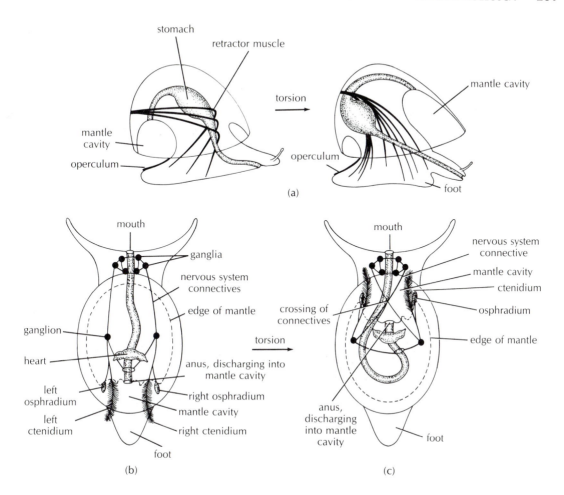

FIGURE 15.4. *(a) Torsion in the free-swimming larva of a primitive prosobranch gastropod,* Patella *sp. Note that the mantle cavity moves from the posterior to the anterior of the larva, providing a space into which the foot of the larva can be withdrawn. Following torsion, the head and foot can be fully retracted into the mantle cavity and the aperature tightly sealed by the rigid operculum. The larval swimming organ has been omitted for clarity. (From J.E. Morton, 1967. Molluscs, 2nd ed. Hutchinson; after Crofts.) The consequences of torsion to the adult gastropod: (b) shows the untorted state of a hypothetical ancestral gastropod-like mollusc; (c) shows the rearrangement of internal anatomy following the evolution of torsion. Note that the primitive gill has leaflets extending from both sides of the central axis. As will be discussed later, this is termed a bipectinate gill. (From Hyman, 1967. The Invertebrates, Vol. VI. McGraw-Hill. Reproduced by permission; after Hescheler.)*

Pulmonata exhibit a marked reduction in the degree to which torsion occurs during development. In some species, an apparent **detorsion** occurs subsequent to torsion. The selective pressures responsible for the evolution of torsion, as with all "why did it happen?" evolutionary questions, are ultimately unknowable. Yet, the phemonenon is so dramatic that it is difficult not to be intrigued by it. Torsion occurs in no other class of mollusc.

Small gastropod species may move largely through the action of cilia located on the ventral surface of the foot, but most species move by means of pedal waves of muscle contraction. Unlike peristaltic waves, pedal waves generally do not involve circular muscles, do not involve muscular contractions of great magnitude, and are restricted to the ventral surface of the animal.

The musculature of the foot is predominantly vertical (dorso-ventral) and transverse. Hemocoelic spaces are interspersed between regions of vertical musculature (Fig. 15.5). At the start of a pedal wave, the dorso-ventral musculature contracts at the anterior portion of the foot. Apparently, the transverse muscles do not relax, preventing the increased hemocoelic pressure from causing the foot to widen. Instead, the foot is extended forward by the increased hydrostatic pressure in the hemocoel. A wave of contraction of the dorso-ventral musculature then follows, moving posteriorly. The edges of the foot are temporarily sealed against the substrate with mucus secreted by glands on the foot, so that a small negative pressure (suction) is generated in the space between the substrate and the raised portion of the foot. If the hemocoelic spaces at the

leading edge of the wave are closed off from the general circulation, as they appear to be, they become deformed as they are squeezed up against the overlying musculature. This increase of pressure within the hemocoel helps bring about the re-extension of the dorso-ventral muscles at the trailing edge of the wave, by exerting downward force when the dorso-ventral muscles relax. The re-extension of the dorso-ventral musculature is also aided by the negative pressure created beneath the raised portion of the foot; this small space thus acts as a hydrostatic skeleton, even though it is external to the body, allowing the musculature at the forward edge of the wave to antagonize that at the trailing edge. In this case, the hydrostatic skeleton operates through a temporary pressure decrease, essentially sucking the raised portion of the foot downward, rather than through a temporary pressure increase.

The above description applies to **retrograde waves** (*retro* = L: back; *grad* = L: step); the wave of muscular contraction travels in the direction opposite that in which the snail is moving. Pedal waves may also be **direct** (i.e., moving in the same direction as the displacement of the animal). The basic principles of pedal wave generation are often greatly embellished by different gastropod species.

Gastropods play an important, although indirect, role in the transmission of several major human diseases, many species serving as obligate intermediate hosts in the life cycles of parasitic flatworms (Phylum Platyhelminthes, Class Trematoda; see Chapter Seven). Indeed, much research on the control of these flatworm parasites has focused on the regulation of snail populations.

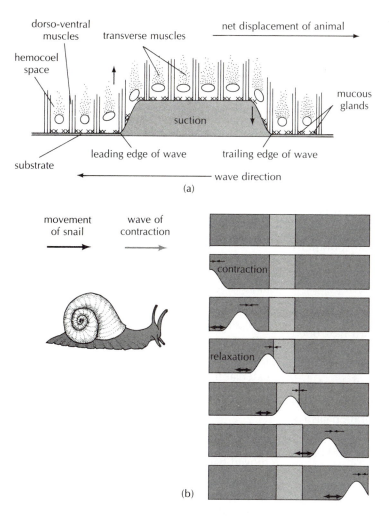

dorso-ventral muscles

transverse muscles

net displacement of animal

hemocoel space

mucous glands

suction

leading edge of wave

trailing edge of wave

substrate

wave direction

(a)

movement of snail

wave of contraction

contraction

relaxation

(b)

FIGURE 15.5. *(a) Diagrammatic longitudinal section through the foot of a moving gastropod. A pedal wave is traversing the foot from anterior to posterior, while the animal advances in the opposite direction. This is called a retrograde pedal wave. Localized contraction of dorso-ventral muscles raises a small portion of the foot away from the substrate, producing a small but measurable suction. This suction helps stretch the dorso-ventral muscles when they relax, pulling the area at the rear of the wave back against the substrate. The compression of hemocoelic spaces at the leading edge of the wave also aids in re-extension of the dorso-ventral muscles. (Modified from Jones and Trueman, 1970. J. Exp. Biol., 52: 201.) (b) Direct waves in the locomotion of a terrestrial snail. The wave moves in the direction of locomotion; i.e., from posterior to anterior. Only the portion of the foot lifted away from the substrate moves forward (note movement of light-shaded region of foot). Many waves travel down the foot simultaneously, as shown in the snail at the left. (After Lissmannn, 1945. J. Exp. Biol., 21: 58.)*

Subclass Prosobranchia

Members of this group, the largest of the three subclasses of the Gastropoda, are mostly marine. A small percentage of prosobranchs live in freshwater or terrestrial environments. Over 50,000 living prosobranch snails have been described. Prosobranchs are generally free-living and mobile, although some species have evolved sessile or even parasitic life-styles. Free-living prosobranchs may be herbivores, carnivores, deposit-feeders, omnivores, or suspension-feeders, depending upon the species.

Prosobranchs are the most primitive of gastropods; that is, the other two gastropod subclasses most likely evolved from prosobranch-like ancestors. Most prosobranch species possess a well-developed shell, mantle cavity, osphradium, and radula, and the foot of most prosobranch gastropods bears a rigid disc of protein (sometimes strengthened with calcium carbonate) called the **operculum.** When the foot is withdrawn into the shell, the operculum may completely seal the shell aperture, thus protecting the snail from physical stresses such as dehydration and from predators (Fig. 15.3a).

The typical prosobranch gill is a **ctenidium,** consisting of a series of flattened triangular sheets lying one adjacent to the next

---

Subclass Proso•branchia
/ /
(G: ANTERIOR GILL)

(Fig. 15.6). Deoxygenated blood moves through an afferent blood vessel from the open system of blood sinuses comprising the hemocoel to the individual sheets of the ctenidium. The blood then moves across each sheet, becomes oxygenated, and is transported to the auricle of the heart by an efferent blood vessel. From the auricle, the blood is pumped into the single associated ventricle and is then distributed to the tissues through a single aorta leading to the blood sinuses of the hemocoel. Primitive species possess a pair of auricles, a pair of efferent blood vessels, and a pair of ctenidia.

Water is drawn into the mantle cavity and across the gill sheets by the movements of gill cilia. In many prosobranch species, a portion of the mantle is drawn out into a cylindrical extension called the **siphon** (Fig. 15.7), and it is through this siphon that water is drawn across the **osphradium** (a chemoreceptor and tactile receptor) and into the mantle cavity (Fig. 15.6a). The muscular siphon is moved back and forth by the snail, enabling the animal to sample water from particular directions. In burrowing species, the siphon is extended through the substrate to the water above. The gastropod siphon is especially well-developed in carnivores and scavengers, and generally reduced or absent in suspension-feeders or deposit-feeders.

The direction of water movement across the gill is counter to the direction of blood flow. This **counter current exchange system** increases the efficiency of gas exchange between the ctenidium and the water. In the alternative situation, in which water and blood move in the same direction (Fig. 15.8a), the magnitude of the oxygen concen-

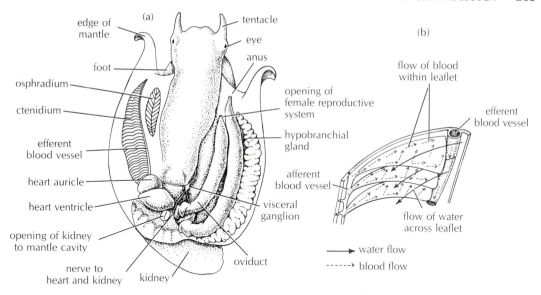

FIGURE 15.6. (a) The periwinkle Littorina littorea, removed from its shell and with its mantle cut mid-dorsally to reveal the arrangement of organs within the mantle cavity. Note the major blood vessel (efferent blood vessel) that conducts blood from the gills to the auricle of the heart. From here, the blood moves to the ventricle and then to the tissues. The hypobranchial gland secretes mucus for binding particles carried by the ctenidial filaments. (From Fretter and Graham, 1976. A Functional Anatomy of Invertebrates. Academic Press.) (b) Detail of ctenidium, showing direction of blood flow across individual gill sheets. (From Fretter and Graham.)

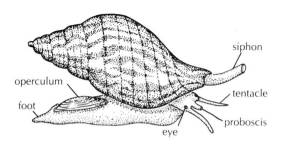

FIGURE 15.7. Prosobranch gastropod (Fasciolaria tulipa), showing extended siphon. (After Niesen.)

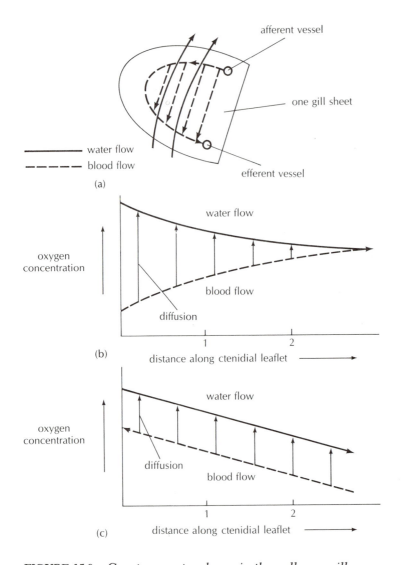

FIGURE 15.8. *Countercurrent exchange in the molluscan gill.
(a) The direction of water flow across the surface of each ctenidial
leaflet is opposite to the direction of blood flow through the gill
capillaries. (See Fig. 15.6b.) The afferent vessel carries deoxygenated
blood to the ctenidial leaflets from the tissues. Oxygenated blood
leaves the gill through the efferent vessel, carrying the blood to the
heart and thence to the tissues. (b,c) Illustration of the principle of
countercurrent exchange. Arrow length signifies magnitude of the
concentration gradient for oxygen between water and blood. In (b),
water and blood flow in the same direction, so the oxygen content
in the two fluids quickly comes to equilibrium. In (c), the
countercurrent situation, equilibrium is never attained; diffusion of
oxygen occurs along the entire gill sheet surface, permitting a greater
percentage of the oxygen in the water to be taken up by the blood.*

tration gradient between the water and the blood continually decreases along the surface of the gill sheet as the water loses oxygen to the adjacent blood, and the rate of gas exchange between water and blood decreases correspondingly. In the countercurrent situation (Fig. 15.8), however, a concentration gradient exists at all points along the gill.

At point 2 in Fig. 15.8b, the water has already given up much of its oxygen to the adjacent blood, and is nearly at the same oxygen concentration as the blood next to it. Essentially, no further gas exchange between the two fluids occurs as the blood continues to move across the gill. In terms of oxygen exchange, the gill surface extending beyond point 2 is essentially wasted. This is not the case with countercurrent exchange. At point 1 in Fig. 15.8c, even though the blood has gained oxygen from the water between points 1 and 2, it continues to come into contact with water of even higher oxygen content, so that diffusion of gas from water to blood continues. In other words, as blood moves from point 2 to point 1 in frame c, it is always in contact with water whose oxygen content is higher than that of the blood, so that diffusion of gases between blood and water occurs along the entire length of the gill.

Much of prosobranch evolution is a story of changes in gill numbers (from two in the more primitive species, contained within the Order Archaeogastropoda, to one in the more advanced species, found within the Meso- and Neogastropoda), and changes in the orientation of gill filaments extending from the ctenidial axis. In the primitive **bipectinate** condition, gill filaments extend from both sides of the ctenidial axis, while in the relatively advanced **monopectinate** con-

dition (Fig. 15.6, 15.8a), the filaments project from only one side of the supporting axis. Regardless of the number of gills and placement of gill filaments in different prosobranch species, however, the principle of countercurrent exchange applies to all. Indeed, the countercurrent principle applies to all ctenidia in the phylum Mollusca, except in the class Cephalopoda.

Considerable anatomical and functional diversity exists within the Prosobranchia. The members of one group, the **heteropods,** show especially striking modifications of the basic prosobranch body plan and life-style. The heteropods are planktonic, voracious carnivores, whose shell is reduced or absent and whose foot is a thin undulating paddle that propels the animal through the water (Fig. 15.9). Except for the viscera, the body is nearly transparent, an excellent adaptation for inconspicuous travel in the water. As illustrated in Fig. 15.9, heteropods swim upside down!

Subclass Opisthobranchia

Members of this subclass, which includes the sea hares, sea slugs, and bubble shells, are almost all marine. About 1100 species have been described. The major characteristics of this group that distinguish its members from the prosobranchs are: (1) a trend toward reduction or loss of the shell; (2) reduction or

---

Subclass Opistho•branchia
/ /
(G: POSTERIOR GILL)

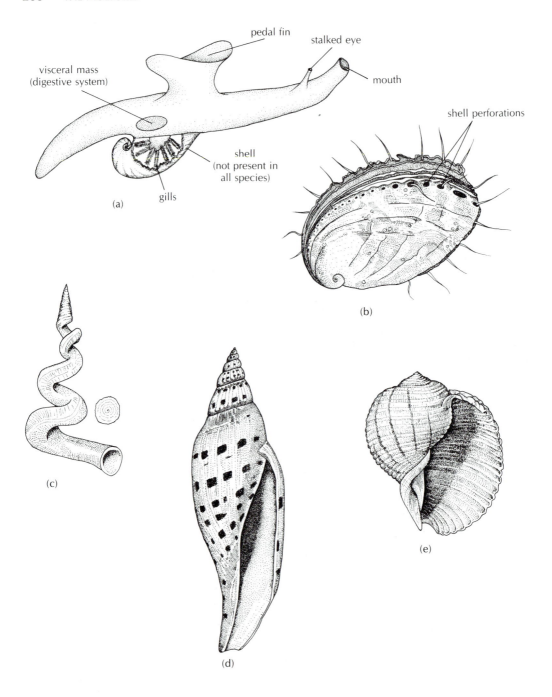

pedal fin

stalked eye

mouth

visceral mass
(digestive system)

shell perforations

shell
(not present in
all species)

gills

(a)

(b)

(c)

(d)

(e)

FIGURE 15.9. *Prosobranch diversity. (a) A heteropod; all species are planktonic. The visceral mass contains the digestive tract, heart, nephridia, and much of the reproductive system. (After Laverack and Dando; after Morton.) (b) The abalone,* Haliotis *sp. Cilia on the gills draw water into the mantle cavity at the anterior end of the body. Water leaves through a series of dorsal shell perforations. The openings of the digestive and excretory systems are located beneath one of these openings. (After Hyman.) (c) A worm-shell snail,* Vermicularia *sp. This is a sessile gastropod that lives attached to solid substrates, including rocks and other shells; the foot is reduced appropriately. As a juvenile, the animal produces a typical, spirally coiled shell, as observed near the apex of the adult shell. As the animal grows, however, the coiling becomes very loose, so that the whorls become disconnected. This corkscrew-shaped shell resembles that secreted by some species of sessile, polychaete worm. (From Hyman, 1967.* The Invertebrates, *Vol. VI. McGraw-Hill. Reproduced by permission.) (d) A volute shell,* Scaphella *sp. All volutes are marine carnivores, feeding on other invertebrates. (After Hardy.) (e) The giant tun shell,* Tonna galea. *Note the conspicuous shell ridges, the low spire, and the large body whorl. The adult shell lacks an operculum. (From Pimentel; after Abbott.)*

loss of the operculum; (3) limited torsion during embryogenesis; (4) reduction or loss of the mantle cavity; and (5) reduction or loss of the ctenidia. Most species that have lost the ctenidia have evolved other respiratory structures that are unrelated to the ancestral gill. For example, in many sea slugs or nudibranchs (order Nudibranchia), gas exchange occurs across brightly colored dorsal projections called **cerata** (Fig. 15.10b). In at least one species, the cerata exhibit rhythmic muscular contractions, apparently serving to move blood through the hemocoelic sinuses.

Cerata function in digestion and defense as well as in respiration. They contain extensions of the digestive system, and may also house unfired nematocysts obtained from ingested cnidarian prey; these nematocysts then function in the defense of the nudibranch against predation. Instead of cerata, many other nudibranchs possess feathery gills arising from the dorsal surface (Fig. 15.10a).

In addition to a pair of tentacles adjacent to the mouth, as in prosobranchs, the opisthobranch head typically bears a second pair of tentacles located dorsally and called **rhinophores**. The rhinophores are believed to be chemosensory, making them analogous with the osphradium of prosobranch gastropods and of those opisthobranchs bearing a mantle cavity.

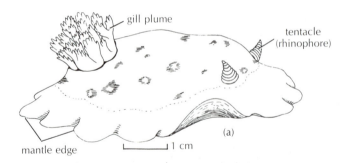

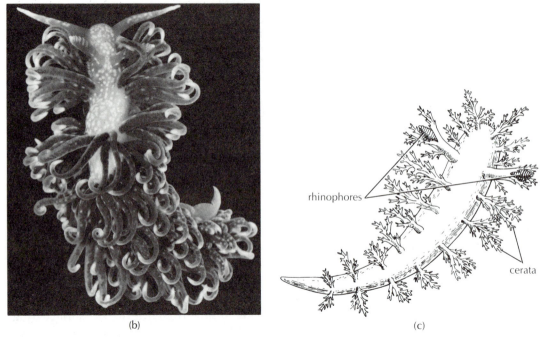

FIGURE 15.10. *(a)* A dorid nudibranch, Dialula sandiegensis. *Note the gills arranged as a plume around the anus. (From McConnaughey, Bayard H., and Zottoli, Robert, 1983.* Introduction to Marine Biology. *St. Louis, MO: The C.V. Mosby Co.) (b) An eolid nudibranch,* Spurilla neapolitana. *The conspicuous dorsal projections are cerata, which serve for gas exchange and also contain outfoldings of the digestive system. (Courtesy of L.S. Eyster.) (c) A dendronotid nudibranch,* Dendronotus arborescens. *Note the elaborate branching of the gills. This and related species are capable of simple swimming by flexing the body from side to side. (After Kingsley.)*

Opisthobranchs show varying degrees of departure from the ancestral, prosobranch-like condition. Adult sea slugs, for example, have no mantle cavity, ctenidia, osphradium, shell, or operculum, and some species show no evidence of torsion as adults. The larvae of sea slugs, on the other hand, have a pronounced mantle cavity, shell, and operculum, indicating a clear affinity with the prosobranchs. Another common opisthobranch group, the sea hares (order Aplysiacea) have a mantle cavity (with gill and osphradium) as adults, and most adults also have a shell. However, the mantle cavity is very small and on the right side of the animal, and the shell is reduced and internal. Clearly, the sea hares are closer to the ancestral, prosobranch-like condition than are the sea slugs. Even so, there is little evidence of torsion in the adult sea hare.

The most primitive of the opisthobranchs are the bubble shells (order Cephalaspidea), which include the greatest number of opisthobranch species. Although shell reduction, internalization, or loss is exhibited by most members of this order, a few species possess a conspicuous, external, spirally coiled shell, with operculum. Moreover, the mantle cavity (containing a gill and osphradium) is well developed.

Although locomotion is generally by means of cilia and pedal waves along the ventral surface of the foot, some opisthobranchs, such as the sea hares, can swim in short spurts by flapping lateral folds of the foot called **parapodia.** In other members of this subclass, the entire foot is drawn out into two thin lobes, also called **parapodia,** which

are used for swimming. These animals are known as pteropods ("wing-footed"). (See Fig. 15.11.) The pteropods may or may not have shells, depending on the species, and are permanent members of the plankton. Pteropods have no specialized respiratory organs, gas exchange being accomplished across the general body surface.

Subclass Pulmonata

In contrast to the prosobranch and opisthobranch gastropods, few of the 20,000 pulmonate species are marine, and those that are occur only intertidally and in estuaries. Most pulmonate species are found in terrestrial or freshwater environments; slugs and "escargot" are terrestrial members of this subclass.[4] A coiled shell is present in some pulmonate species, but the shell is reduced, internalized, or completely lost in others (the slugs) (Fig. 15.12). Only a few species have an operculum on the foot. Most pulmonates possess a long radula, in keeping with their generally herbivorous diet, and the head commonly bears two pairs of tentacles. Torsion is limited to about 90°, so that the nervous system is not so greatly twisted and the mantle cavity opens on the right side of the body, as in many opisthobranchs.

---

4 See Topics for Discussion, No. 10.

Subclass Pulmo • nata
/
(L: LUNG)

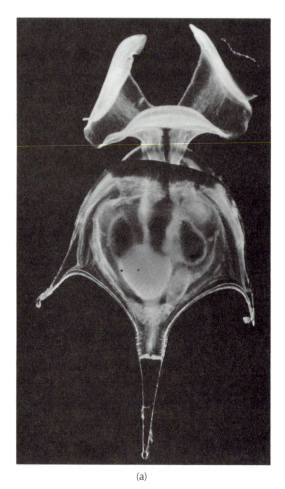

(a)

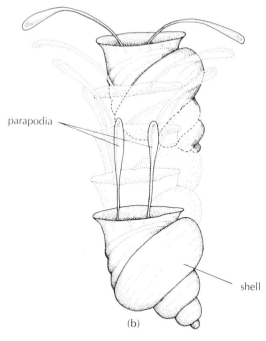

parapodia

shell

(b)

FIGURE 15.11. *Pteropods. (a)* Cavolina *sp., with parapodia exposed. (Courtesy of R.W. Gilmer.) (b)* Spiratella *sp., swimming. The animal is shown executing its power stroke, in which the parapodia are swept downward forcefully, generating forward thrust. The muscular parapodia are elaborations of the ancestral foot. (After Morton.)*

A major feature distinguishing the pulmonates from members of the other two gastropod subclasses is that the mantle cavity has become highly vascularized and now functions as a lung. Downward movement of the floor of the mantle cavity increases the volume of the cavity so that air, or in some cases water, is drawn into the mantle cavity for respiration. The fluid is then expelled by decreasing the volume of the mantle cavity. Flow of the air or water into and out of the lung occurs through a single small opening called the **pneumostome** (*pneumo* = G: lung; *stoma* = G: mouth). Although ctenidia are lacking in pulmonates, a gill has been secondarily evolved in some species. This gill takes the form of folds of mantle tissue near the pneumostome.

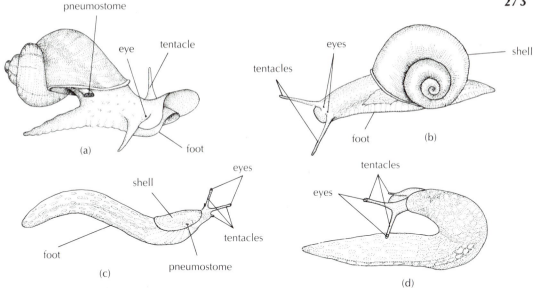

FIGURE 15.12. *Pulmonate diversity. (a) Lymnaea sp., with conspicuous pneumostome. Members of this genus are common inhabitants of freshwater lakes and ponds throughout the world, and serve as intermediate hosts for a variety of trematodes and cestodes (Platyhelminthes). (From Pennak,* Fresh-Water Invertebrates of the United States *2nd ed. John Wiley & Sons; after Baker.) (b) Helisoma sp., another freshwater pulmonate. Note the planospiral shell, with all the whorls lying in a single plane. (From Pennak; after Baker.) (c) A terrestrial slug,* Arion fuscus, *with a reduced, external shell. (After Kingsley.) (d)* Limax flavus, *a terrestrial slug with a small internal shell (not shown). (After Kingsley.)*

CLASS BIVALVIA (= Pelecypoda)

There are over 25,000 living bivalve species, including clams, scallops, mussels, and oysters. On the basis of gill structure, the bivalves may be divided into two major subclasses, the Protobranchia and the Lamellibranchia, and one very small subclass, the Septibranchia.[5] Major bivalve characteristics include: (1) a hinged shell, the two valves (left and right sides) of which are joined together by a springy ligament that, uncontested, acts to spring the shell valves

> *Class Bi • valvia (= Pelecy • poda)*
>
> **(L = TWO VALVED) (G = HATCHET FOOT)**

---

5 A more complex classification system, based largely upon morphological characteristics of the hinge area, is also widely used, as proposed in the *Treatise of Invertebrate Paleontology*, 1957–71. (Numerous volumes) R.C. Moore (Ed.). Geological Society of America.

apart; (2) lateral compression of the body and foot; (3) poorly developed head and associated sensory structures; (4) a spacious mantle cavity, relative to that found in other molluscan subclasses; (5) a sedentary life-style. A radula/odontophore complex is never found in the Bivalvia. Bivalves are primarily marine, but about 10–15% of all species occur in fresh water. No species are terrestrial.

The hinged portion of the bivalved shell is dorsal (Fig. 15.13). The shell valves, then, are on the left and right sides of the animal. The shell opens ventrally. On the dorsal surface, adjacent to the hinge, one frequently sees a conspicuous bulge in the shell. This bulge, termed the **umbo,** is comprised of the earliest shell material deposited by the ani-

mal. Distinct **growth lines** typically run parallel to the outer margins of the shell, as illustrated in Fig. 15.14. The foot projects ventrally and anteriorly, in the direction of movement, and the siphons, when present, project posteriorly.

Subclass Protobranchia

Members of this subclass retain what biologists regard as a morphologically primitive state of bivalve organization. The group is

Subclass Proto•branchia
(G: FIRST GILL)

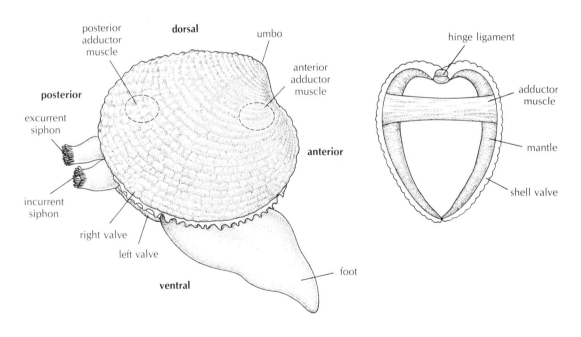

FIGURE 15.13. *A bivalve, indicating the orientation of the body within the shell valves. Note that the hinge is located dorsally, and the valves open ventrally. The siphons protrude posteriorly.*

FIGURE 15.14. *Growth lines in the shell of the bivalve* Arctica islandica. *The age of the shell can be determined by counting growth lines, since the patterns of shell growth are seasonal. The most recent growth lines are near the outer margin of the shell valve. This individual was estimated to be 149 years old. (Courtesy of Douglas S. Jones.)*

entirely marine, and all species live in soft substrates. One pair of gills is present in the mantle cavity. The gill filaments may be flattened triangular sheets (Fig. 15.15b), as in the typical gastropod ctenidium, or they may be finger-like structures (Fig. 15.15c) which are rather circular in cross section. The gills of protobranch bivalves are always **bi-pectinate** (the gill sheets extend from opposite sides of a central axis). The gill filaments hang down into the mantle cavity, dividing it into an **incurrent** (ventral) and an

excurrent (dorsal) chamber. Thus, water generally enters the mantle cavity ventrally, passes between gill filaments, and exits dorsally. The protobranch gill functions primarily in gas exchange, although it may also play some role in food collecton by filtering particles from the water.

Most of the food collection in protobranch bivalves is accomplished by **palp proboscides,** which are muscular, tentacle-like extensions of the tissue surrounding the mouth (Fig. 15.15a). The palp proboscides protrude

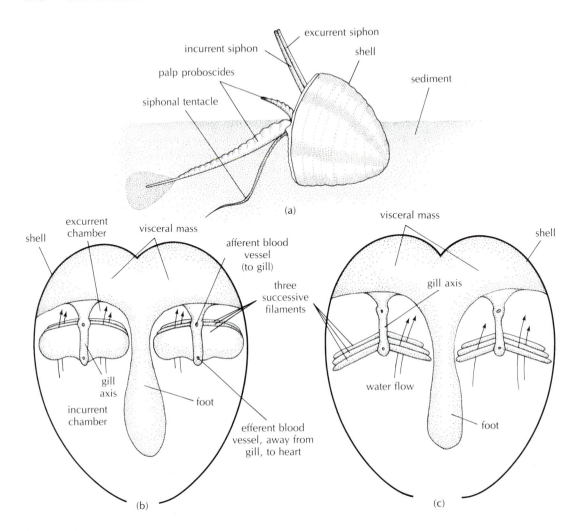

FIGURE 15.15. *(a) A protobranch bivalve,* Yoldia limatula, *with its foot and the lower part of its shell buried in the sediment. (After Meglitsch; after Drew.) (b,c) Two types of protobranch gills, with the more primitive gill type on the left. The gills function primarily in gas exchange. (After Russell-Hunter.)*

between the shell valves and probe the surrounding mud substrate, entangling particles in mucus. The mucus is then transported into the mantle cavity by cilia. Flattened structures (**labial palps**) attached to the palp proboscides at their bases sort the particles by ciliary action, transporting light particles to the mouth for ingestion and transferring heavy particles (which are less likely to be nutritional) to the margins of the labial palps, where these particles are ejected into the mantle cavity and expelled. This rejected material is termed **pseudofeces,** since it is waste material that has never been ingested. Protobranch bivalves are clearly restricted to soft sediments by their mode of feeding. This type of feeding, in which sediment is taken in and the organic fraction is digested, is called **deposit-feeding,** and is quite commonly encountered among invertebrates. The earthworm is a well-known terrestrial example of an annelid deposit-feeder.

Although protobranchs are present in shallow water, they are much better represented in deep water. Protobranch bivalves may account for about $\frac{3}{4}$ of all bivalves collected in sediment sampled from depths of one thousand meters or more.

Subclass Lamellibranchia

Most bivalves are lamellibranchs. Although the majority of lamellibranchs are marine, the only freshwater bivalve species are also members of this subclass. The freshwater species are mostly contained within a single family, the Unionidae.

Among the bivalves, water typically enters the mantle cavity through an **incurrent siphon** and exits through an **excurrent siphon** after passing between adjacent gill filaments. The siphons, where present, are tubular extensions of mantle tissue that can be protruded far beyond the posterior shell margins

Subclass Lamelli • branchia

(G: PLATE GILL)

(Fig. 15.16c). The siphons of some lamellibranchs can be extended approximately 15 cm beyond the shell. As in the prosobranch gastropods and the protobranch bivalves, water currents are generated by the action of the gill cilia, and the gill is the primary gas exchange surface. But the gill of lamellibranchs has another important role to play as well.

The lamellibranch ctenidium is modified to provide an enormous surface area for the collection, sorting, and transport of suspended particles. The individual gill filaments are thin and greatly elongated. The mantle cavity is divided by the gill filaments into an incurrent and an excurrent chamber, so that all water entering the mantle cavity must pass between the gill filaments before exiting (Fig. 15.17). Most significantly, the gill filaments are often bent to form food grooves along the ventral surface, and the surfaces of the gill filaments facing into the incurrent chamber of the mantle cavity show complex patterns of ciliation. Particles are carried to the food grooves by some of these cilia, and are then transported from gill filament to gill filament along the food groove to the labial palps, where sorting of particles occurs as in the Protobranchia.

Periodically, the lamellibranch bivalve closes its valves forcefully. This results in the expulsion of water and unwanted particles (pseudofeces) from the incurrent siphon.

After food particles are ingested at the mouth and passed down the esophagus, entangled in strings of mucus, they are drawn into the stomach, stirred, and, in part, digested by the action of a rotating translucent rod known as the **crystalline style.** The crystalline style is composed of structural protein

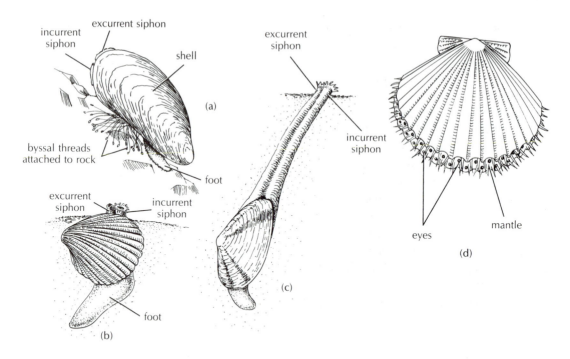

FIGURE 15.16. *Lamellibranch diversity. (a) A mussel attached to a rock by byssal threads. These threads are proteinaceous, and are secreted by a gland located at the base of the foot. (b) A cockle,* Laevicardium substriatum, *in its normal feeding position. (c) The soft-shell clam,* Mya arenaria, *in its normal feeding position. The fused siphons may extend more than 30 cm from the shell, permitting this clam to live well below the surface of the sediment. (d) A scallop,* Pecten sp. *Numerous light receptors are present along the edge of the mantle. The scallop can swim by forcefully contracting the well-developed posterior adductor muscle, ejecting water through openings near the shell hinge. Scallops lack siphons and have lost the anterior adductor muscle. It is the large, well-developed posterior adductor muscle that is served when one orders "scallops" in a restaurant. (All after Niesen.)*

and several digestive enzymes. One end of the rod lies in a **style sac**, a pouch of the intestine lined by cilia (Fig. 15.18b). The activity of these cilia causes the rod to rotate. The end of the style protruding into the stomach abrades against a chitinous **gastric shield** as the rod rotates. This abrasion causes the rod to slowly degrade at the end, releasing digestive enzymes into the stomach. Additions to the crystalline style are made in the style sac. A morphologically and functionally similar crystalline style appara-

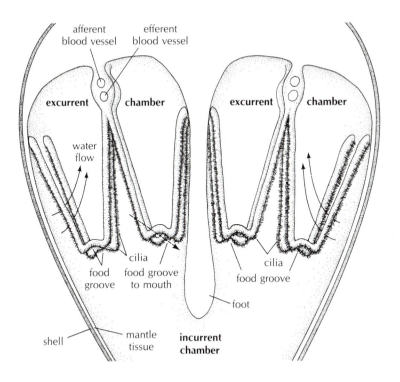

FIGURE 15.17. *Typical lamellibranch gill. (Reprinted with permission of Macmillan Publishing Company from* A Life of Invertebrates *by W.D. Russell-Hunter. Copyright © 1979 by W.D. Russell-Hunter.*

tus is also found in the stomachs of some suspension-feeding gastropod species. In contrast, protobranch bivalves possess a style sac containing mucus and digestive enzymes, but do not have a crystalline style. As in most other molluscs, the bivalve stomach connects with larger **digestive glands (digestive diverticula)**, which serve as the major sites of digestion and absorption.[6]

Some lamellibranchs, such as mussels, live attached to hard substrates by means of pro-

teinaceous secretions called **byssal threads.** A few lamellibranch species (some scallops) live unanchored atop the substrate and are capable of short bursts of "swimming," achieved by repeatedly clapping the shell valves together; rapid expulsion of water from the mantle cavity results in a quick displacement of the animal. Most bivalves, however, live wihin a burrow. Burrowing into a soft substrate (e.g., sand or mud) is accomplished by the muscular foot (Fig. 15.19). The foot is initially extended into the substratum by hydrostatic means, through

6 See Topics for Discussion, No. 9.

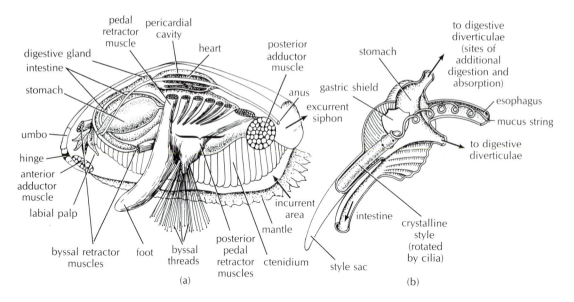

FIGURE 15.18. *(a) Internal anatomy of a lamellibranch bivalve, the mussel* Mytilus edulis. *(After Turner.) (b) The digestive system of a suspension-feeding lamellibranch bivalve, such as the blue mussel or soft-shell clam. The crystalline style aids digestion; rotated by cilia in the style sac, digestive enzymes are released as the tip of the rod is abraded against the gastric shield. A crystalline style apparatus is also commonly encountered among prosobranch gastropods. (After Morton.)*

contractions of the appropriate musculature in the foot. The **adductor muscles,** which attach the animal to the shell in all bivalve species, are relaxed at this time. This relaxation of the adductor muscles permits the release of energy that was stored in the compressed hinge ligament, so that the shell valves now press tightly against the surrounding substrate. This laterally directed force of the shell valves provides anchorage for the shell as the foot extends down into the substratum. Thus, downward extension of the foot need not eject the animal from the burrow. The adductor muscles then con-

tract, drawing the shell valves toward each other. This action releases the shell anchor. Contraction of the adductor muscles also pumps blood into the foot, which then dilates at the tip to form another anchor.

In addition to swelling the foot, abrupt closure of the shell valves forces water out of the mantle cavity, blowing away some of the sediment adjacent to the opening between the shell valves. With the foot firmly anchored in the substrate, the shell can now be pulled downward by contraction of the **pedal retractor muscles** from the foot to the shell. The pedal retractor muscles do not contract

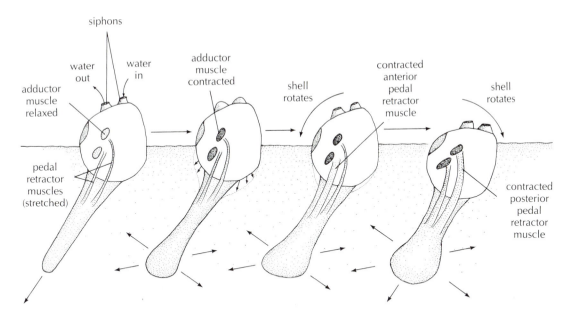

FIGURE 15.19. *Burrowing by bivalves. The foot serves both to penetrate into the substrate and to anchor the animal as the body is pulled downward. See text for discussion. (After Trueman.)*

simultaneously, but rather sequentially, so that the bivalve shell "rocks" forward and then backward as it progresses, slicing into the substrate.

A good many marine bivalve species burrow into wood, shell, coral, and other hard substrates as newly metamorphosed juveniles. The shell plays a major role in burrow formation by such species, actually carving into the substrate. As the animal grows, the body remains hidden within, and protected by, the substrate into which the burrow has been formed, with only the siphons protruding to the outside. The damage done by wood-boring bivalves can be considerable, as any marina-owner or boat-owner knows all too well. Wood-boring bivalves have been found in all marine habitats, including the deep sea at several thousand meters.[7]

A major group of wood-boring lamellibranch bivalves are misleadingly referred to as "shipworms" (Figs. 15.20, 15.21). Among the shipworms, filtering food particles from suspension in typical lamellibranch fashion may actually play a minor role in the feeding biology. Shipworms appear to meet many of their nutritional requirements by the digestion of wood and by the activities of symbiotic, nitrogen-fixing bacteria. Wood digestion is brought about by symbiotic bacteria (and possibly protozoans), which are

---

7 See Topics for Discussion, No. 11.

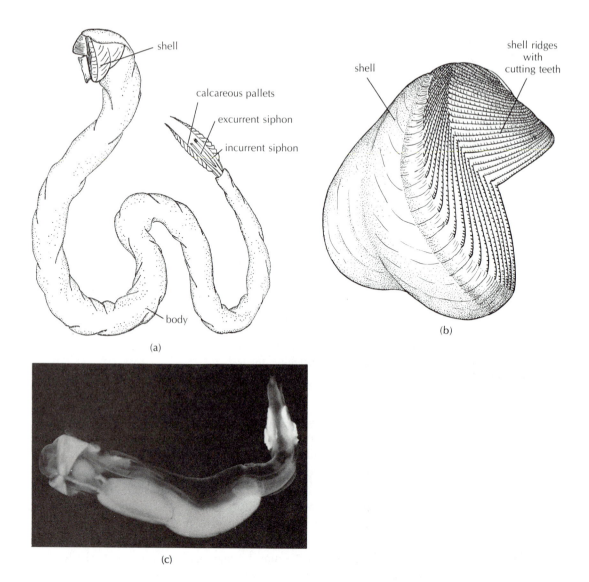

(a)

(b)

(c)

FIGURE 15.20. *(a) External morphology of a shipworm removed from its wood burrow. The shell is used to excavate the burrow. The pallets seal off the opening of the burrow when the animal withdraws. The pallets are thus analogous to the operculum of gastropods. (b) Detail of the shell, showing the cutting teeth. (c) Photograph of the shipworm* Teredora malleolus. *[(a,b) Courtesy of R.D. Turner, from Turner, 1966. A Survey and Illustrated Catalogue of the Teredinidae.* Museum of Comparative Zoology, Harvard University. *(c) Courtesy of C. Bradford Calloway.]*

(a)   (b)

FIGURE 15.21. *(a) A piece of wood riddled with wood-boring bivalves. (b) X-ray photo of a piece of wood containing shipworms. (Courtesy of R.D. Turner.)*

found in high densities within a specialized portion of the shipworm stomach called the **wood-storing caecum.**

Subclass Septibranchia

The septibranchs are a small group of carnivorous bivalves that feed on small planktonic animals (zooplankton) and on pieces of decomposing animal tissue. All species are marine, and most species are found in very deep

Subclass Septi • branchia
(G: FENCE GILL)

water. Only seven genera of septibranch bivalves have ever been described.

The ctenidium of septibranchs is highly modified, forming a muscular septum. The septum divides the mantle cavity into ventral and dorsal chambers, as in other bivalves, but is perforated by a series of ciliated openings. One-way valves regulate water flow through these openings in some species. The septum can be moved forcefully upward within the mantle cavity, drawing water in through the incurrent siphon and simultaneously expelling water through the excurrent siphon. Septibranchs thus feed as organic vacuum-cleaners, commonly sucking in small crustaceans and annelids. The stomach is lined with hardened chitin, serving to grind up ingested food. Labial palps, although present, are quite small and do not have any sorting function. A very reduced style sac is found, but never a crystalline style.

Class Polyplacophora

The 800 species in this class of molluscs are known as the chitons (not to be confused with the polysaccharide, chitin!). Chitons are generally found close to shore, particularly in the intertidal zone, and are restricted to living on hard substrate. The most distinctive external feature of chitons is the shell, which occurs as a series of eight overlapping and articulating elongated plates covering the dorsal surface (Fig. 15.22a). These plates are partially or largely embedded in the mantle tissue from which they are secreted. Because the shell is multisectioned, the body can bend to conform to a wide variety of underlying substrate configurations. The thick lateral mantle of chitons is called the **girdle.**

The mantle cavity of chitons takes the form of two lateral grooves on either side of the body (Fig. 15.22b). Up to about 80 bipectinate ctenidia hang down from the roof of each groove to the foot, dividing each elongated mantle cavity into incurrent and excurrent chambers. Water is drawn into the incurrent chamber by the action of the gill cilia. The flow of water is anterior to poste-

rior, so that waste products are discharged posteriorly in the excurrent stream. The flow of blood through the individual gill lamellae is opposite in direction to the flow of water, so that the principle of countercurrent exchange applies.

The foot extends along the entire ventral surface of the animal and is completely covered by the overlying shell and girdle. Locomotion is accomplished by means of pedal waves, as in the gastropods. When disturbed, the chiton can press the girdle tightly against the substrate. By then lifting up the central portion of the foot (and the inner margin of the mantle tissue as well, if required), while retaining a tight seal against the substrate along the entire outer margin of the foot (and girdle), a suction is generated which binds the animal tightly against the substrate. The secretion of mucus along the girdle helps maintain the suction. This ability to cling tightly to the substrate is a particularly effective adaptation for life in areas of heavy wave action.

The nervous system of chitons lacks ganglia in many species, and ganglia are only poorly developed in others. Sensory systems are also reduced; adult chitons lack statocysts, tentacles, and eyes on the head. However, **aesthetes** (organs derived from mantle tissue and extending through holes in the shell plates) apparently function as general light receptors.

The mouth and anus are at opposite ends of the body (Fig. 15.22b). Food particles are scraped from the substrate by a radula/odontophore complex. Despite the fact that chitons feed on small particles, most of which are algal, a crystalline style is not a component of the digestive system.

Class Poly • placo • phora

(G: MANY PLATE BEARING)

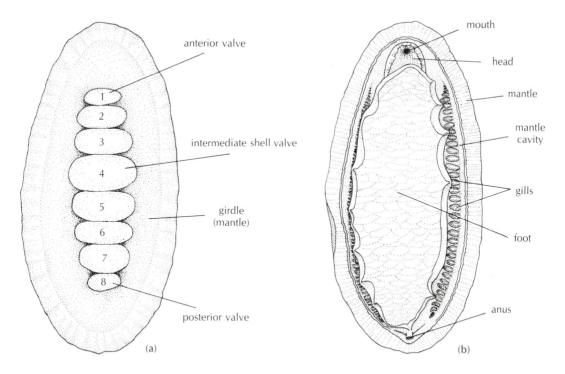

FIGURE 15.22. *The polyplacophoran* Katharina tunicata: *(a) dorsal view; (b) ventral view. All chitons are dorso-ventrally flattened, with the foot forming a large suction-cup for clinging to firm substrates. The shell is composed of eight articulating plates, permitting the entire animal to curl and therefore conform to the topography of the underlying surface. (Modified from Beck and Braithwaite, 1968.* Invertebrate Zoology Laboratory Workbook, *3rd ed. Burgess Publishing Company, Minneapolis, Minnesota.)*

CLASS SCAPHOPODA

The 300–400 species in this class are all marine and live in sand or mud substrates, mostly in deep water. Scaphopods possess these "typical" molluscan features: foot; mantle tissue; mantle cavity; radula; and shell. The scaphopod shell is never spirally wound, but rather grows linearly as a hollow, curved tube. Hence the common names "tooth shell" and "tusk shell." The shell has an opening at each end (Fig. 15.23). Water enters at the narrower end, which protrudes above the substrate. Inflow of water is due to the action of ciliated cells restricted to ridges of mantle tissue. Periodically, water is expelled through this same opening by a sudden contraction of the foot musculature.

Unlike many other molluscs, scaphopods possess no ctenidia. A heart and circulatory

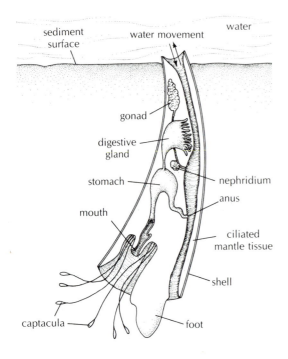

FIGURE 15.23. Dentalium *sp., a scaphopod, in its normal feeding orientation. Food particles are captured by the captacula. (After Borradaile; after Naef.)*

system are also lacking; instead, the blood circulates through the various large sinuses of the hemocoel as a consequence of the rhythmic movements of the foot.

Burrowing into soft substrate is accomplished by the foot, essentially as described previously for the bivalves. Scaphopods capture small food particles, including foraminiferans (Protozoa), from the surrounding sediment and water using specialized, thin tentacles known as **captacula**. The food is then transported to the mouth by captacular cilia, or, in the case of large particles, by muscular movements of the tentacles themselves.

CLASS CEPHALOPODA

It is truly wondrous that animals such as the squid and octopus are to be found in the same phylum as the clams, snails, scaphopods, and chitons. Unlike most molluscs, cephalopods are generally fast-moving, active carnivores with a very impressive amount of behavioral complexity. Members of this class are exclusively marine.

Ctenidia and a radula are present in all cephalopod species. A mantle cavity and foot are present as well, but they do not generally function in "typical" molluscan fashion. The head and associated sensory organs of cephalopod molluscs are extremely well developed. As you will see, cephalopods are the supreme testament to the impressive plasticity of the basic molluscan body plan.

Most of the 600 or so extant cephalopod species lack an external shell. The only living cephalopods to possess an external shell are species in the genus *Nautilus*. The shell of *Nautilus* is spiral, but unlike that of gastropods, it is divided by **septa** into a series of compartments (Figs. 15.24, 15.25). The living animal is found in the largest, outermost chamber. The septa are penetrated by a calcified tube of tissue (the **siphuncle**), which spirals through the shell from the visceral mass. Liquid may be transported to and from the shell chambers through the siphuncle, gas diffusing into or out of the chambers as the fluid volumes are altered. As the gas content of each chamber is changed by this

Class Cephalo•poda

(G: HEAD FOOT)

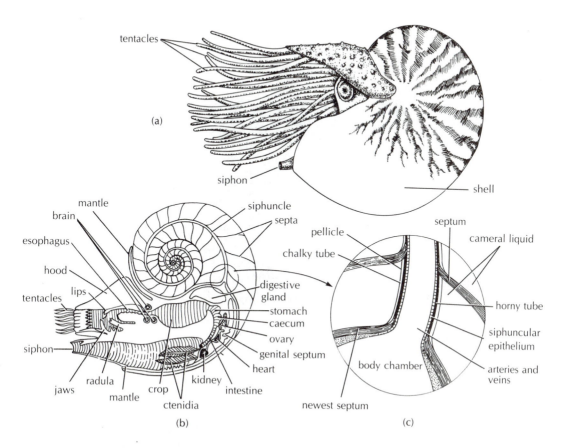

FIGURE 15.24. *(a) Nautilus sp., in its normal swimming position. (b) Longitudinal section through the shell of the same individual, showing the septa isolating the different shell compartments, the siphuncle, and the internal anatomy. The entire body of the animal is found in the outermost chamber of the shell. (After Engemann and Hegner; after Borradaile and Potts.) (c) Detail of the newest and second most recent shell chambers. (From Ward et al., 1980. The buoyancy of the chambered nautilus. Copyright © 1980 by Scientific American, Inc. All rights reserved.)*

mechanism, the buoyancy of the shell, and therefore of the animal, is changed as well. Locomotion is accomplished by jet propulsion; i.e., by the expulsion of water resulting from contractions of a flexible, hollow tube called the **siphon** or **funnel.** The funnel, which is derived from the foot of the "typical" mollusc, may be pointed in a variety of directions, providing good directional control for locomotion.

FIGURE 15.25. *Section through the shell of* Nautilus, *showing the large body chamber (BC), siphuncle (S), and smaller chambers separated by septa. Holes have been drilled through the shell in several of the most recent chambers for experimental purposes; they are not natural. (Courtesy of L. Greenwald.)*

Other cephalopods also move by jet propulsion, but the expulsion of water occurs far more forcefully and a much greater volume of water is expelled. In contrast to the nautiloids, an external shell is lacking and the mantle plays the major role in movement (Fig. 15.26a). The mantle tissue is thick, and replete with both circular and radial musculature. Contraction of the radial musculature of the mantle tissue, while the circular muscle fibers are relaxed, causes the volume of the mantle cavity to increase. Water then enters the mantle cavity all along the mantle margin. This causes the animal to be drawn forward slightly. When the radial muscles relax and the circular muscles then contract, the margins of the mantle tissue form a tight seal against the head, and a large volume of water is forcefully expelled entirely through the flexible, hollow funnel. The expulsion of a large volume of water through the funnel at great velocity enables these cephalopods to move at high speeds.

Some modern cephalopods (e.g., the European cuttlefish) retain a chambered shell internally; the shells of such species are involved in regulation of buoyancy, as in *Nautilus*.[8] The shell of squids is also internal, but is little more than a stiff proteinaceous rod called the **pen**. In *Octopus*, no vestiges of a

8 See Topics for Discussion, No. 2.

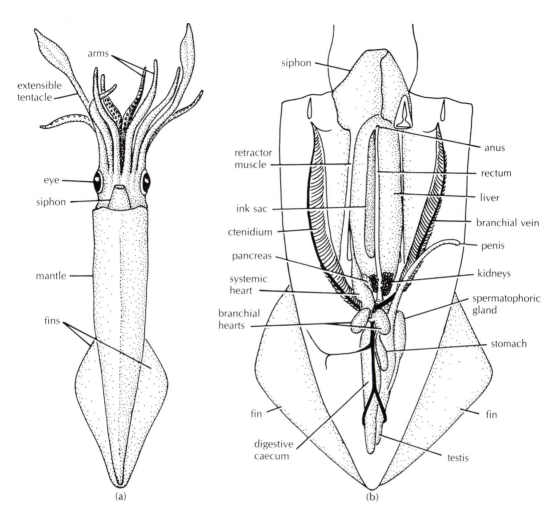

FIGURE 15.26. *(a) External anatomy of the squid,* Loligo *sp. (b) The internal anatomy of the squid. Digestion is initiated in the stomach and completed in the large associated caecum. (Modified after several sources.)*

shell remain. Loss of the shell has clearly been a major trend in cephalopod evolution; over 7000 species of shelled cephalopod are known from fossils.

The mantle cavity of cephalopods generally contains a pair of ctenidia (Fig. 15.26b), but, unlike the situation in other molluscs, blood and water flow in the same direction, rather than in counter-current fashion. Moreover, the ctenidia of cephalopods are not ciliated. Water circulation is maintained by the continual emptying and refilling of the mantle cavity, accomplished through contraction of the mantle musculature.

Cephalopods are also unique among molluscs in having a completely closed circulatory system, in which blood flows entirely through a system of arteries, veins, and capillaries. The blood sinuses found in other molluscs are not present in cephalopods. In addition to a **systemic heart,** which receives oxygenated blood from the gills and sends it back to the tissues, an accessory (**branchial**) heart is associated with each gill. The branchial hearts increase the blood pressure, helping to push blood through the capillaries of the gills.

In addition to forming the siphon, derivatives of the molluscan foot form the muscular **arms** and extensible **tentacles** of cephalopods (Fig. 15.27). The mouth thus lies in the center of the "foot"! The total number of arms and tentacles is usually either 8 or 10, depending on the species, although nautiloids possess nearly 40. All cephalopod tentacles are provided with small suction cups, which are used for clinging to substrates or to objects, including potential food items. The arms are additionally studded with receptors sensitive to touch.

The degree of **cephalization** (concentration of sensory and nervous tissue at the anterior end of an animal) found among the Cephalopoda exceeds that found in any other invertebrate. A large, complex, highly differentiated brain is found in cephalopods (Fig. 15.28a). The brain of the common octopus, *Octopus vulgaris,* is composed of over ten distinct lobes. Studies of cephalopod behavior indicate a clear capability for memory and learning.

Each cephalopod has two eyes, which are incredibly similar in gross morphology to those of vertebrates (Fig. 15.28b); the eyes of cephalopods and vertebrates are among the most beautiful examples of convergent evolution encountered among animals. Like the vertebrate eye, the cephalopod eye possesses a cornea, lens, iris, diaphragm, and retina. The cephalopod eye is also focusable and image-forming, although the process of image formation in vertebrates and cephalopods differs in detail. In the vertebrate eye, light is focused by altering the shape of the lens. In contrast, light is focused within the cephalopod eye by moving the lens toward or away from the retina.

The skin of cephalopods contains several layers of tiny colored cells called **chromatophores,** which overlay reflective cells called **iridocytes** (Fig. 15.29). Expansions and contractions of the chromatophores are mediated by muscle elements in the skin, and are under direct nervous control. Thus, the coloration of cephalopod skin may change extremely quickly. Defensive, camouflaging, and courtship-related changes of color have been described.[9]

Most cephalopods have an **ink sac** associated with the digestive system. The dark-pigmented fluid secreted by the ink sac may be deliberately discharged through the anus, forming a cloud that presumably confuses potential predators, and which may also act as a mild narcotic.

Clearly, the story of cephalopod evolution is one of developing adaptations for an active carnivorous life-style: reduction and elimi-

---

9 See Topics for Discussion, No. 5.

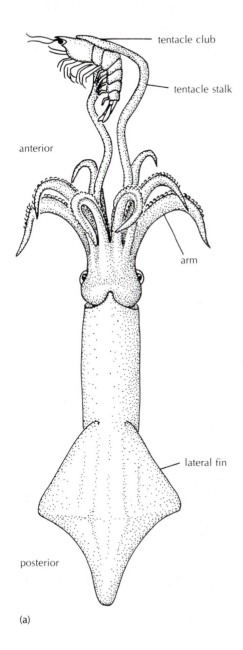

FIGURE 15.27. (a) The squid, Loligo pealei, capturing prey with its suckered, extensible tentacles. (From Kier, 1982. J. Morphol., 172: 179.) (b) In this sequence, the crustacean prey was dropped in front of the squid at T_0. The events shown occurred within 70 milliseconds. (Courtesy of W.M. Kier.)

tentacle club

tentacle stalk

anterior

arm

lateral fin

posterior

(a)

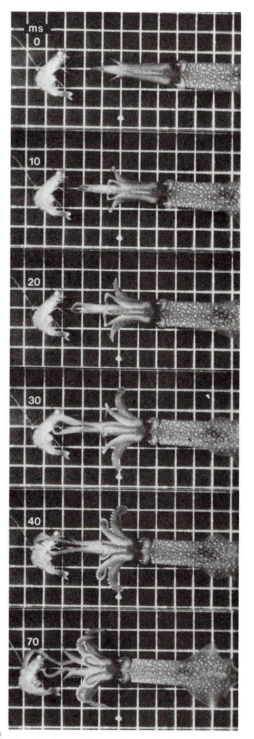

ms
0

10

20

30

40

70

(b)

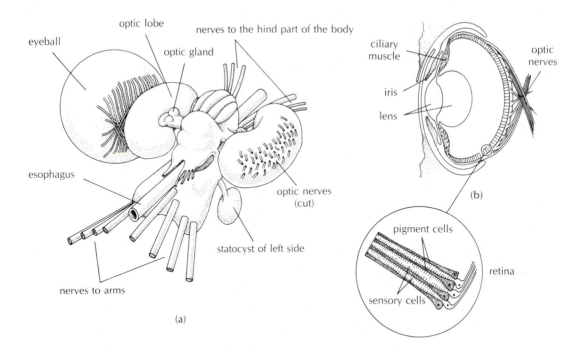

FIGURE 15.28. *(a) A cephalopod brain. Cephalopods possess the most complex brains found among invertebrates. (From Wells, 1968. Lower Animals. McGraw-Hill. Reproduced by permission.) (b) The eye of* Octopus, *demonstrating convergent evolution with the eye of vertebrates. Unlike the eye of vertebrates, cephalopod eyes are focused by moving the lens back and forth, rather than by altering the shape of the lens. (From Wells. Reproduced by permission.)*

nation of the shell; replacement of ciliary activity by muscular activity for locomotion and respiration; modification of the foot, mantle, and blood circulatory system; extensive development of the head, sensory organs, brain, and nervous system.

CLASS MONOPLACOPHORA

Prior to 1952, this class was known only from the fossil record. Since that time, seven living species have been described, all marine and all collected from very deep water. A single, unhinged shell is present, as in many gastropods (Fig. 15.30). The shell of adult monoplacophorans is flattened rather than spirally wound, although the larval shell is spiral. The monoplacophoran foot is

> *Class Mono•placo•phora*
> **(G: ONE SHELL BEARING)**

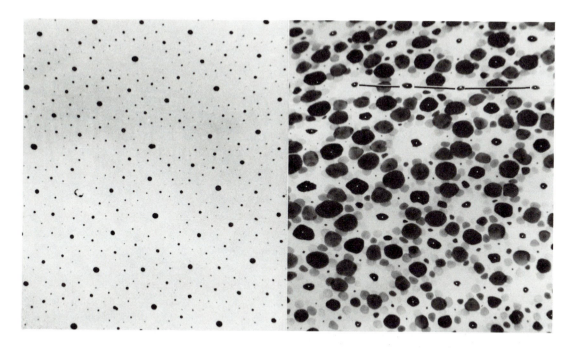

FIGURE 15.29. *Chromatophores photographed from a living squid. The chromatophores are contracted at the left, and expanded at the right. (Courtesy of R. Hanlon, 1982.* Malacologia, *23: 89.)*

flattened, as in gastropods and poly-placophorans.

The mantle cavity takes the form of two lateral grooves, as in polyplacophorans; five or six pairs of gills hang down within the mantle grooves. Whether or not these gills are homologous with the typical molluscan ctenidium is, however, uncertain. In addition to the gills, the pedal retractor muscles, auricles and ventricles of the heart, gonads, and nephridia come in multiple copies. Both a radula and a crystalline style are present, and the gut is linear, with the mouth being anterior and the anus posterior.

CLASS APLACOPHORA

About 200 species of aplacophorans have been described to date, but many species are known from only one or two specimens. Like the cephalopods, scaphopods, mono-placophorans, and chitons, aplacophorans are entirely marine. They occur primarily in deep water sediments, especially at high lati-

Class A • placo • phora
(G: NOT SHELL BEARING)

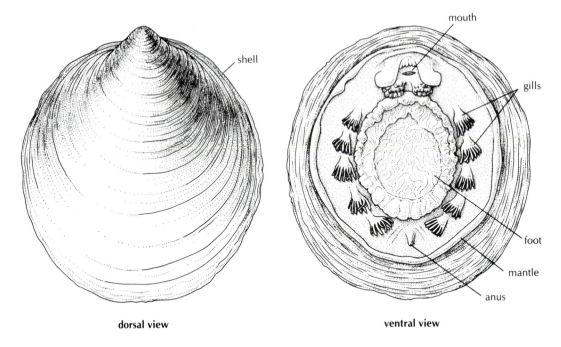

dorsal view   ventral view

FIGURE 15.30. *The monoplacophoran* Neopilina galatheae. *This animal was known only as a fossil until 1952, when it was dredged off the Pacific coast of Mexico from 5000 meters depth. The adult shell is about 3 cm in length. (After Lemche.)*

tudes. The body is vermiform, with no indication of segmentation (Fig. 15.31). Aplacophorans lack a shell and lack a well-defined foot. On the other hand, some species possess external calcareous spicules, a radula, a style sac, ctenidia, and a posterior mantle cavity. Vestiges of a foot are also found in some species. Even with these features, aplacophorans are so different from other molluscs that some workers have considered them to belong to a different phylum altogether, the Platyhelminthes.

OTHER FEATURES OF MOLLUSCAN BIOLOGY

1. Reproduction and Development

Although some gastropods are parthenogenetic, reproduction is usually sexual among molluscs. Most species are **dioecious** (i.e., sexes are separate), but there are a large number of exceptions to this generalization. A number of prosobranch gastropods, opisthobranch gastropods, and lamellibranch bi-

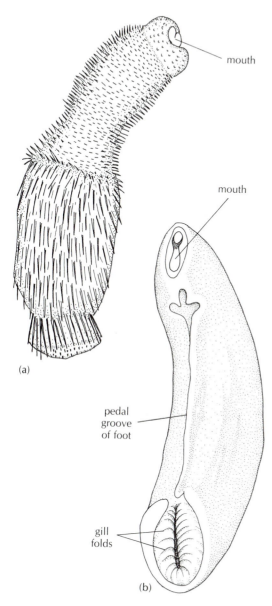

(a)

mouth

mouth

pedal groove of foot

gill folds

(b)

FIGURE 15.31. *Members of the Aplacophora:* (a) Falcidens cf. recisum (*Schwabl, 1963*). *(Courtesy of A.H. Scheltema.)* (b) *Neomenia carinata. (From Hyman.)*

valves, for example, are **protandric hermaphrodites.** That is, the sex of a single individual changes from male to female with age. All pulmonates, and those opisthobranchs that are not protandric hermaphrodites, are **simultaneous hermaphrodites,** a single individual producing both eggs and sperm simultaneously. Indeed, the gonad of such an individual is termed an **ovotestis.** Simultaneous hermaphrodites often have reciprocal copulation, resulting in a mutual exchange of sperm.

Generally, the molluscan genital ducts are associated with a portion, or a modified portion, of the excretory system. Fertilization of eggs is exclusively external in the Scaphopoda, and probably in the Monoplacophora as well. External fertilization is quite common among the bivalves, chitons, and aplacophorans, less so among the gastropods, and nonexistent among the cephalopods. Terrestrial and freshwater molluscs (i.e., some gastropods and bivalves) fertilize only internally, as adaptations to stressful conditions that would otherwise be imposed upon the gametes and embryos by the environment (see Chapter 1). Cephalopods show a particularly distinctive set of adaptations for achieving internal fertilization. In particular, one arm of the male is modified as a copulatory organ (Fig. 15.32). This modified arm, called a **hectocotylus** (*hecto* = G: one hundred; *cotylo* = G: sucker), transfers packets of sperm (**spermatophores**) to the female and is often left behind by the male after the sperm transfer is accomplished.

Not surprisingly, free-living larval stages are associated with the development of most

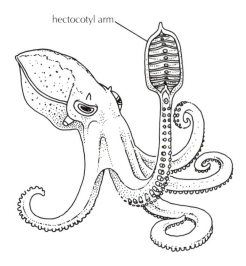

hectocotyl arm

FIGURE 15.32. Octopus lentus *male with hectocotylized arm. The arm is turned up to show where the spermatophores will be carried. (After Huxley; after Verrill.)*

species that fertilize their eggs externally in the surrounding sea water. The embryo passes through a conspicuous **trochophore** stage, resembling that of polychaete annelids (Fig. 15.33a). In gastropods and bivalves, the prototroch of the trochophore stage gradually becomes outfolded into a distinctive, ciliated organ known as the **velum**. A larva with a velum is called a **veliger** (Fig. 15.33b, c). The velum may be used for locomotion, food (phytoplankton) collection, and gas exchange. The velum is lost upon metamorphosis to adult form.

Even when fertilization is internal, a free-swimming dispersive larva often occurs at some point during development. In the more advanced species, particularly among the gastropods, the free-living larval stage is often suppressed. Development to the juvenile stage may take place entirely within a jelly mass, an egg capsule (Fig. 15.33d), or a specialized brood chamber of the female. Only among the cephalopods is a free-living, morphologically/ecologically distinct larval stage entirely absent from the life cycle. Cephalopod eggs develop within gelatinous masses, often attentively protected and aerated by the mother until the young hatch out as fully formed miniatures of the adult.

2. Circulation, Gas Exchange, and Blood Pigments

All molluscs have a blood circulatory system (Fig. 15.34). In cephalopods, the system is completely closed; all blood flow occurs through arteries, veins, and capillaries. In all other molluscs, the circulatory system is largely an open one, blood moving through a series of large sinuses comprising the hemocoel. In gastropods, the turgor of the tentacles and foot is dependent upon the amount of blood in the sinuses of these tissues. In most molluscs, including the apparently primitive monoplacophorans, the blood is pumped by a heart. However, in scaphopods, which lack a heart, muscular contractions of the foot have primary responsibility for moving the blood through the large sinuses.

The bloods of many mollusc species lack a specialized oxygen transport pigment. Where one is found, it is either hemoglobin, as in some bivalves, or hemocyanin, as in pulmonates, prosobranchs, and cephalopods. **Hemocyanin** is a pigment structurally similar to hemoglobin, except that the heme group contains copper rather than iron.

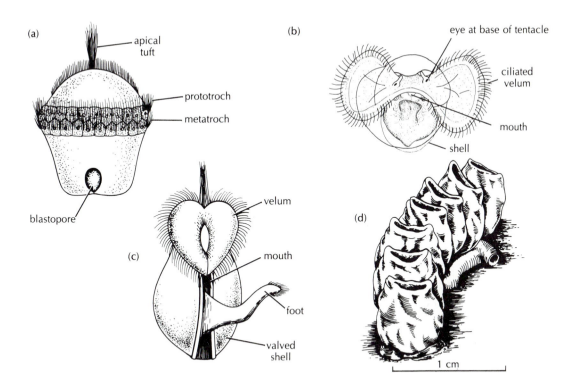

FIGURE 15.33. *Molluscan free-living larvae. (a) The trochophore larva of* Patella, *a primitive prosobranch. (From Hyman.) (b) A gastropod veliger in frontal view. (Courtesy of R. S. Scheltema.) (c) A bivalve veliger. (After Harmer and Shipley.) (d) Egg capsules of the marine prosobranch gastropod* Conus abbreviatus. *Each capsule contains numerous embryos. (From Kohn, 1961.* Pacific Sci., *15: 163.)*

Gas exchange may take place across gills housed within a mantle cavity (e.g., in cephalopods, bivalves, chitons, and prosobranch gastropods); external gills (e.g., in some opisthobranch gastropods); or other vascularized tissues (e.g., the cerata of opisthobranch gastropods and the tissue lining the mantle cavity of pulmonates and scaphopods).

3. Nervous System

The degree of development of the nervous system corresponds to the activity level of the possessor. The ganglia of molluscs range from being nonexistent in many chitons to being extremely well-developed in the cephalopods. Among the cephalopods, the ganglia

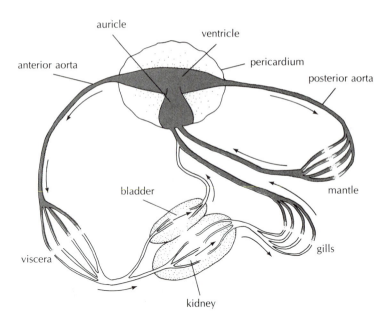

FIGURE 15.34. *The pattern of blood circulation in the lamellibranch* Anodonta *sp. The shaded areas indicate the route of oxygenated blood. (From Hickman, Cleveland P., 1973. Biology of the Invertebrates, 2nd ed. St. Louis, MO: The C.V. Mosby Co.)*

form a true brain. In other molluscs, the major ganglia are generally paired and are connected by nerve fibers, forming a ring through which the esophagus passes. Major nerve cords may run to the mantle, foot, gills and osphradium, viscera, and radula. The ganglia associated with innervation of these tissues are named, respectively, the pleural, pedal, parietal, visceral, and buccal ganglia. Details vary among members of the different classes.

In a number of molluscs, certain neurons are specialized for very rapid impulse conduction. These unusually wide, **giant fibers** are especially prominent among cephalopods, connecting the cerebral ganglia with the musculature of the mantle (Fig. 15.35). Giant fibers are extremely important in syn-

chronizing contractions of the mantle musculature for jet propulsion, and make rapid escape responses possible. The fibers have also been important to humans; our present knowledge of how nerve impulses are generated and transmitted comes largely from studies made on these cephalopod giant fibers.

4. Digestive System

All molluscs have a complete digestive system with a separate mouth and anus. The mouth leads into a short esophagus, which in turn leads to a stomach. Associated with the stomach are one or more **digestive glands** or **digestive caeca.** Digestive enzymes are secreted into the lumen of these glands. Addi-

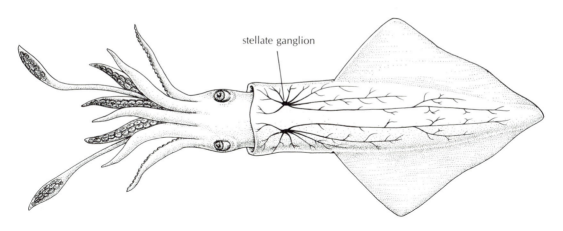

stellate ganglion

FIGURE 15.35. *The nervous system of a cephalopod (squid), showing the system of giant fibers. Individual axons may be as large as one millimeter in diameter. (From Keynes, 1958. The nerve impulse and the squid. Copyright © 1958 by Scientific American, Inc. All rights reserved.)*

tional extracellular digestion takes place in the stomach proper. In cephalopods, digestion is entirely extracellular. In most other molluscs, the terminal stages of digestion are completed intracellularly, within the tissue of the digestive glands. The absorbed nutrients enter the blood circulatory system for distribution throughout the body, or are stored in the digestive glands for later use. Undigested wastes pass through an intestine and out through the anus.

Other aspects of food collection and processing have already been discussed where appropriate for each group.

5. Excretory System

A urine is typically formed as the coelomic fluid passes through one or more pairs of metanephridia (often called "kidneys"). As in annelids, coelomic fluid generally enters the nephridium through a nephrostome. Recall that the molluscan coelom is little more than a small cavity (**pericardium**) surrounding the heart (*peri* = G: around; *cardio* = G: heart) (Fig. 15.34). Thus, the coelomic fluid appears to be largely a filtrate of blood, containing small waste molecules such as ammonia that are forced across the heart wall with the filtrate. Additional wastes are actively secreted into the coelomic fluid by glands lining the pericardium. As in annelids, the primary urine is further modified by selective resorption and secretion as it travels through the metanephridial tubules. The penultimate urine is then discharged into the mantle cavity through a **nephridiopore** and carried away by water currents.

TAXONOMIC SUMMARY

PHYLUM MOLLUSCA

  CLASS GASTROPODA—SNAILS AND SLUGS
    SUBCLASS PROSOBRANCHIA
    SUBCLASS OPISTHOBRANCHIA
    SUBCLASS PULMONATA

  CLASS BIVALVIA—CLAMS, MUSSELS, AND SHIPWORMS
    SUBCLASS PROTOBRANCHIA
    SUBCLASS LAMELLIBRANCHIA
    SUBCLASS SEPTIBRANCHIA

  CLASS POLYPLACOPHORA—CHITONS

  CLASS SCAPHOPODA—TOOTH SHELLS

  CLASS CEPHALOPODA—SQUID, OCTOPUS, AND
      CUTTLEFISH

  CLASS MONOPLACOPHORA

  CLASS APLACOPHORA

TOPICS FOR FURTHER DISCUSSION AND INVESTIGATION

1. What is the functional significance of differences in shell morphology among gastropods and bivalves?

Bertness, M.D., and C. Cunningham, 1981. Crab shell-crushing predation and gastropod architectural defense. *J. Exp. Marine Biol. Ecol.*, 50: 213.

Bottjer, D.J., 1981. Periostracum of the gastropod *Fusitriton oregonensis*: natural inhibitor of boring and encrusting organisms. *Bull. Marine Sci.*, 31: 916.

Bottjer, D.J., and J.G. Carter, 1980. Functional and phylogenetic significance of projecting periostracal structures in the Bivalvia (Mollusca). *J. Paleontol.*, 54: 200.

Conover, M.R., 1979. Effect of gastropod shell characteristics and hermit crabs on shell epifauna. *J. Exp. Marine Biol. Ecol.*, 40: 81.

Currey, J.D., and J.D. Taylor, 1974. The mechanical behavior of some mollusc hard tissues. *J. Zool., London*, 173: 395.

Palmer, A.R., 1977. Function of shell sculpture in marine gastropods: hydrodynamic destabilization in *Ceratostoma foliatum*. *Science*, 197: 1293.

Palmer, A.R., 1979. Fish predation and the evolution of gastropod shell sculpture: experimental and geographical evidence. *Evolution*, 33: 697.

Schmitt, R.J., 1982. Consequences of dissimilar defenses against predation in a subtidal marine community. *Ecology*, 63: 1588.

Stanley, S.M., 1969. Bivalve mollusk burrowing aided by discordant shell ornamentation. *Science*, 166: 634.

Vermeij, G.J., 1979. Shell architecture and causes of death of micronesian reef snails. *Evolution*, 33: 686.

Vermeij, G.J., and A.P. Covich, 1978. Co-evolution of freshwater gastropods and their predators. *Amer. Nat.* 112: 833.

Vermeij, G.J., and J.D. Currey, 1980. Geographical variation in the strength of thaidid snail shells. *Biol. Bull.*, 158: 383.

2. A number of cephalopod species retain shells as adults, either internally or externally. How do these shells play a role in the regulation of buoyancy?

Clarke, M.R., E.J. Denton, and J.B. Gilpin-Brown, 1979. On the use of ammonium for buoyancy in squids. *J. Marine Biol. Assoc. U.K.*, 59: 259.

Denton, E.J., 1974. On buoyancy and the lives of modern and fossil cephalopods. *Proc. R. Soc. London B*, 185: 273.

Greenwald, L., C.B. Cook, and P.D. Ward, 1982. The structure of the chambered nautilus siphuncle: the siphuncular epithelium. *J. Morphol.*, 172: 5.

Greenwald, L., P.D. Ward, and O.E. Greenwald, 1980. Cameral liquid transport and buoyancy in chambered nautilus (*Nautilus macromphalus*). *Nature*, 286: 55.

Ward, P.D., and L. Greenwald, 1982. Chamber refilling in *Nautilus*. *J. Marine Biol. Assoc. U.K.*, 62: 469.

3. Discuss the adaptive significance of gastropod torsion.

Crofts, D.R., 1955. Muscle morphogenesis in primitive gastropods and its relation to torsion. *Proc. Zool. Soc. London*, 125: 711.

Fretter, V., 1967. The prosobranch veliger. *Proc. Malac. Soc. London*, 37: 357.

Fretter, V., 1969. Aspects of metamorphosis in prosobranch gastropods. *Proc. Malac. Soc. London*, 38: 375.

Ghiselin, M.T., 1966. The adaptive significance of gastropod torsion. *Evolution*, 20: 337.

Hardy, A.C., 1959. "Pelagic Larval Forms," in *The Open Sea: Its Natural History*, Boston: Houghton Mifflin.

Kriegstein, A.R., 1977. Stages in the post-hatching development of *Aplysia californica*. *J. Exp. Zool.*, 199: 275.

Thompson, T.E., 1967. Adaptive significance of gastropod torsion. *Malacologia*, 5: 423.

Underwood, A.J., 1972. Spawning, larval development and settlement behavior of *Gibbula cineraria* (Gastropoda: Prosobranchia) with a reappraisal of torsion in gastropods. *Marine Biol.*, 17: 341.

4. A number of molluscs demonstrate a symbiotic relationship with photosynthetic algae that is reminiscent of that between zooxanthellae and cnidarians. How are the photosynthetic symbionts acquired by the molluscs, and what benefits are obtained by the hosts?

Clark, K.B., and K.R. Jensen, 1982. Effects of temperature on carbon fixation and carbon budget partitioning in the zooxanthellal symbiosis of *Aiptasia pallida* (Verrill). *J. Exp. Marine Biol. Ecol.*, 64: 215.

Fitt, W.K., and R.K. Trench, 1981. Spawning, development, and acquisition of zooxanthellae by *Tridacna squamosa* (Mollusca, Bivalvia). *Biol. Bull.*, 161: 213.

Gallop, A., J. Bartrop, and D.C. Smith, 1980. The biology of chloroplast acquisition by *Elysia ciridis*. *Proc. R. Soc. London B*, 207: 335.

Trench, R.K., D.S. Wethey, and J.W. Porter, 1981. Observations on the symbiosis with zooxanthellae among the Tridacnidae (Mollusca: Bivalvia). *Biol. Bull.*, 161: 180.

5. Investigate the behavioral and chemical defenses of gastropods, bivalves, or cephalopods against predation.

Bullock, T.H., 1953. Predator recognition and escape responses of some intertidal gastropods in the presence of starfish. *Behavior*, 5: 130.

Cimino, G., S. De Rosa, S. De Stefano, and G. Sodano, 1982. The chemical defense of four Mediterranean nudibranchs. *Comp. Biochem. Physiol.*, 73B: 471.

Feder, H.M., 1963. Gastropod defensive responses and their effectiveness in reducing predation by starfishes. *Ecology*, 44: 505.

Feder, H.M., 1972. Escape responses in marine invertebrates. *Sci. Amer.*, 227: 92.

Fishlyn, D.A., and D.W. Phillips, 1980. Chemical camouflaging and behavioral defenses against a predatory seastar by three species of gastropods from the surf grass *Phyllospadix* community. *Biol. Bull.*, 158: 34.

Garrity, S.D., and S.C. Levings, 1983. Homing to scars as a defense against predators in the pulmonate limpet *Siphonaria gigas* (Gastropoda). *Marine Biol.*, 72: 319.

Geller, J.B., 1982. Chemically mediated avoidance response of a gastropod, *Tegula funebralis* (A. Adams), to a predatory crab, *Cancer antennarius* (Stimpson). *J. Exp. Marine Biol. Ecol.*, 65: 19.

Greenwood, P.G., and R.N. Mariscal, 1984. Immature nematocyst incorporation by the aeolid nudibranch *Spurilla neapolitana*. *Marine Biol.*, 80: 35.

Hadlock, R.P., 1980. Alarm response of the intertidal snail *Littorina littorea* (L.) to predation by the crab *Carcinus maenas* (L.). *Biol. Bull.*, 159: 269.

Margolin, A.S., 1964. A running response of *Acmaea* to seastars. *Ecology*, 45: 191.

Messenger, J.B., 1974. Reflecting elements in cephalopod skin and their importance for camouflage. *J. Zool. London*, 174: 387.

Packard, A., and G.D. Sanders, 1971. Body patterns of *Octopus vulgaris* and maturation of the response to disturbance. *Anim. Behav.*, 19: 780.

Parsons, S.W., and D.L. Macmillan, 1979. The escape responses of abalone (Mollusca, Prosobranchia, Haliotidae) to predatory gastropods. *Marine Behav. Physiol.*, 6: 65.

Phillips, D.W., 1975. Distance chemoreception-triggered avoidance behavior of the limpets *Acmaea* (*Collisella*) *limatula* and *Acmaea* (*Notoacmaea*) *scutum* to the predatory starfish *Pisaster ochraceus*. *J. Exp. Zool.*, 191: 199.

Prior, D.J., A.M. Schneiderman, and S.I. Greene, 1979. Size-dependent variation in the evasive behaviour of the bivalve mollusc *Spisula solidissima*. *J. Exp. Biol.*, 78: 59.

Young, R.E., C.F.E. Roper, and J.F. Walters, 1979. Eyes and extraocular photoreceptors in midwater cephalopods and fishes: their role in detecting downwelling light for counter-illumination. *Marine Biol.*, 51: 371.

6. Gastropods continue to have pronounced impact on their communities even after death. Investigate the ways in which empty gastropod shells influence the populations of other invertebrates living in the same locality.

Conover, M.R., 1979. Effect of gastropod shell characteristics and hermit crabs on shell epifauna. *J. Exp. Marine Biol. Ecol.*, 40: 81.

McLean, R., 1983. Gastropod shells: a dynamic resource that helps shape benthic community structure. *J. Exp. Marine Biol. Ecol.*, 69: 151.

Spight, T.M., 1977. Availability and use of shells by intertidal hermit crabs. *Biol. Bull.*, 152: 120.

Vance, R., 1972. Competition and mechanism of coexistence in three sympatric species of intertidal hermit crabs. *Ecology*, 53: 1062.

Wilber, T.P., Jr., and W.F. Herrnkind, 1984. Predaceous gastropods regulate new-shell supply to salt marsh hermit crabs. *Marine Biol.*, 79: 145.

7. Do molluscs show an immune response?

Cheng, T.C., K.H. Howland, and J.T. Sullivan, 1983. Enhanced reduction of T4D and T7 coliphage titres from *Biomphalaria glabrata* (Mollusca) hemolymph induced by previous homologous challenge. *Biol. Bull.*, 164: 418.

Tripp, M.R., 1960. Mechanisms of removal of injected microorganisms from the American oyster,

Crassostrea virginica (Gmelin). *Biol. Bull.*, 119: 273.

8. Investigate adaptations for carnivorous behavior in prosobranch gastropods.

Carr, W.E.S., 1967. Chemoreception in the mud snail, *Nassarius obsoletus*. I. Properties of stimulating substances extracted from shrimp. *Biol. Bull.*, 133: 90.

Carriker, M.R., D. van Zandt, and T.J. Grant, 1978. Penetration of molluscan and non-molluscan minerals by the boring gastropod *Urosalpinx cinerea*. *Biol. Bull.*, 155: 511.

Rittschoff, D., L.G. Williams, B. Brown, and M.R. Carriker, 1983. Chemical attraction of newly hatched oyster drills. *Biol. Bull.*, 164: 493.

9. Investigate the cycles of digestive activity encountered among molluscs and prosobranch gastropods.

Curtis, L.A., 1980. Daily cycling of the crystalline style in the omnivorous, deposit-feeding estuarine snail *Ilyanassa obsoleta*. *Marine Biol.*, 59: 137.

Hawkins, A.J.S., B.L. Bayne, and K.R. Clarke, 1983. Coordinated rhythms of digestion, absorption and excretion in *Mytilus edulis* (Bivalvia: Mollusca). *Marine Biol.*, 74: 41.

Morton, J.E., 1956. The tidal rhythm and the action of the digestive system of the lamellibranch *Lasaea rubra*. *J. Marine Biol. Assoc. U.K.*, 35: 563.

Palmer, R.E., 1979. Histological and histochemical study of digestion in the bivalve *Arctica islandica* L. *Biol. Bull.*, 156: 115.

Robinson, W.E., and R.W. Langton, 1980. Digestion in a subtidal population of *Mercenaria mercenaria* (Bivalvia). *Marine Biol.*, 58: 173.

Yonge, C.M., 1926. Structure and physiology of the organs of feeding and digestion in *Ostrea edulis*. *J. Marine Biol. Assoc. U.K.*, 14: 295.

10. Investigate the adaptations of gastropods for terrestrial existence.

Boss, K.J., 1974. Oblomovism in the mollusca. *Trans. Amer. Microsc. Soc.*, 93: 460.

Howes, N.H., and G.P. Wells, 1934. The water relations of snails and slugs. I. Weight rhythms in *Helix pomatia* L. *J. Exp. Biol.*, 11: 327.

Sloan, W.C., 1964. The accumulation of nitrogenous compounds in terrestrial and aquatic eggs of prosobranch snails. *Biol. Bull.*, 126: 302.

Verderber, G.W., S.B. Cook, and C.B. Cook, 1983. The role of the home scar in reducing water loss during aerial exposure of the pulmonate limpet, *Siphonaria alternata* (Say). *Veliger*, 25: 235.

Wells, G.P., 1944. The water relations of snails and slugs. III. Factors determining activity in *Helix pomatia* L. *J. Exp. Biol.*, 20: 79.

11. Discuss the ecological significance of wood-boring bivalves in the deep sea.

Turner, R.D., 1973. Wood-boring bivalves, opportunistic species in the deep sea. *Science*, 180: 1377.

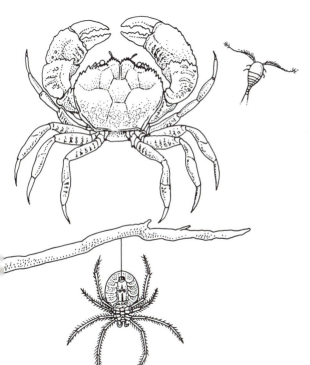

16

THE ARTHROPODS

Phylum *Arthro · poda*
(G: JOINTED FOOT)

INTRODUCTION AND GENERAL CHARACTERISTICS

At least 75% of all animal species described to date belong to this phylum, making the arthropod body plan by far the best represented in the animal kingdom. Like the annelid stock from which they are most likely derived, arthropods are basically metameric. In most modern members of the phylum, however, the underlying metameric, serial repetition of like segments is masked by the fusion and modification of different regions of the body for highly specialized functions. This specialization of groups of segments, known as **tagmatization,** is also seen to some extent in polychaete annelids, but reaches its greatest extent in the Arthropoda. Two of the major arthropod classes (Insecta and Crustacea) have three distinct tagmata: head, thorax, and abdomen.

Arthropods have one conspicuous feature in common with members of the Mollusca. Individuals of both groups generally have a

305

hard, external, protective covering—but here the similarity ends. The external coverings in the two phyla are deposited through entirely different mechanisms; differ greatly in chemical composition; have distinctly different physical properties; and perform largely different functions. Whereas the molluscan shell generally functions exclusively as protection for the soft parts within, the arthropod integument functions additionally as a locomotory skeleton.

The outermost layer (the **epicuticle**) of the arthropod exoskeleton (Fig. 16.1) is secreted by epidermal cells and is generally waxy, being composed of a firm lipoprotein layer underlain by layers of lipid. The cuticle is thus water-impermeable. The outer body surface can therefore not serve for gas exchange. On the other hand, the arthropod is more resistant to water loss by dehydration compared with most other invertebrates. The epicuticle is quite thin, constituting perhaps only 3% of the total thickness of the exoskeleton. The bulk of the exoskeleton is made up of the **procuticle,** composed largely of a polysaccharide (**chitin**) in association with a number of proteins. The procuticle has hardening elements added for strength. In the Crustacea, this hardening is partly achieved through the deposition of calcium carbonate in some layers of the procuticle. Hardening is also accomplished by "tanning" the protein component of the procuticle. The tanning process, also called **sclerotization,** involves the formation of crosslinkages between protein chains, and contributes to the hardening of the cuticle in all arthropods. In the insects, hardening of the

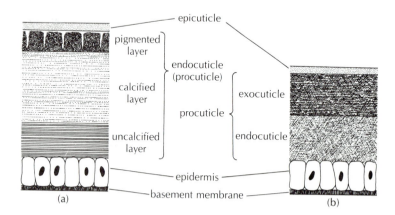

FIGURE 16.1. *The cuticle of (a) crustaceans and (b) insects. The cuticles of both groups of animals are secreted by the underlying epidermis. (Reprinted with permission of Macmillan Publishing Company from* A Life of Invertebrates *by W.D. Russell-Hunter. Copyright ©1977 by W.D. Russell-Hunter.)*

cuticle is accomplished entirely through tanning. A similar process commonly occurs among other phyla, by the way. Sclerotization is involved in forming the hinges and byssal threads of bivalves; strengthening the periostracum of the molluscan shell; hardening the molluscan radula; and hardening the jaws of some annelids, the egg coverings of some platyhelminths, and the trophi of rotifers.

The arthropod procuticle is not hardened uniformly over the entire body, and therein lies its major functional significance. In many regions of the body, the procuticle is thin and flexible in certain directions, forming joints (Fig. 16.2). Through the presence of appropriate musculature, the arthropods thus have a jointed skeleton that functions in much the same way as does the vertebrate skeleton; pairs of muscles antagonize each other through a system of rigid levers. The development of a jointed, flexible exoskeleton is the essence of arthropod success, and as will be discussed later, has opened up a lifestyle inaccessible to any other invertebrate group: flight.

The coelom can play no major role in the locomotion of an animal encased in a suit of rigid plates, and the arthropod coelom is greatly reduced in extent accordingly. The main body cavity is instead a **hemocoel,** part of the blood circulatory system, as in the Mollusca.

In contrast to shell growth among molluscs, the outer protective covering of arthropods is not added gradually at a growing edge. Instead, it is secreted over all regions of the body simultaneously. Once the hardening process is completed, the arthropod is literally encased in its armor, except where

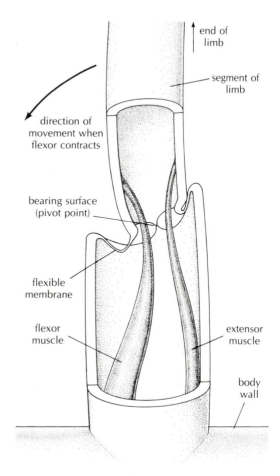

FIGURE 16.2. *The articulation of a crustacean limb. The procuticle is hardened everywhere except at the joints, as indicated. (After Russell-Hunter.)*

the armor is pierced by sensory hairs and gland openings. Major regions of the foregut and hindgut are also lined with cuticle. In order to increase in body size, the arthropod must shed the cuticle—including that lining the gut—grow larger, and then harden a new cuticle to fit the larger (and, in certain instances, morphologically altered) body. The old cuticle is partially degraded by enzymatic secretions and is split prior to its removal.

The splitting of the old cuticle is accomplished by uptake of water and air, and an increase in blood pressure, causing the body to swell. The process of removing the existing exoskeleton is called **ecdysis,** from the Greek word meaning "an escape" or "a slipping out of." In practice, the new cuticle is actually secreted before the old one is shed, which may partially explain why the arthropod does not become totally nonfunctional during the molting process. High blood pressure within the hemocoel may also play a role in maintaining body form and function prior to hardening of the new cuticle. Of course, potential collapse of the body during molting is more of a problem in air than in water, since air is relatively unsupportive. This may be a factor in explaining why terrestrial arthropods are smaller than aquatic species. Hardening of the cuticle does not occur until morphological alterations, if any, and the increase in body size have taken place. Thus, the time between ecdysis and hardening of the new cuticle is a period of increased vulnerability to predators; most arthropods seek protective shelter during ecdysis.

It is important to realize that although increases in size are discontinous in arthropods, growth of tissue (**biomass**) is actually a continuous process. The number of cells in the epidermis, for example, increases continuously, the additional tissue often becoming folded or pleated until the old cuticle is shed and the increase in body size can take place.

The process of ecdysis and formation of the new exoskeleton is under hormonal control. Basically, one gland (the **Y-organ,** located in the head of crustaceans, for example, or the **prothoracic glands,** located in the thorax of insects) produces a hormone called **ecdysone,** which stimulates molting to occur. However, between molts, ecdysone production seems to be inhibited by the production of a second hormone. In crustaceans, this second hormone is produced by a gland (the **X-organ**) located in the eyestalks. When this second organ ceases effective production of its hormone, inhibition of the Y-organ is removed and ecdysone may be produced. Alternatively, it may be that the X-organ secretion does not turn the Y-organ off but rather inhibits the action of ecdysone directly. In either case, ecdysis cannot occur until the X-organ ceases production of its inhibitory hormone (Fig. 16.3); surgical removal of crustacean eyestalks results in premature ecdysis. A number of other important arthropod functions are known to be under hormonal control as well, including regulation of the reproductive cycle; regulation of body fluid osmotic concentration; migration of eye pigments; and the functioning of the **chromatophores,** which are used to vary body color in some arthropod groups.

The arthropod nervous system merits special mention, since it is operationally quite different from the nervous systems of both vertebrates and other invertebrates. In vertebrates, one muscle fiber is innervated by a single neuron. The strength of muscle contraction depends upon the number of fibers contracting, and the number of fibers contracting in a given muscle is dependent upon the number of axons fired. In arthropod muscle, in contrast, the strength of con-

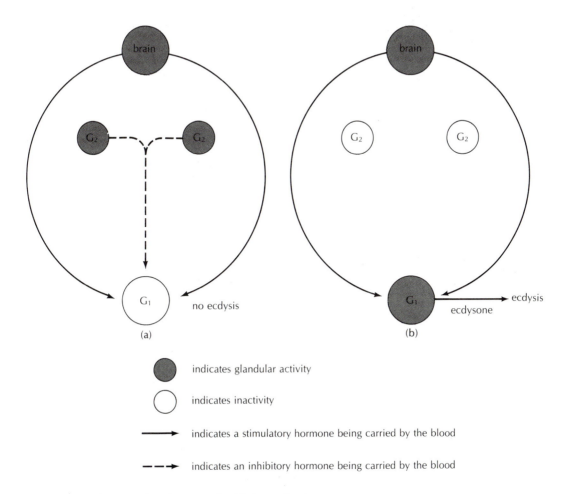

FIGURE 16.3. *Schematic diagram of a likely mechanism regulating ecdysis in arthropods. Glands associated with molting are indicated by G_1 and G_2. (a) The hormone produced by G_2 inhibits ecdysone production by G_1 and no molting occurs. (b) Gland G_2 is inactive; the brain hormone stimulates ecdysone production by G_1 and the animal molts.*

traction depends upon the rate of delivery of nerve impulses to the fibers. Moreover, a single muscle fiber may be innervated by as many as five different types of neurons (Fig. 16.4). The type of contraction (fast but brief versus slow and sustained) depends in part upon the source of the stimulation to the muscle. In addition, some of the neurons are inhibitory; action potentials delivered to such inhibitory neurons can alter the outcome of signals delivered down other axons to the same muscle fiber. As yet another compli-

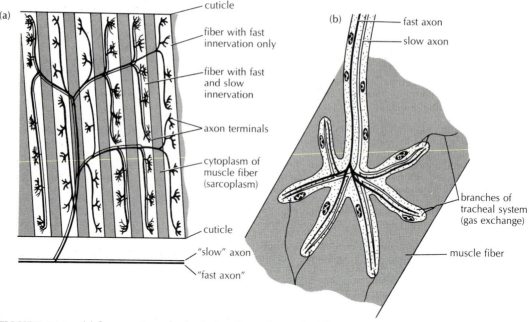

FIGURE 16.4. (a) Innervation of a typical arthropod muscle. Note that only some of the muscle fibers receive innervation from the "slow" axon. (b) Neuromuscular junction, in detail. (Both from Hoyle, 1974. In M. Rockstein, ed., The Physiology of the Insecta, Vol. 4. Academic Press. Reprinted by permission.)

cation, arthropods have several physiological and functional types of muscle fiber. That is, the rate of contraction is partly a property of the individual muscle fiber itself. Fine control of arthropod movement is therefore dependent upon the types of muscle fiber stimulated and the interaction of several types of neurons terminating on a single muscle fiber. Finally, it should be noted that a single arthropod neuron may innervate a large number of muscle fibers, so that a given muscle may be innervated by very few neurons (2–3, in some cases). In contrast, a given vertebrate muscle may receive innervation from hundreds of millions of neurons.

The musculature of arthropods differs significantly from that of other invertebrate groups. Arthropod muscle is entirely striated, whereas most other invertebrates possess primarily (or entirely) smooth muscle. The functional ramifications of this difference are considerable, in that striated muscle is far more responsive than smooth muscle tissue. That is, the time required for striated muscle to complete a contraction is generally far less than that required for smooth muscle (Table 16.1). Without striated muscle, arthropods would never have achieved flight.

The arthropod circulatory system is of interest as well, in that blood leaves the heart

TABLE 16.1. Contraction Times of Invertebrate Muscle (From various sources).

SOURCE	CONTRACTION TIME (SECONDS)
Anthozoa	
sphincter muscle	5.000
circular muscle	60–180
Scyphozoa	0.500–1.000
Annelida	
earthworm circular muscle	0.300–0.500
Bivalvia	
anterior byssus retractor	
muscle	1.000
Gastropoda	
tentacle retractor muscle	2.500
Arthropoda	
Limulus abdominal muscle	0.195
insect flight muscle	0.025

through closed vessels in most species, but enters the heart directly from the hemocoel through perforations in the heart wall. These openings into the heart are called **ostia** (Fig. 16.5). The circulatory system is thus an open one, with the oxygenated blood moving through a series of sinuses and finally being drawn back into the heart through the ostia as the heart expands. A "heart with ostia" is one of the diagnostic features of the Arthropoda, although gas exchange is achieved by radically different means in a number of arthropod groups.

In the classification scheme adopted here, there are 10 classes, 17 subclasses, and at least 85 orders contained within the phylum Arthropoda. In contrast, the phylum Mollusca contains 7 classes, 9 subclasses, and about 30 orders. Even more impressively, the phylum Arthropoda contains at least ten times the number of species found in the phylum Mollusca. Despite the large number

of major and minor taxonomic groupings found within the Arthropoda, most adult arthropods show only minor deviations from the general body plan. Taxonomic distinctions among arthropods depend largely on the number, distribution, embryological origin, form, and function of appendages.

Since the focus of this book is on the basic vocabulary and grammar of invertebrate zoology, rather than on the diversity of form and function encountered within each group, I will include only some of the arthropod classes and subclasses in the following discussion. Selecting the "essential" arthropod groups is not an easy task. I have been careful to include those groups having representatives that are generally familiar to the average person and/or those groups having major ecological and/or evolutionary significance. A taxonomic breakdown of arthropods follows. The groups selected for detailed discussion are indicated by *.

(a)

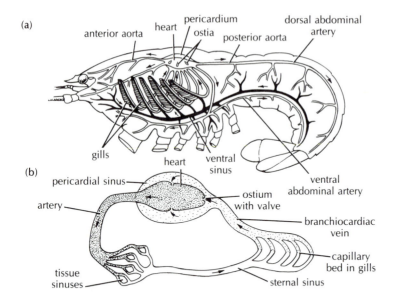

(b)

FIGURE 16.5. *(a) Circulatory system of a typical arthropod, demonstrating the heart with ostia. The animal illustrated is a lobster. (After Engemann and Hegner; after Gegenbauer.) (b) Diagrammatic illustration of the blood circulatory pattern. Oxygenated blood is transported through the unshaded vessels to hemocoelic channels, in which gas exchange between blood and tissues occurs. Deoxygenated blood (shaded vessels) collects in a series of ventral venous sinuses. From here, the blood moves to the gills, is oxygenated, and returns to the pericardial sinus surrounding the heart.* (From Hickman, Cleveland P., 1973. Biology of the Invertebrates, 2nd ed. St. Louis, MO: The C.V. Mosby Co.)

Phylum Arthropoda
  Subphylum Trilobitomorpha*
    Class Trilobita*
  Subphylum Chelicerata*
    Class Merostomata*—horseshoe crabs
    Class Arachnida*—spiders, mites, ticks, scorpions
    Class Pycnogonida—sea spiders
  Subphylum Mandibulata*
    Class Chilopoda*—centipedes
    Class Diplopoda*—millipedes
    Class Pauropoda

    Class Symphyla
    Class Insecta*
      Subclass Apterygota (A = G: without; *pterygo* = G: a wing)
    Order Protura
    Order Collembola—springtails
    Order Diplura
    Order Thysanura—bristletails, silverfish
    (The above four orders are characterized by absence of wings and lack of an abrupt metamorphosis during de-

velopment. Members of these orders most likely represent the most primitive insect condition of about 310 million years ago. Most groups of winged insects do not appear in the fossil record until at least 30 million years later, with major groups such as the Lepidoptera (moths and butterflies) and Diptera (flies) not appearing until about 150 million years ago.)

Subclass Pterygota (i.e., winged insects)

Order Orthoptera—grasshoppers, praying mantids, cockroaches, walking sticks, locusts, crickets

Order Dermaptera—earwigs

Order Plecoptera—stoneflies

Order Isoptera—termites

Order Embioptera

Order Odonata—dragonflies, damselflies

Order Ephemeroptera—mayflies

Order Mallophaga—biting lice, hair lice

Order Anoplura—sucking lice

Order Psocoptera—book lice

Order Thysanoptera—thrips

Order Zoraptera

Order Hemiptera—the true bugs

Order Homoptera—cicadas, aphids, spittle bugs, mealybugs, leafhoppers

Order Mecoptera—scorpion flies

Order Trichoptera—caddis flies

Order Neuroptera—ant lions, dobsonflies, lacewings

Order Lepidoptera—butterflies, moths, silkworms

Order Diptera—flies, mosquitos, gnats, midges

Order Siphonaptera—fleas

Order Coleoptera—beetles, weevils

Order Strepsiptera

Order Hymenoptera—ants, bees, wasps (The above orders are characterized by presence of wings, although the wings are secondarily reduced or lost in some species, as in the Siphonaptera (fleas). Members of the above orders pass through several mor-

phologically distinct developmental stages; i.e., all species show some degree of metamorphosis during ontogeny. Metamorphosis is generally associated with the transformation of a wingless developmental stage into a winged adult.)

Class Crustacea*

Subclass Cephalocarida

Subclass Branchiopoda*—brine shrimp, fairy shrimp, water fleas

Subclass Ostracoda

Subclass Copepoda*—copepods

Subclass Mystacocarida

Subclass Branchiura—fish lice

Subclass Cirripedia*—barnacles

Subclass Malacostraca*

Order Leptostraca

Order Anaspidacea

Order Bathynellacea

Order Stygocaridacea

Order Stomatopoda

Order Mysidacea—opossum shrimps

Order Cumacea

Order Tanaidacea

Order Isopoda*—pill bugs, wood lice

Order Amphipoda*—sand fleas

Order Euphausiacea—krill

Order Decapoda*—crabs, lobster, crayfish, shrimp, hermit crabs

Three classes in the above list (Arachnida, Insecta, and Crustacea) include well in excess of 95% of all arthropod species. I should point out that the classification scheme presented is only one of several that have been proposed. Although the Crustacea, Merostomata (including horseshoe crabs and spiders), and Trilobita appear to be quite closely related to each other, the nature of the relationship between these groups and the terrestrial arthropods is obscure. Some zoologists consider that members of the classes Chilopoda, Diplopoda, Symphyla, Pauropoda, and Insecta have had a distinct evolutionary origin from the Crustacea, and suggest their placement within a new taxonomic group, the Uniramia. In the given classification scheme, the Uniramia would form a separate subphylum, as would the Crustacea.[1]

I emphasize that the following discussion omits even some major taxonomic groups within the Arthropoda. However, consideration of the groups chosen should be sufficient to present the major principles and vocabulary of arthropod architecture and biology. Please note that termites are not ants, and that centipedes, millipedes, and spiders are not insects!

Subphylum Trilobitomorpha

CLASS TRILOBITA

There are no living representatives of this class, although approximately 4000 species have been described from the fossil record. The trilobites were especially common approximately 500 million years ago, but were extinct by about 250 million years ago, at the end of the Paleozoic era. Morphological differences apparent among the fossilized remains imply that a significant ecological diversity existed within the group at one time, varying from burrowing to walking and swimming forms.

The body was flattened dorso-ventrally, and divided into three sections (Fig. 16.6). Sections I and III were covered by a continuous unjointed sheet of exoskeleton (a carapace), so that the underlying metameric segmentation was not visible when viewed from the dorsal surface. A pair of **compound eyes** were found laterally on the first body section. Each eye was composed of many, elongated light-receiving chambers (**ommatidia**); hence the term "compound eye."

Adjacent to the mouth, located on the ventral surface of one of the segments of body region I, was a chitinous lip, the **labrum**. Each body segment posterior to the mouth bore a pair of 2-branched (**biramous**) appendages. The innermost branch was devoid of long setae and presumably functioned in walking. The segments of the outer branch bore long filaments. These may represent gill filaments, or may simply have been setae used for swimming, filtering food, or digging in loose substrate. This serial repetition of identical biramous appendages along the entire length of the body is clearly the primitive arthropod condition. More advanced groups show increasing specialization of appendages for specific tasks. This specialization has often involved the reduction or complete loss of one of the two branches of each primitive biramous limb.

Subphylum Chelicerata

Chelicerates are characterized by being the only group of arthropods that lack even a single pair of antennae. The first anterior

Subphylum Chelicerata
(G: CLAW)

---

1 See Topics for Discussion, No. 2.

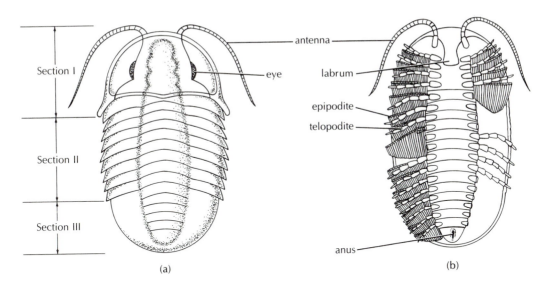

FIGURE 16.6.   *Trilobite as seen in (a) dorsal view and (b) ventral view. Note the pair of biramous appendages associated with each segment, and that the appendages are very uniform in structure all along the body. For a given appendage, the epipodite is the branch that bears the long filaments, the function of which remains uncertain. (From Beck and Braithwaite, 1968.* Invertebrate Zoology—Laboratory Workbook, *3rd ed.* Burgess Publishing Company, Minneapolis, Minnesota.)

segment bears no appendages at all. The second anterior segment bears a pair of clawed appendages (**chelicerae**) adjacent to the mouth. These appendages are used for grabbing and shredding food. Members of the Chelicerata also lack mandibles, appendages found adjacent to the mouth in many other arthropod groups and used for chewing and grinding food during ingestion.

CLASS MEROSTOMATA

This group is composed primarily of extinct species. Only four species are currently living, including the so-called horseshoe crab,

Limulus polyphemus (Fig 16.7). Note that despite the common name, these animals are not true crabs. Horseshoe crabs make their livings burrowing through the surface layers of muddy substrate and ingesting smaller animals that they come across in the process. All members of the Merostomata are marine. Curiously, living representatives are found only in the waters of the Eastern United States and Indonesia.

Characteristically, the head and thorax are fused into a single functional unit, the **prosoma** (*pro* = G: forward; *soma* = G: body), or **cephalothorax,** and are covered with a single unjointed sheet of exoskeleton (the **carapace**) (see fig. 16.7). A pair of compound

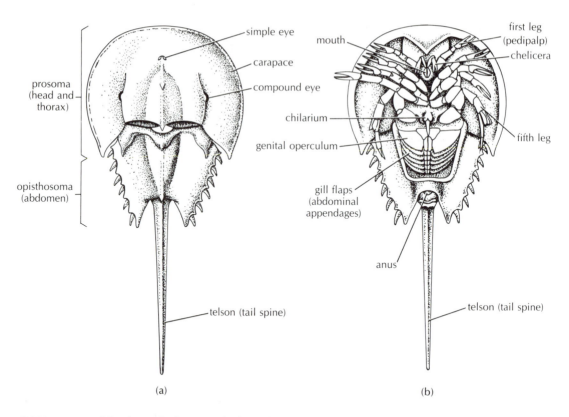

FIGURE 16.7. Limulus polyphemus, *the horseshoe crab. (a) Dorsal view, showing major divisions of the body. (b) Ventral view, showing appendages. Note that the abdominal appendages are modified as flattened sheets for gas exchange. (Modified from several sources.)*

eyes is present laterally on the dorsal surface of the prosoma.

The first pair of appendages found ventrally on the prosoma are the **chelicerae** (*cheli* = G: claw). These are followed by five pairs of like appendages, the **walking legs**, all but the last of which bear claws. The first pair of walking legs (i.e., the appendages on the third anterior segment) are called **pedipalps**, but are morphologically and functionally indistinguishable from the other walking legs.

The last pair of walking legs is structurally modified for removal of mud during burrowing. A pair of small, hairy appendages (**chilaria**) is found on the last segment of the prosoma; these may be involved in crushing food prior to ingestion.

The abdomen (**opisthosoma:** *opistho* = G: behind; *soma* = G: body) bears six pairs of appendages. The first pair are modified for reproduction, and the subsequent five pairs are modified to serve as gills. The underside of each gill flap bears approximately 150 leaf-like gas exchange surfaces, through which blood circulates.

CLASS ARACHNIDA

Although the earliest members of this group were undoubtedly marine, the nearly 50,000 living arachnid species so far described are primarily terrestrial (Fig. 16.8). Moreover, the relatively few living aquatic species are derived from terrestrial forms. The arachnids are thus clearly a group that, in the course of their evolution, have left the sea. This class includes many familiar but generally unpopular organisms, including spiders, mites, ticks, and scorpions. All species are carnivorous or parasitic.[2] The mites and ticks are of particular economic importance, acting as parasites of humans and of the plants and animals on which we depend for food. In addition, many arachnid species produce a potent venom that can, in some species, be quite toxic and even deadly to humans.

Like the Merostomata, the head and thorax of arachnids are fused to form a **prosoma,** which is covered by a carapace (Fig. 16.9). From zero to four pairs of eyes are found on the prosoma, with four pairs being most common. The anterior-most pair of appendages borne by the prosoma are **chelicerae,** which generally tear apart food prior to ingestion. The next pair of appendages are the **pedipalps,** which are variously modified for grabbing, killing, or reproducing, and in some species may have a sensory function as well. The basal segment of each pedipalp

forms a **maxilla,** which, like the chelicerae, aids in the mechanical preparation of food. The pedipalps are followed by four pairs of **walking legs.** The **abdomen,** or **opisthosoma,** is generally distinct from the prosoma; in some arachnids, including the spiders, the two divisions are connected by a narrow stalk called a **pedicel.** The presence of a pedicel increases the range of movement of the abdomen, facilitating the precise placement of silk threads in web-building and prey capture. In a few arachnid groups, notably the ticks and mites, the prosoma and opisthosoma have fused together, and the entire dorsal surface is covered by a single carapace (Fig. 16.9).

Respiration in the more primitive forms is by means of pairs of modified, internalized book gills, now known as **book lungs.** These flattened respiratory surfaces on the abdomen are connected to the outside by means of openings called **spiracles.** The spiracles of some species can be closed between "breaths," to limit water loss. In more advanced species, the spiracles may lead into a system of tubules known as **tracheae.** The tracheae form a system of branching tubules that ultimately terminate directly on the tissues. Gas exchange therefore occurs without use of the blood circulatory system. In some arachnid species, both book lungs and tracheae are present.

Some arachnids, the spiders, bear 3–4 pairs of small abdominal appendages called **spinnerets** (Fig. 16.9). These appendages are located ventrally and posteriorly, near the anus, and connect to internal glands that secrete proteins. The proteinaceous secretions of these silk glands are extruded through the spinnerets to form silk, which may be used to

---

2 See Topics for Discussion, No. 5.

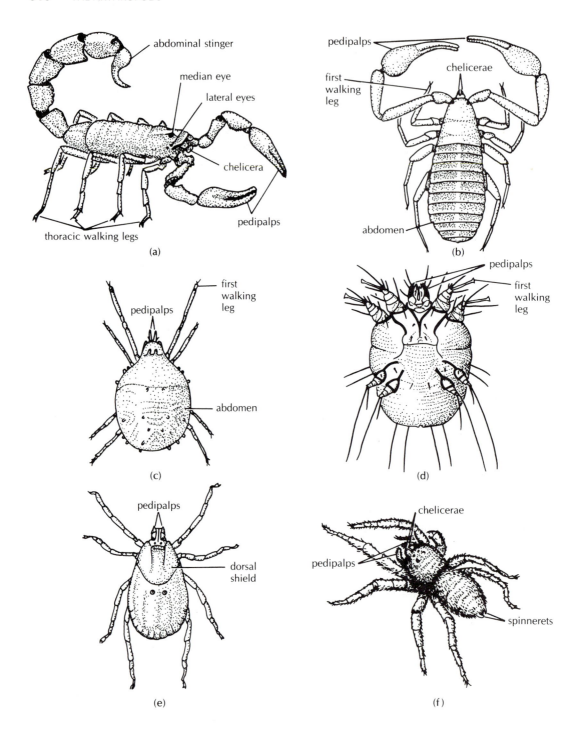

(a)

abdominal stinger

median eye

lateral eyes

chelicera

pedipalps

thoracic walking legs

(b)

pedipalps

chelicerae

first walking leg

abdomen

(c)

pedipalps

first walking leg

abdomen

(d)

pedipalps

first walking leg

(e)

pedipalps

dorsal shield

(f)

chelicerae

pedipalps

spinnerets

form safety lines during climbing, form webs for trapping prey, build homes, mate, or protect developing young.[3] Humans have also put arachnid silk to use, notably as cross hairs in optical equipment. The chemical composition of silk varies even within an individual, according to the manner in which the silk is to be used. Some insects and centipedes also house silk-producing glands. In fact, the silk used to make silk clothing comes from the cocoons of lepidopterans, especially the silkworm, *Bombyx mori.* Silkworms have been used for this purpose for more than 4000 years.

Subphylum Mandibulata

CLASSES CHILOPODA AND DIPLOPODA

The Chilopoda and Diplopoda contain the centipedes ("one hundred feet") and millipedes ("one thousand feet"), respectively. Because of their many legs (Fig. 16.10), the members of both groups are often referred to together as myriapods ("many feet"). Chilopods are generally fast-moving carnivores, living in soil, humus, under logs, and occa-

3 See Topics for Discussion, No. 6.

sionally in people's homes. Although most of the 3000 species are terrestrial, some are marine. The body is covered by a cuticle, but the cuticle is unwaxed. Moreover, respiration is accomplished by tracheae, but the spiracles cannot be closed. Thus, centipedes are generally restricted to moist environments (or moist microenvironments) because of difficulty in restricting water loss. Many species conserve water by being nocturnal, i.e., by avoiding the heat of the day and becoming active only at night.

The head bears a single pair of antennae. In addition, we find a pair of mandibles for chewing, a pair of first and second **maxillae,** and a pair of **maxillipeds.** The maxillipeds are modified for subduing prey. They contain poison glands and resemble fangs. Eyes are often lacking; when present, light receptors are generally simple ocelli.

The head is followed by 15 or more leg-bearing segments. Some species have **repugnatorial glands** on the ventral surface of each segment or on some of the legs themselves. These glands function in discouraging predation by producing an adhesive ejaculate. A number of species have spinneret glands, with which silk can be produced. Although most species of centipede are long-legged runners, some species are adapted for burrowing through soil. In these species, the legs are reduced and thrust is

FIGURE 16.8. *Arachnid diversity. (a) Scorpion. (b) Pseudoscorpion. (c) Red spider mite. (d) The human itch mite,* Sarcoptes scabei. *(e) A tick,* Dermacentor andersoni. *This tick is a vector in the transmission of Rocky Mountain Spotted Fever. (f) Spider. (After several sources.)*

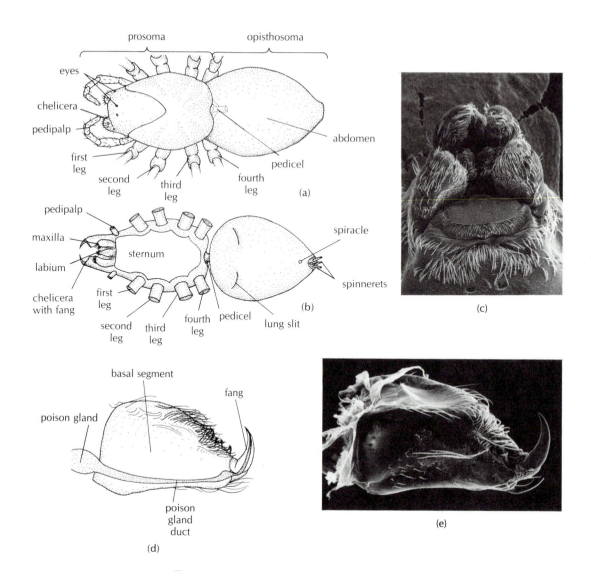

FIGURE 16.9. *(a) Typical spider, in dorsal view. (After Sherman and Sherman.) (b) Diagrammatic ventral view of spider. The legs have been removed for clarity. (After the Kastons.)*
(c) Posterior-ventral view of spider abdomen, showing three pairs of jointed, flexible spinnerets. Silk is extruded through the lumen of the hollow bristles. (Courtesy of Brent Opell.) (d) Detail of an arachnid chelicera that has become modified to form a fang. (From Foelix, 1982. Biology of Spiders. Boston: Harvard Univ. Press. Reprinted by permission.) (e) Scanning electron micrograph of the chelicera of Zoxis geniculatus. *This species belongs to one of only three families lacking poison glands; the fang is used primarily for grooming rather than prey capture. Prey are subdued by quickly and thoroughly wrapping them with silk. (Courtesy of Brent Opell.)*

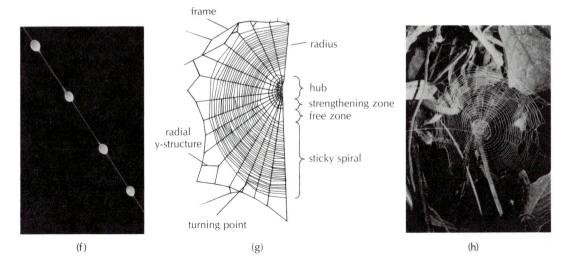

FIGURE 16.9 (cont.). *(f) A silk strand with adhesive droplets, from the spiral of an orb web built by* Mangora *sp. This silk can be stretched to nearly three times its resting length before it breaks. (Courtesy of Brent Opell.) (g) Structure of an orb web. (From Foelix, 1982.* Biology of Spiders. *Harvard University Press.) (h) Photo of an orb web. (Courtesy of Brent Opell.)*

generated by exploiting the properties of a hydrostatic skeleton, earthworm-style.

There are approximately 7500 millipede species, about twice the number of known centipede species. In contrast to the chilopods, the diplopods are primarily slow-moving, deposit-feeding animals that make their living plowing through soil and decaying organic material. Some carnivorous species also exist. Pairs of segments have become fused in the millipedes, so that each new segment (a diplosegment) bears two pairs of legs, as well as two pairs of spiracles and ventral ganglia. In most species, the integument (body covering) is impregnated with calcium salts, as in the Crustacea. The

covering of millipedes is therefore more protective against abrasion and predation than that of the centipedes. As in the centipedes, however, the cuticle is not waxy. Although many species lack eyes, as many as 80 ocelli are found on the heads of some species. As with the centipedes, compound eyes are absent. The head appendages consist of a pair of uniramous (single-branched) antennae, a pair of mandibles, and a pair of first maxillae. Second maxillae are lacking among millipedes. Most species have an abundance of repugnatorial glands, which eject a variety of toxic, repellent secretions.

Both chilopods and diplopods are generally small animals, often only a few mm, or at

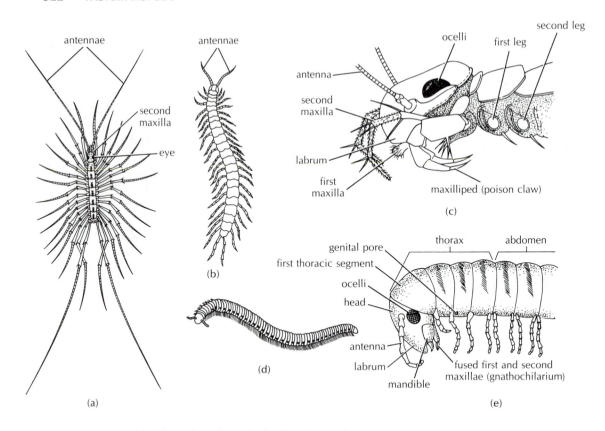

FIGURE 16.10. (a) A long-legged centipede, Scutigera coleoptrata, capable of especially rapid locomotion. Members of this species are seldom greater than 3 cm long; they are often encountered in moist areas (e.g., bathrooms) in buildings. (After Pimentel.) (b) Scolopendra sp., a tropical centipede that may grow to a length of 25 cm. (After Huxley.) (c) Detail of centipede head, Scutigera coleoptrata, in lateral view. Note the compound eye—actually a dense and organized cluster of ocelli—and the conspicuous poison claw. A poison-secreting gland is housed within the maxilliped. (After Snodgrass.) (d) Millipede, dorsal view. (After Huxley.) (e) Detail of millipede head. Note that two pairs of legs are borne by each diplosegment of the abdomen. (Reprinted with permission of Macmillan Publishing Company from The Invertebrates: Function and Form, 2nd ed., by Irwin W. Sherman and Vilia G. Sherman. Copyright ©1976 by Irwin W. Sherman and Vilia G. Sherman.)

most 1 cm in length. Some species of both groups have been reported to attain lengths of nearly 30 cm in the tropics, however! Myriapods are believed to be close relatives of the insects.

CLASS INSECTA

Insects have been reported from nearly every habitat except the deep sea. Although most species are terrestrial, many species live, either as adults or as larvae, in fresh water, or in saltwater marshes (Fig. 16.11). A few species (the ocean-striders) live on the surface waters of the open ocean. Nearly one million species of insects have been described so far, with at least three times that number of species probably awaiting description. This tremendous number of species is in large part attributable to the feeding specializations, dispersal capabilities, and predator-avoidance possibilities associated with the evolution of flight. No other invertebrates, and relatively few vertebrate species, have evolved this capability. Indeed, when insects first evolved flight, they achieved access to a life-style previously unexploited by any other organism. Their adaptive radiation was thus unhindered by competition from other groups of animals.

Insects are among the best-studied of invertebrates, in large part because of their omnipresent impact on humans. The majority of flowering plants, including many of agricultural importance, depend upon insects for pollination. The study of the mutual interdependence of plants and insects, and the evolution of these interactions, is a bur-geoning field. Insects are also significant vectors of disease (e.g., malaria, bubonic plague, typhoid fever, yellow fever) and are major threats to agriculture, especially as larvae. A few insect species (e.g., bees, wasps, and some beetles) produce secretions that are toxic to humans. Alternatively, other secretions (e.g., silk, honey) are of commercial importance. Finally, the social interactions[4] and division of labor seen among some insect groups are magnificently complex, and have long occupied the attention of animal behaviorists.

The insect body is divided into three conspicuous tagmata: head, thorax, and abdomen. The thorax generally bears three pairs of legs directed ventrally, giving rise to the alternate name applied to insects, the Hexapoda (*Hexa* = G: six; *pod* = G: foot). Two pairs of wings are generally carried dorsally. The wings are outfoldings of the thoracic integument, and consist of two thin, chitinous sheets. The legs may be modified for walking, jumping, swimming, digging, or grasping, and are generally studded with a variety of sensory receptors, including receptors for taste, smell, and touch. Such receptors are also found on the mouthparts and elsewhere on the body. The main light receptors are a pair of **compound eyes,** but single unit eyes (**ocelli**) are usually present on the head as well.

An **ocellus** is simply a small cup with a light-sensitive surface backed by light-absorbing pigment. Such simple photoreceptors are found in many phyla, including the

---

4 See Topics for Discussion, No. 10.

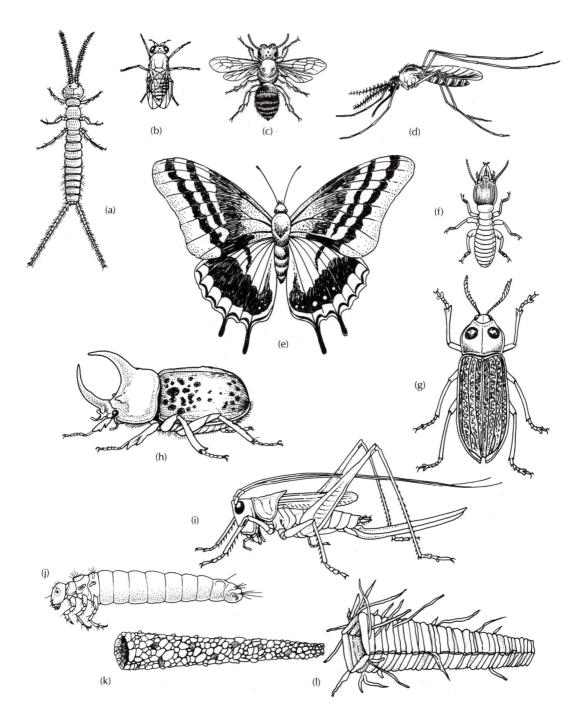

(a)

(b)

(c)

(d)

(e)

(f)

(g)

(h)

(i)

(j)

(k)

(l)

Platyhelminthes, Annelida, and Mollusca. The cup is often covered by a lens. As in all known visual systems, the photosensitive pigment of the ocellus is a vitamin A derivative in combination with a protein. Stimulus by light causes a chemical change in the photoreceptor pigment, generating action potentials, and these are then carried by nerve fibers to be interpreted elsewhere. Each insect head typically bears three ocelli. Ocelli are generally not image forming.

Compound eyes, however, are capable of forming images. For any eye to form an image, light must be focused on the receptor surfaces, and the animal must be able to examine individual components of the image independently. The animal must also possess a nervous system of sufficient sophistication to reconstruct the image detected by the sensory system. Insects satisfy all of these criteria.

In the human eye, light enters through a single lens and is focused on the retina. The components of the resulting inverted image are sampled by myriad closely packed receptor cells in the retina. Nerve impulses from the individual receptor cells are then integrated and interpreted by the brain. With a compound eye, the image is broken up into its components before it reaches the receptor cells, i.e., as it enters the eye.

The compound eye is composed of many individual units, called **ommatidia** (*ommato* = G: eye; *ium* = G: little). The compound eyes of some insect species contain many thousands of ommatidia, each oriented in a slightly different direction from the others as a result of the convex shape of the eye (Fig. 16.12). The visual field of such a multifaceted, convex eye is very wide, as anyone who has tried to surprise a fly will know. It is interesting to note that some species of polychaete annelid and bivalve mollusc have evolved photoreceptors that operate on similar principles.

Each ommatidium consists of two fixed-focus lenses (usually a 6 or 8-sided outer facet called the **cornea** and an underlying gelatinous **crystalline cone**); a series of up to eight light-sensitive bodies (**retinular cells**); collars containing shielding pigments, which can migrate within fixed tracts in response to changing light intensities; and nerve fibers to carry action potentials to the

FIGURE 16.11. *Insect diversity. (a) An apterygote.* Campodea staphylinus. *These, like the related silverfish, bristle-tails, and springtails, are primitive, wingless insects descended from wingless ancestors. (After Huxley.) (b) Fruit fly,* Drosophila sp. *(After Pimentel.) (c) Leaf-cutter bee. (d) Mosquito. (e) Butterfly. (f) Termite soldier. (After Romoser.) (g, h) Beetles. (i) Grasshopper. (After Pimentel.) (j) Caddisfly larva. (k, l) Protective cases made by the larvae of two caddisfly species. (j, k, and l from McCafferty, 1981.* Aquatic Entomology. *Science Books International.)*

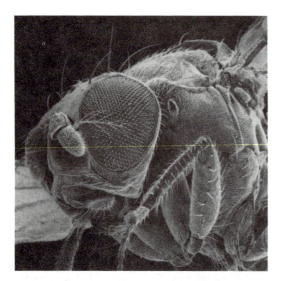

FIGURE 16.12. *Compound eye of a fly, containing hundreds of ommatidia. Note the convex shape of the eye; no two ommatidia are oriented in precisely the same direction. (Courtesy of T. Eisner, Cornell University.)*

optic ganglia for processing (Fig. 16.13). The light-sensitive pigment is actually contained within tens of thousands of **rhabdomeres**, which are fine, microvillar outfoldings of the retinular cell walls. The rhabdomeres within an ommatidium form a discrete, ordered association (often a central shaft) called a **rhabdom**. The rhabdom, in other words, contains the light-sensitive pigment of the retinular cells. In addition, the rhabdom includes a nonpigmented, central area in some species. The collars of the ommatidium function as an iris; pigment migration within these collars is under hormonal control.

The sharpness of the image formed by a compound eye depends upon a number of factors: the extent to which the light impinging on the rhabdomeres of a single om-

matidium enters along a pathway parallel to the optic axis (i.e., the long axis) of that ommatidium (increased resolution); the extent to which light from adjacent ommatidia impinges upon the receptor pigment of an ommatidium (decreased resolution); the amount of difference in direction in which adjacent ommatidia are oriented (decreased angle gives increased resolution); the number of ommatidia per eye (increased number gives increased resolving potential); and the complexity of the information center (i.e., brain) receiving and processing the impulses sent from the ommatidia.

The insect head bears four pairs of appendages: one pair of antennae (which are always uniramous, i.e., single-branched), and three pairs of mouthparts (Fig. 16.14). In sequence, the mouthparts are the **mandibles,** the **maxillae,** and finally a pair of **second maxillae** that have fused to form a single appendage called the **labium.** The mandibles are shielded anteriorly by a downward extension of the head called the **labrum.** The precise morphology of these mouthparts varies considerably according to the feeding biology of the insect (Fig. 16.15). The abdomen lacks appendages, except for a pair of sensory **cerci** borne on the last abdominal segment. The abdomen may also house receptors that monitor the degree to which the body wall is stretched during feeding.

In keeping with a largely terrestrial lifestyle, the gas exchange surfaces of insects have been internalized. Gas exchange is generally accomplished by means of a **tracheal system** (Fig. 16.16). Although resembling the tracheal system found in more advanced arachnids, the insect tracheal system is

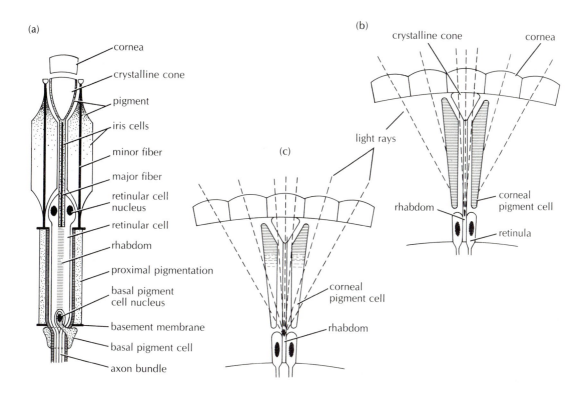

(a)

cornea
crystalline cone
pigment
iris cells
minor fiber
major fiber
retinular cell nucleus
retinular cell
rhabdom
proximal pigmentation
basal pigment cell nucleus
basement membrane
basal pigment cell
axon bundle

(b)

crystalline cone
cornea
light rays
corneal pigment cell
rhabdom
retinula

(c)

corneal pigment cell
rhabdom

FIGURE 16.13. (a) Structure of a single ommatidium in a butterfly. The light-sensitive pigment is found in the rhabdom. (Modified from S.L. Swihart, 1969. J. Insect Physiol., 15: 1347.) (b, c) A compound eye in the (b) light-adapted condition and (c) dark-adapted condition. Note that in the dark-adapted condition, light entering through the lenses of several adjacent ommatidia impinges upon a single rhabdom. Migration of pigment within the pigment collars prevents this from happening in the light-adapted eye, improving visual acuity and directional sensitivity. (Reprinted with permission of Macmillan Publishing Company from The Science of Entomology by W.S. Romoser. Copyright ©1973 by W.S. Romoser.)

thought to have been independently evolved; i.e., the tracheal systems in the two groups are convergent, evolving independently in different ancestors. One pair of **spiracles** opening into the tracheal system is found on the thorax, with additional pairs of spiracles located on many of the abdominal segments. The spiracles of most species can be closed, deterring evaporative loss of water. The tracheae are lined by cuticle, which is shed and

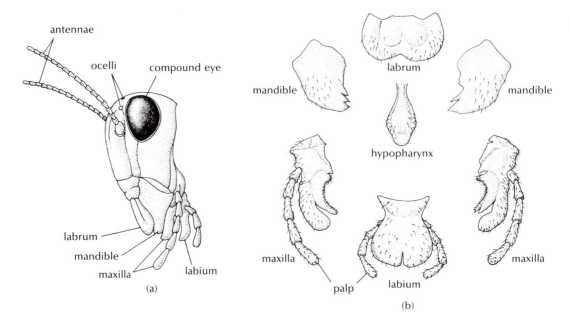

FIGURE 16.14. *(a) Insect head, showing the various appendages and eyes. (After Snodgrass.) (b) Detail of appendages, drawn in proper orientation viewed anteriorly. The morphology of the mouthparts differs widely among species, and correlates with feeding biology. In butterflies and moths, for example, the maxillae are greatly elongated for taking up nectar. (After James and Harwood.)*

resecreted by the underlying epidermis each time the animal molts. The tracheal tubules are kept from collapsing by means of chitinous rings embedded in the walls. The tracheae branch to form a network of smaller tubules called **tracheoles,** which are less than 1 μm in diameter. These branch again and terminate directly on the tissues of the animal. Thus, gas exchange between the tissues and the environment is accomplished directly, without the involvement of the blood circulatory system. Some insect species lack tracheae, either as adults or during development. Gas exchange in such animals must occur across general body surfaces. Such surfaces obviously cannot be waxy, and the animals are thus restricted to life in moist habitats.

Water conservation is another correlate of a terrestrial life-style.[5] Uric acid is the primary end product of protein metabolism

5 See Topics for Discussion, No. 3.

FIGURE 16.15. *Scanning electron micrograph of an insect head,* Manduca sexta, *the tobacco hornworm. (Courtesy of N. Milburn.)*

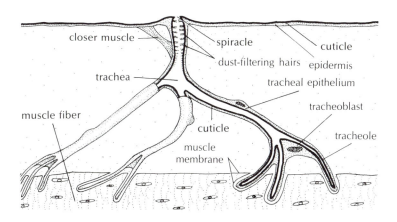

FIGURE 16.16. *The insect tracheal system. The spiracles of terrestrial insects can generally be closed by contraction of appropriate musculature, to regulate water loss. The finest tubes of the tracheal system, the tracheoles, develop from the tracheoblast cell. (After Chapman; after Meglitsch.)*

among insects; this nontoxic nitrogenous compound is excreted in nearly dry, solid form. The major excretory organs are called **Malpighian tubules** (Fig. 16.17), of which up to 250 pairs are found in the insect hemocoel. Waste products, notably a soluble derivative of uric acid, are actively transported from the blood into the distal portion of the Malphigian tubules. Increased acidity in the proximal portion of the tubules causes the uric acid to precipitate out of solution. Most of the water contained in the urine is then resorbed during its passage through the rectum.

One feature that sets the insects apart from all other invertebrates, and most vertebrates as well, is the ability to fly. The insect wing is a lateral outfolding of the body wall. These outfoldings are very thin and lightweight, and are structurally supported by a characteristic network of veins connecting with the blood circulatory system; by tracheal tubules; and by rows of pleats radiating outward from the base of the wings to the tips. The evolution of wings has opened up to insects a life-style that is virtually inaccessible to other animals.

Flight requires the generation of both lift and thrust. Thrust is something we have an intuitive feel for; it is the force we exert in one direction that creates motion in the opposite direction (every action produces an equal and opposite reaction). The generation of lift is more mysterious. In fact, I hesitate to discuss it, for fear that you'll never want to fly again. The secret is contained in the following equation, modified from the original of Daniel Bernoulli (1700–1782). When dealing with flight, the equation pertains to air moving across a solid surface, such as a wing.

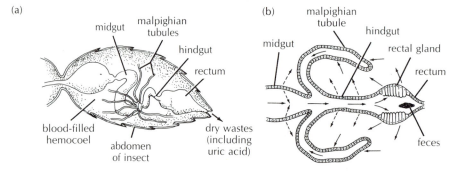

FIGURE 16.17. *(a) Malpighian tubules in the insect abdomen. (From Purves and Orians, 1983. Life: The Science of Biology. Sinaner Associates/Willard Grant Press.) (b) Diagrammatic illustration of the relationship between the Malpighian tubules and the posterior portion of the digestive tract. Fluid moves from the hemocoel into the tubules, where it joins wastes moving toward the anus. The arrows indicate the extensive reclamation of water that occurs in the hindgut and rectum (After Wilson; after Wigglesworth.)*

$$\frac{1}{2}\,dv^2 + p + dgh = \text{a constant}$$

where d = density of the air; v^2 = the square of the velocity of the air relative to the wing; p = air pressure at the surface of the wing; g = the gravitational constant; and h = the height of the air above the wing surface. The term dgh is related to potential energy and the $1/2\,dv^2$ term is related to the expression for kinetic energy. When you fly in an airplane, the principles embodied within this equation are what keep you aloft. In this situation, the air we are talking about does not change density and is always the same height above the wing, so d and dgh are constants. The equation states, then, that if the velocity of the air moving across the wing increases, the pressure above the wing surface must decrease; this must occur if the sum total of the three expressions on the left side of the equation is to remain constant. This pressure decrease above the wing produces lift; that is, the pressure below exceeds the pressure above, and the body rises. You can prove to yourself that lift is generated by differential air flow over the upper and lower surfaces of an object by blowing along the length of a strip of paper. Try a strip about 1″ wide and 6–8″ long, holding one end of the paper just below your mouth. If you blow hard enough along the length of the strip, the end of the paper will rise.

Obviously, an insect does not blow over its wings to generate lift. Neither does the propeller or jet engine of an airplane function by blowing air over the upper surfaces of the wings. Instead, the beating of an insect wing, like the spinning of a propeller, serves to move the wing through the air. Wing beat frequencies vary from less than 10 to more than 1000 beats per second in different insect species. The wings are shaped so that when air moves over their surfaces, the air moving over the upper surface of the wing has to travel farther to reach the back of the wing than the air moving over the lower surface. Such a surface is termed an **airfoil** (Fig. 16.18). Since the entire wing moves a given distance per unit of time, the shape of the

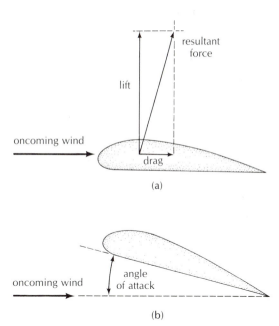

FIGURE 16.18. *(a) Diagrammatic illustration of an insect wing in cross section. The shape illustrated is called an airfoil. As air travels more rapidly over the upper surface of the wing, a net lift is produced, as illustrated. (b) Flight direction can be altered by changing the angle of attack. (From Vogel, 1981.* Life in Moving Fluids. *Willard Grant Press.)*

wing ensures that the air moves faster across the upper surface than across the lower surface, generating lift according to Bernoulli's principle. By varying the angle at which the wing moves through the air, the direction of the net forces generated can be altered by the insect, producing movement in the direction of choice (even backwards in many species).

Most insects have two pairs of wings, although many species have only a single pair. Insect species differ considerably with respect to wing morphology and the manner in which the wings are operated by the thoracic musculature. Much of the modification in wing structure and function encountered among different groups of insects seems to reflect selection for increased energy efficiency and increased fine directional control during flight. Many insects can hover, and even fly backward, a great advantage for mating and egg laying "on the wing." Other morphological modifications serve to protect the insect body or the wings themselves.

Although I will not here go into the details of insect flight mechanics,[6] it is worthwhile to list the major characteristics that make flight possible in the Insecta:

(1) abundance of striated muscle specialized for rapid, strong contractions;
(2) muscle antagonism by means of a lightweight, jointed skeleton, permitting a great amount of movement to be generated from relatively short changes in muscle length;
(3) small body size;
(4) water-impermeable outer body covering, preventing dehydration;
(5) efficient systems for exchange of gases and for storage and distribution of nutrients to the musculature;

(6) highly developed nervous and sensory systems for steering, navigating, sensing wind direction, and locating food and mates in flight.

In many of the faster flying insects, the flight muscle is highly specialized. In these species the muscle fibers are capable of contracting many times following stimulation by a single nerve impulse. This is termed **asynchronous flight**, since wing beat frequency does not correspond to the frequency of nerve impulse generation. The wing beat frequencies of over 1000 per second that have been recorded in mosquitos are made possible by this unique insect invention.

CLASS CRUSTACEA

Subclass Malacostraca

This subclass contains nearly 75% of all described crustacean species. Malacostracans are among the most familiar of crustaceans, and include such forms as the crabs, hermit crabs, shrimp, and lobsters. The basic body plan is tripartite, consisting of a **head, thorax,** and **abdomen** (Fig. 16.19). The head and thorax may be covered by a **carapace,** extending posteriorly from the head, and therefore may function as a single unit, the **cephalothorax.** In some species, the carapace bears a prominent anterior projection called

Subclass Malaco•straca
(G: SOFT A SHELL)

---

6 See Topics for Discussion, No. 7.

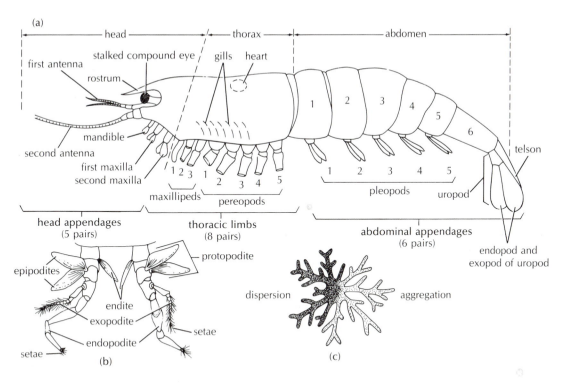

FIGURE 16.19. *(a) General external anatomy of a crustacean, showing the head, thorax, abdomen, and associated appendages. (Modified after Russell-Hunter and other sources.) (b) Illustration of biramous appendages. (c) Patterns of pigment dispersion in crustacean chromatophores. In the dispersed configuration, the cuticle becomes dark; pigment aggregation within the chromatophores causes the cuticle to become lighter in color. Movement of pigment within the chromatophores is under hormonal control in response to changing light intensities. (From Weber, 1983. Amer. Zool., 23: 495.)*

the **rostrum.** Large, stalked **compound eyes** are conspicuous, as are two pairs of head appendages, the first and second **antennae.** Often in the zoological literature, the first pair of antennae are known as the "antennules," and the second pair of antennae is simply referred to as the antennae. In malacostracans, both pairs of antennae are

primarily sensory. In other groups, the second antennae may also play roles in feeding, locomotion, and mating. In addition to the two pairs of antennae, the malacostracan head bears three pairs of smaller appendages, which are involved in feeding or in the generation of respiratory currents. These appendages are, in sequence beginning from the mouth and moving posteriorly: the **mandibles** (crush food), and the first and second **maxillae** (generate water currents and manipulate food). The next eight segments of the cephalothorax are thoracic segments, commonly bearing in sequence the first, second, and third **maxillipeds** (food manipulation) and five pairs of thoracic **walking legs,** commonly known as **pereopods.** The first three pairs of pereopods may be chelate (claw-bearing), in which case they also function in feeding and in defense.

Each of the six abdominal segments bears a pair of appendages as well. The first five pairs of abdominal appendages are referred to as **pleopods;** these function primarily in the generation of respiratory currents, and, in females, in the brooding of eggs and developing young. The last pair of abdominal appendages are the **uropods.** These flat appendages lie on either side of the telson, forming a tail.

Malacostracan appendages are generally **biramous;** i.e., they have two branches (*bi* = L: two; *rami* = L: branch). The portion of the limb proximal to the branch point is the **protopodite** (*proto* = G: first). The inner and outer branches are the **endopodite** and **exopodite,** respectively (Fig. 16.19b). The expodite is often less well developed than is the endopodite. Frequently, lateral protuberances occur on the protopodites themselves. These are termed **endites** or **epipodites,** depending on whether they project inward or outward, respectively. Epipodites commonly function as gills. Some crustacean appendages no longer have both the endopodite and exopodite, and are, therefore, **uniramous** (one-branched) append-

ages. The first antennae and the maxillae of lobster, for example, are biramous, whereas the second antennae and the thoracic appendages (**pereopods**) are uniramous. The abdominal appendages are biramous, a uniquely malacostracan characteristic.

The body surface of many malacostracan species is covered with **chromatophores** (Fig. 16.19c). Chromatophores are highly branched cells containing pigment granules. The pigments come in a variety of colors, including red, black, yellow, and blue. More than one pigment may be found within a single chromatophore, and the distribution of pigments differs among the chromatophores of a single animal. By varying the distribution of pigments in the different chromatophores, the animal can alter the color of its body considerably. The migration of pigment granules within chromatophores is under hormonal control.[7] The hormones are manufactured by the so-called X-organ located in the eyestalks, and transported a short distance to the **sinus gland** for storage. From here, the hormone is transported as needed through the bloodstream. Because chromatophore operation is under hormonal control rather than under direct nervous control, arthropod color changes never occur as rapidly as do those of cephalopods.

The above description of a typical malacostracan is most applicable to members of the order Decapoda (*deca* = G: ten; *pod* = G: foot; in reference to the total number of legs on the thorax plus abdomen), which includes about 8500 species and is the

---

7 See Topics for Discussion, No. 13.

largest of the malacostracan orders. All of the following are decapods: lobsters, crayfish, hermit crabs, true crabs, shrimp (Fig. 16.20 a–b).

There are only about one-half as many species in the order Isopoda. Most isopods are marine, although both freshwater and terrestrial species (including the familiar "pill bugs" or "sow bugs") occur. Unlike the decapods, isopods have no carapace. Moreover, they have only a single pair of maxillipeds, in contrast to the three pairs found in decapods, and have uniramous first antennae as opposed to the biramous first antennae characteristic of decapods. Isopods tend to be small, about 0.5–3 cm in length. Compound eyes, if present at all, are not on moveable stalks. Isopods are characteristically flattened dorso-ventrally (Fig. 16.20e). Gas exchange in isopods is accomplished by means of flattened pleopods. Thus, respiratory appendages are associated with the abdomen. A number of terrestrial isopod species possess a system of tracheae. This is another example of convergent evolution, in which two or more groups of animals have independently evolved similar adaptations in response to similar selective pressures. In this case, as in the insects and arachnids, selection has favored internalization of the respiratory system as a means of deterring water loss in a terrestrial environment.

In contrast to the isopods, members of the order Amphipoda tend to be flattened laterally (Fig. 16.20g, h). Again, no carapace is present; individuals possess only a single pair of maxillipeds; and the compound eyes are sessile. More than 5000 species of amphipods have been described, mostly from salt water.

However, both freshwater and terrestrial species exist. In contrast to the isopods, amphipod gills are found on the thorax, attached to the pereopods. In some species, a portion of the protopodite (the **coxa**) of each of several pairs of anterior appendages is elaborated into a large, flattened sheet, contributing significantly to the flattened appearance of the amphipod body.

Subclass Branchiopoda

Branchiopods are a diverse group of small, primarily freshwater crustaceans. On each thoracic appendage, the **coxa,** one of the basal segments of the crustacean appendage, is modified to form a large, flattened paddle; this paddle functions in gas exchange and locomotion, giving rise to the name of the subclass (*branchio* = G: a gill; *pod* = G: foot). Most species are filter-feeders, although a few species are carnivorous. The bodies of most branchiopods are at least partially enclosed in a bivalved carapace (Fig. 16.21a–c). However, some species (the fairy shrimp) completely lack a carapace (Fig. 16.21d). This group includes the well-known brine shrimp (*Artemia salina*), which are found in waters whose salinities range from about 1/10 to 10 times the salt concentration of open-ocean sea water. The fertilized eggs of this species are commonly marketed as "sea monkeys."

> ### *Subclass Branchio•poda*
> (G: GILL FOOT)

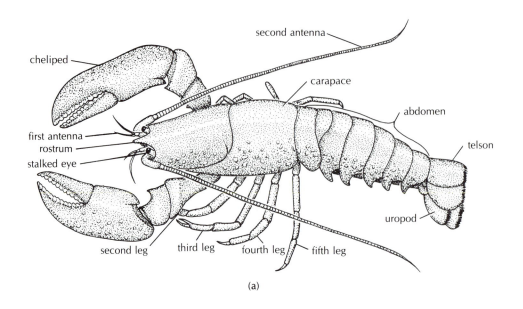

(a)

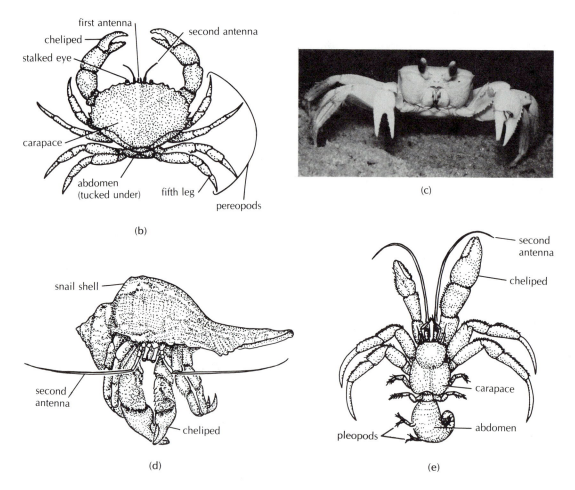

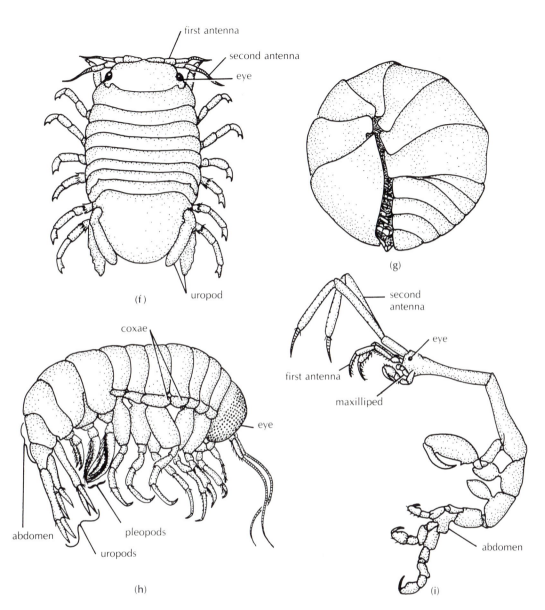

first antenna

second antenna

eye

(f)

uropod

(g)

second antenna

eye

first antenna

maxilliped

abdomen

(i)

coxae

eye

abdomen

pleopods

uropods

(h)

FIGURE 16.20. *Malacostracan diversity. (a) A lobster,* Homarus americanus. *(Modified after several sources.) (b–e) Various crabs, including (d) a hermit crab in an empty gastropod shell and (e) the same individual removed from the shell. Note that in other crabs, the abdomen is tucked up under the rest of the body. (Photo courtesy of W. Lang.) (f–g) Isopods: (f)* Sphaeroma quadridentatum. *Note the dorso-ventral flattening of the body, the uniramous antennae, and the absence of a carapace. The eyes are not stalked. (After*

Harger.) (g) Terrestrial isopod (pillbug), curled to form a ball. (After Pimentel.) (h–i) Amphipods. (h) Hyperia gaudichaudii *in side view. Note the lateral compression of the body, the large, unstalked compound eyes (only one is shown), and the absence of a carapace. Some species reach 90 mm in length, although most are less than 1 cm. (After Stebbing.) (i)* Caprella equilibra, *a species highly modified for clinging to algae, hydroids, and other substrates. Caprellid amphipods range between 1 and 32 mm in length. (After Light.)*

(a)

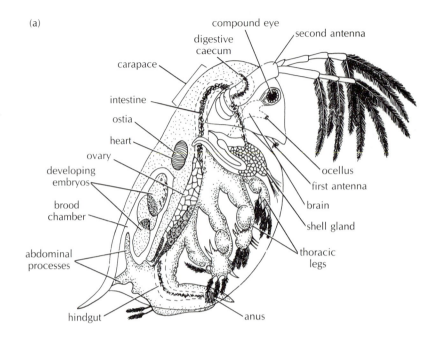

compound eye
digestive caecum
second antenna
carapace
intestine
ostia
heart
ovary
developing embryos
brood chamber
ocellus
first antenna
brain
shell gland
abdominal processes
thoracic legs
hindgut
anus

(b)

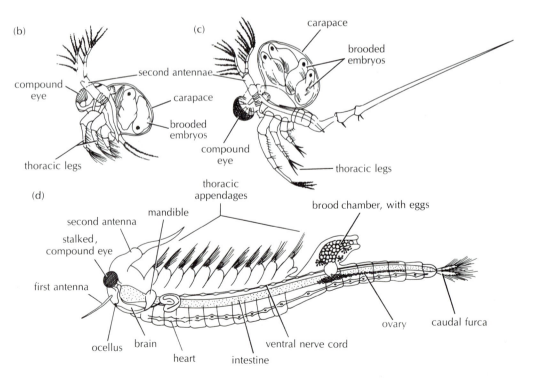

compound eye
second antennae
carapace
brooded embryos
thoracic legs

(c)

carapace
brooded embryos
compound eye
thoracic legs

(d)

thoracic appendages
mandible
brood chamber, with eggs
second antenna
stalked, compound eye
first antenna
ocellus
brain
heart
intestine
ventral nerve cord
ovary
caudal furca

Of the approximately 1000 branchiopod species so far described, at least 50% are contained within the order Cladocera, the water fleas. Cladocerans, including the familiar genus *Daphnia*, dominate the zooplankton of freshwater lakes. A few species are marine. All species are microscopic. Most of the body is contained within a bivalved carapace. Protruding from this carapace is the head, bearing a pair of large, biramous second antennae that are used to propel the animal in the water. The first antennae and the second maxillae are much reduced. The thoracic region bears 5–6 pairs of appendages, which generate feeding and respiratory currents. Food particles are filtered from the water by fine setae on these appendages. There are no abdominal appendages. In addition to the second antennae, the head also bears a single huge compound eye, formed by the fusion of the ancestral compound eyes from each side of the head. The eye is not stalked, but can be rotated in various directions by associated musculature. Other branchiopod species (tadpole shrimp) completely lack a second antenna; only a vestige remains. Clearly, the body plan of branchiopods is rather plastic.

Subclass Copepoda

There are at least 7500 species in the subclass Copepoda, most of which are marine and most of which feed on unicellular, free-floating plants (**phytoplankton;** *phyto* = G: plant; *plankton* = G: that which is forced to wander). Copepods are invariably small, usually less than 1–2 mm long. Some copepod species occur in freshwater lakes and ponds, and there are terrestrial species living in soil or in moist surface films in humid environments. Perhaps two-thirds of all copepod species are planktonic in the ocean. Because of their small size, such copepods are, to a

Subclass Cope•poda
(G: OAR FOOT)

FIGURE 16.21. *Branchiopod diversity. (a)* Daphnia pulex *(female), the freshwater water flea. Except for the head, most of the body lies within a laterally flattened, bivalve-like carapace. The long, biramous second antennae are the primary locomotory appendages of cladocerans. The compound eyes are paired and sessile (not stalked). (After Pennak; after Claus.) Other cladocerans: (b)* Podon intermedius, *and (c)* Bythotrephes longimanus. *The carapace is greatly reduced in size, serving primarily as a brood chamber. Cladocerans range from a few hundred microns to nearly 2 cm in length. (From Hutchinson, 1967.* A Treatise on Limnology, *vol. 2. John Wiley & Sons.) (d)* Artemia salina, *the brine shrimp, in normal swimming orientation. A pair of stalked, compound eyes are present. Brine shrimp may attain lengths of about 1 cm. (After Brown; after Lochhead.)*

large extent, at the mercy of currents. Together with other animals of limited locomotory capability (relative to movement of water currents), these copepods form a major component of the **zooplankton** (*zoo* = G: animal). Most other copepod species are specialized for life in or on substrates, forming a major component of the **meiobenthos** (i.e., the community of small animals living in association with sediment). Locomotion of planktonic copepods is accomplished primarily by the actions of a pair of biramous second antennae. Benthic species use their thoracic appendages to walk over surfaces.

In terms of numbers of individuals, copepods are among the most abundant animals on earth, and are among the most important herbivores of the ocean. In part, they collect phytoplankton through the activities of the first and second maxillae (Fig. 16.22a), although the details of food capture are complex and incompletely understood.[8] Copepods are at the base of the food chain in the ocean in another respect as well: they are a major source of food for primary carnivores.

Most free-living copepods have a single,

median eye on the head. The eye usually consists of three lens-bearing ocelli; two of the units look forward and upward, and the third ocellus is directed downward. Yet, exceptions do exist; some species have a pair of eyes, placed laterally and with conspicuous lenses, while other species lack eyes entirely. Compound eyes are never encountered among the Copepoda.

In contrast to many other crustaceans, copepods lack gills and abdominal appendages. The other tagma (head and thorax) do bear appendages (Fig. 16.23). The structure and function of copepod appendages vary substantially with species, life-style, and, often, with sex as well. One or both first antennae may be hinged in the male (Fig. 16.22), functioning in the capture of females for mating, in addition to the standard sensory function. In addition, one or both fifth thoracic appendages of the male may terminate in a claw (Fig. 16.22), which is used to hold the female during mating.

Although a free-living, suspension-feeding existence is the norm for planktonic copepods, a number of planktonic species are carnivorous. Many benthic species, living on sediment or in the spaces between sand grains, scrape off food particles from the sub-

---

8 See Topics for Discussion, No. 12.

FIGURE 16.22. *Diversity among the free-living copepods.*
(a) Euchaeta prestandreae, *a typical, actively swimming copepod.* (After McConnaughey and Zottoli; after Brady, 1883. Challenger Reports, Vol. 8.) *The fifth right thoracic legs of a female (1) and a male (2) copepod,* Centropages typicus, *illustrating the clawed appendage of the male. The right first antenna of the male (3) illustrates the hinge.* (From C. B. Wilson, 1932. The Copepods of the Woods Hole Region. Smithsonian Institution Bulletin 158.) (b) *Free-living copepod carrying two clusters of fertilized eggs on the abdomen.* (From Pimentel.)

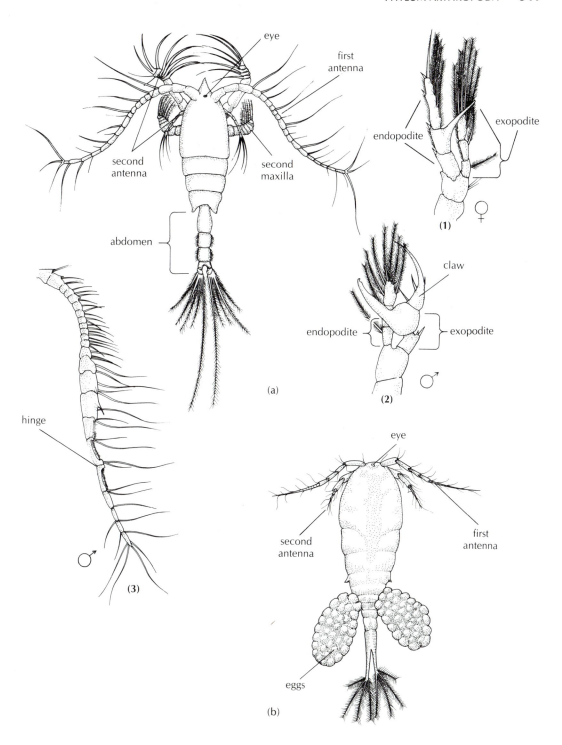

eye

first antenna

second antenna

second maxilla

abdomen

endopodite

exopodite

♀

(1)

claw

endopodite

exopodite

♂

(2)

(a)

hinge

♂

(3)

eye

second antenna

first antenna

eggs

(b)

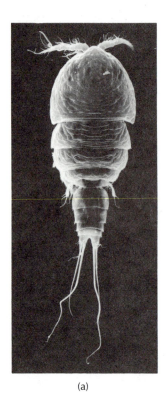

(a)

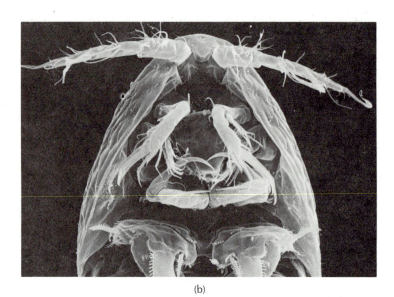

(b)

FIGURE 16.23. *(a) Scanning electron micrograph of a free-living cyclopoid copepod in dorsal view. The total length is about one mm. (b) Ventral view of same animal, showing head appendages. (Courtesy of C. Bradford Calloway.)*

strate. Still other copepods, about 25% of all described species, are parasites on a variety of vertebrates and invertebrates. As you might expect, the head appendages and other body parts are modified, sometimes extravagantly, as reflections of these different life-styles (Fig. 16.24).

Subclass Cirripedia

Cirripedes are more commonly billed as the barnacles. The approximately 1000 species in this subclass are exclusively marine, and

> *Subclass Cirri•pedia*
> L: HAIRY FOOT

show a greater departure from the basic crustacean body plan than the members of any other subclass. Unlike most other crustaceans, cirripedes are exclusively sedentary organisms, permanently affixed to, or burrowed into, living (including whales) or nonliving substrates. Barnacles attached to moving or floating substrates are often conspicuously stalked (Fig. 16.25b). Other species may be cemented directly to the substrate at the "basis." In keeping with a non-motile existence, the head is greatly reduced, the first antennae are much reduced, and the second antennae are completely absent. Most species live within a thick calcium carbonate protective shell (Fig. 16.25), which they secrete. Because of this shell, barnacles were classified as molluscs until about 150 years ago. Indeed, the barnacle shell is se-

342

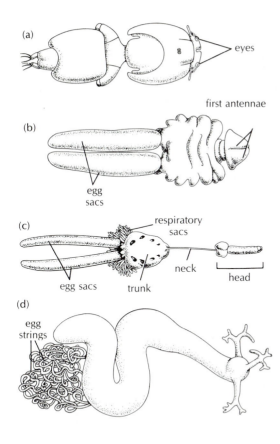

(a)

eyes

first antennae

(b)

egg
sacs

(c)

respiratory
sacs

egg sacs   trunk   neck   head

(d)

egg
strings

FIGURE 16.24. *Parasitic copepods. (a) Caligus curtus, female, found on the outside surface of various marine fish, including cod, pollock, and halibut. Total length is 8–12 mm. (b) Chondracanthopsis nodosus, female, taken from the gills of redfish. The body is about 7 mm long. (c) Rebelula bouvieri, female, taken from the flesh of a marine fish. The egg sacs alone are 30–40 mm long. (d) Lernaea branchialis, a parasite in the gills of flounder. The body of the copepod may reach 40 mm in length, while the egg strings may be several hundred millimeters long. (All from C.B. Wilson, 1932. The Copepods of the Woods Hole Region. Smithsonian Institution Bulletin 158.)*

Thus, the scuta and terga can occlude the opening at the top of the shell when the animal is withdrawn, and move apart when the feeding appendages are to be protruded. The feeding appendages, called **cirri,** are modified thoracic appendages (Fig. 16.25a). The food of free-living barnacles is collected by filtering particles from the water, using the cirri. Barnacles lack abdominal segments, gills, and a heart; the blood is circulated through sinuses entirely by movements of the body. The circulatory system is an open one, as it is in all arthropods.

A number of cirripedes live in burrows within calcareous substrates (e.g., shell, coral), or as parasites within the bodies of other animals, including other crustaceans. As might be expected, such specializations in habitat and life-style are reflected by major morphological modifications, particularly by the presence of root-like absorptive structures (Fig. 16.25c,d).[9]

creted by the "mantle" tissue, and the space between the inner wall of the shell and the animal itself is called the "mantle cavity" as in molluscs.

Barnacles have attracted a reasonable amount of attention as foulers of ship bottoms; even a moderate encrustation of barnacles can cause a great reduction in ship speed and fuel efficiency. The happy consequence is that research on the biology of cirripedes has not been difficult to justify.

The shell of barnacles is composed of numerous plates, including the **carina, rostrum, scuta,** and **terga.** The rostrum represents the side of the shell at which the body of the barnacle attaches to the mantle. Certain of the scuta and the terga are movable.

9 See Topics for Discussion, No. 5.

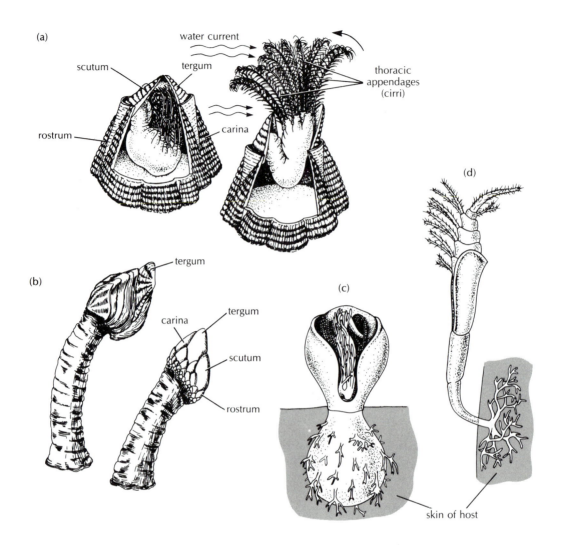

FIGURE 16.25. *(a) The free-living barnacle* Balanus *sp., an "acorn barnacle." The scutum and tergum of the barnacle shell can move apart, allowing the animal to extend its cirri in the water currents and filter out food particles. The scuta and terga form a type of operculum, protecting the animal from predators, salinity stress, and desiccation stress once the thoracic appendages have been withdrawn. (From Wells, 1968. Lower Animals. McGraw-Hill. Reproduced with permission.) (b) Stalked gooseneck barnacles:* Lepas *sp. on the left and* Pollicipes *sp. on the right. Barnacles encrust a variety of substrates, including ship bottoms, turtle shells, and whales. (After Pimentel.) (c, d) Parasitic barnacles, shown in the skin of a shark (c) and an annelid (d) . (From Baer, 1951. Ecology of Animal Parasites. University of Illinois Press.)*

OTHER FEATURES OF ARTHROPOD BIOLOGY

1. Reproduction and Development

Sexual reproduction is the rule among arthropods. Fertilization is internal in most species, but is external in some. Most species are **dioecious** (i.e., have separate sexes). However, there are exceptions. Sedentary and/or parasitic arthropod species are often hermaphroditic, and some free-living species exhibit various degrees of asexual reproduction. **Parthenogenesis,** i.e., production of offspring from unfertilized eggs, is commonly encountered among the Insecta and Branchiopoda, and in some freshwater copepods. Indeed, males have never been found in some species of these groups.

Marine species often have a free-living larval stage in the life history (Fig. 16.26). A **nauplius** larva is typical of several diverse groups of crustaceans, including the copepods, branchiopods, and cirripedes. Although the adult body plan of barnacles has become highly modified from the basic crustacean pattern, the typical nauplius larva remains unchanged. Even highly modified parasitic species produce typical nauplius larvae. In all species, the nauplius has a characteristic triangular shape, a good crustacean carapace, and a single median eye composed of three ocelli. Periodically, the nauplius larva molts, and subsequently adds or modifies appendages and gets larger. After going through several naupliar stages, the nauplius metamorphoses into a morphologically distinct larval stage. Among the Copepoda, the final naupliar stage undergoes a transition to a **copepodite** form. The individual goes through five copepodite stages before the final adult body plan is attained. Among the barnacles, the larva proceeds through several naupliar stages and then metamorphoses into a remarkable **cypris** stage. The cypris larva, or cyprid, is housed in a bivalved, noncalcified carapace. The thoracic appendages with which the larva swims become the filtering appendages of the adult barnacle. A pair of sensory antennae are conspicuous at the anterior end of the animal. When a suitable substrate is located by the cyprid, it secretes a glue from anterior cement glands and the animal then becomes permanently attached to the substrate by its head. Dramatic internal reorganization and production of the adult body enclosure ensues.

Not all crustaceans produce nauplius larvae. Decapod crabs, for instance, produce larvae called **zoeae** (Fig. 16.26d). The zoeal stages are characterized by a pair of large compound eyes and a carapace from which two spines project ventrally and dorsally. Decapods typically go through a number of zoeal stages before metamorphosing into a **megalopa** stage, which looks much like a crab except that the abdomen is not tucked up under the thorax. Other crustacean groups produce other types of larval stages, always recognizable as being arthropod. Freshwater crustaceans often lack free-living larval stages, probably for reasons discussed in the first chapter; freshwater copepods, however, often do have free-swimming naupliar stages in the life history.

A number of marine groups, i.e., isopods and amphipods, also lack larval stages.

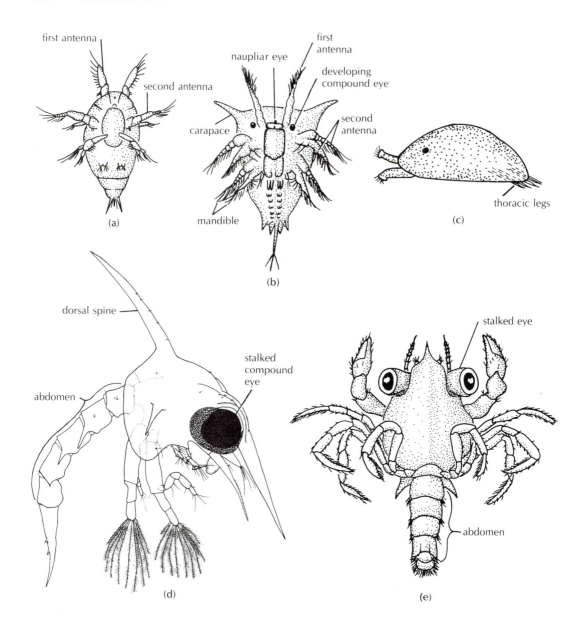

FIGURE 16.26. *Crustacean larval stages. (a) Copepod nauplius. (b) Barnacle nauplius. (After Korschelt.) (c) Barnacle cyprid. (d) Decapod zoea: Portunis sayi, stage IV. (Courtesy of I.P. Williams.) (e) Decapod megalopa larva. (After Hardy.)*

Young emerge as miniatures of the adult, after a period of protection by the adult female. The lack of free-living larval stages may be a pre-adaptation for a terrestrial existence. Not surprisingly, isopods and amphipods are the only groups of malacostracans to have accomplished major radiations into terrestrial or semiterrestrial habitats.

Although internal fertilization is sporadically distributed among marine arthropods, it is the rule among terrestrial species. Indeed, internal fertilization must have been a prerequisite in marine forms, a pre-adaptation, for the invasion of land. The eggs of terrestrial species require some form of protection, especially from desiccation, and are often provided with sufficient food to fuel most or all of their pre-juvenile development. The nutritional requirements of the fertilized egg of terrestrial species can often be met only if the female has access to a high-protein diet during the period of egg formation(**oogenesis**; *oo* = G: egg; *genesis* = G: birth). Hence, many female insects require a blood meal prior to **oviposition** (i.e., discharge and placement of eggs). A number of insects, notably the wasps, meet the nutritional needs of their larvae by placing their eggs in or adjacent to the eggs of other insect species, or within the bodies of other adult insects, which are then devoured by the developing young from the inside out.[10] Some insects deposit their eggs in plants, which respond by forming protective galls. The eggs are inserted into these various substrates through a long tube, called an **ovipositor,** protruding from the ab-

domen. A number of arachnid species are similarly equipped, and for a similar purpose.

During development, insects pass through several larval stages, called **instars.** This is conspicuously true for insect species that undergo a **metamorphosis** from a larval to a distinctly different adult body plan. In some species, this transition is gradual, and the different instars are called **nymphs** (Fig. 16.27a, b). Aquatic nymphs are sometimes referred to as **naiads.** Dragonflies and cockroaches develop in this manner, for example, and are said to be **hemimetabolous** (*hemi* = G: half; *metabolo* = G: change) (Fig. 16.28a). In other insect species, the change to adult form is radical and abrupt: **holometabolous** development (*holo* = G: whole; *metabolo* = G: change). The feeding, immature stages are termed **larvae** (Fig. 16.27c, d). After passing through several larval instars of ever-increasing size, a morphologically distinct, nonfeeding pupal stage is formed. The **pupa** then undergoes extensive internal and external reorganization to form the adult morph. Butterflies provide what is probably the most familiar example of holometabolous development (Fig. 16.28b).

Adults of **apterygote** species, such as silverfish, lack wings (*a* = G: without; *ptero* = G: wing) and so do not pass through any real metamorphosis as they develop. Instead, immatures simply get larger with each succeeding molt, and the body plan resembles that of the final adult at each stage. Such development is termed **ametabolous** (i.e., without change).

Holometabolous development is particularly characterized by a distinctly insect phenomenon, the formation of **imaginal discs.** At the completion of cleavage, small groups

---

10 See Topics for Discussion, No. 5.

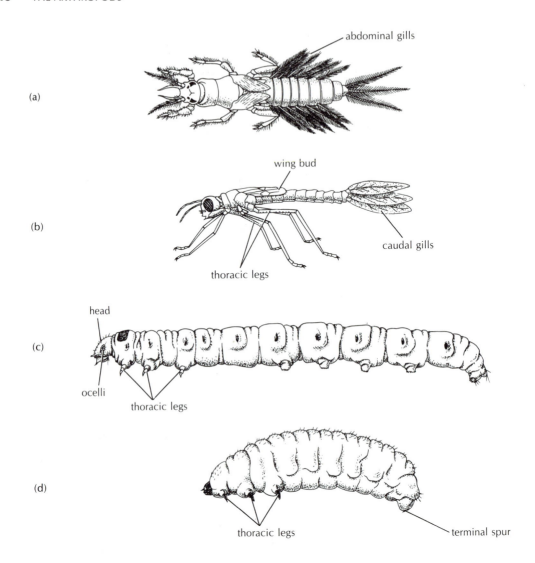

FIGURE 16.27. *Insect development. The nymphs show a gradual transition to the adult form, exhibiting hemimetabolous development. On the other hand, the larval stages of moths, beetles, and related species show little resemblance to the adult; development to adulthood is radical and abrupt, and is termed holometabolous. (a) Mayfly nymph,* Ephemera varia. *(After Pennak; after Needham.) (b) Damselfly nymph. (After Sherman and Sherman.) (c) Moth larva,* Bellura *sp. (After McCafferty.) (d) Beetle larva* Donacia *sp. (After McCafferty.)*

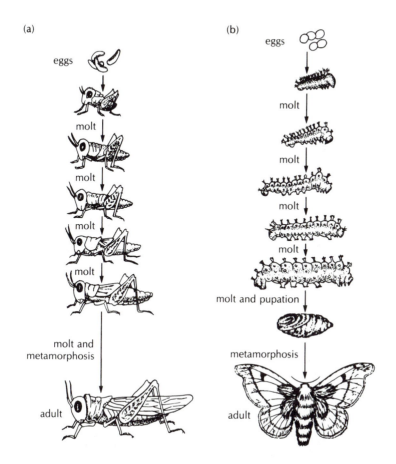

FIGURE 16.28. *(a) Hemimetabolous development of a grasshopper. (b) Holometabolous development in the silkworm moth. (From C.D. Turner, 1955. General Endocrinology. Saunders.)*

of up to about 30 cells give rise to discrete "discs" (spheres, actually), which are destined to differentiate into very well-defined adult epidermal structures, such as the eyes, the antennae, or the wings. Throughout larval development, however, these imaginal discs remain quiescent, or divide and grow at a very slow rate compared to the rest of the larva. The cells of one disc are distinguish-able from those of other discs found elsewhere in the body only by their position and a small number of structural details. Nevertheless, the eventual fate of the cells in a disc is fixed, and at metamorphosis a particular disc will always give rise to the same structure, even if transplanted elsewhere on the body. The orientation of the structure within the body may be incorrect following disc transplantation, but the structure itself will be perfectly formed. Imaginal discs have long been utilized by developmental biologists to probe the manner in which the expression of genes in individual cells is controlled.

Molting and metamorphosis during insect development is under complex environmental and hormonal control. Neurosecretory cells in the brain secrete a **prothoracicotropic hormone** (PTTH, also currently known as the "brain hormone"), a polypeptide that activates a pair of glands in the anterior portion of the thorax (**prothoracic glands**) (Fig. 16.29). The pro-

thoracic glands (PG), in turn, secrete the steroid **ecdysone,** which triggers the molting process—in particular, the resorption of some of the old cuticle and the development of new cuticle. The PG thus seem to perform the same role as the Y-organ in crustaceans. The extent to which morphological differentiation occurs during molting is a function of the amount of another hormone, **juvenile hormone,** present in the blood at certain critical periods (**gates**). Juvenile hormone (JH), a sesquiterpene lipid, is produced by yet another gland, the paired **corpora allata,** located just behind the brain. It has been experimentally demonstrated that the presence

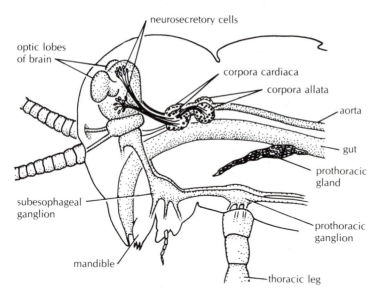

FIGURE 16.29. *Anterior end of an insect, showing the location of the brain hormone, juvenile hormone, and ecdysone secretory centers. Neurosecretory cells in the brain secrete a brain hormone (PTTH) that stimulates the prothoracic glands to secrete ecdysone. The corpora allata secrete juvenile hormone. The corpora cardiaca innervate the corpora allata and are also neurosecretory. One of their major roles is the regulation of heart beat rate. (From Wells, 1968. Lower Animals. McGraw-Hill. Reproduced with permission. (After Jenkins.)*

of high quantities of JH in the blood generally inhibits differentiation. Normal development seems to depend upon either a gradually diminishing concentration of JH with each succeeding stage, or upon the magnitude of the pulses of JH secretion being critically timed to coincide with gates of sensitivity of target tissues to the hormone. JH also has other important roles to play during development; for example, in caste determination in social insects and in stimulating yolk deposition during oogenesis in adult females. A number of intriguing behavioral patterns are also known to be under hormonal control. The more we learn, the more complex and fascinating the role of hormones in the lives of insects becomes.

A number of insects enter into a resting state (**diapause**) at some point in their development, as an adaptation for withstanding adverse conditions (e.g., cold winters). Entrance into the diapause state is under hormonal control, and is generally triggered (and released) by changing day length (**photoperiod**) and/or changing temperature. The stage at which diapause occurs depends upon the species in question. In some species, it is an early embryonic stage that enters diapause. In other species, either a larval instar, the pupal instar, or even the adult typically enters diapause. Diapause is also commonly encountered among copepods and branchiopods.

2. Digestion

The arthropod gut is divisible into three areas: foregut, midgut, and hindgut. All species exhibit a separate mouth and anus, and in all species food must be moved through the digestive tract by muscular activity rather than ciliary activity, since the lumen of the foregut and hindgut is lined with cuticle. Digestion is generally extracellular. Nutrients are distributed to the tissues through the hemal system.

a. Class Merostomata

The members of this class are carnivorous. The mouth leads into an esophagus and thence into a **gizzard.** Both the esophagus and the gizzard are lined with cuticle, which is molted periodically along with the rest of the exoskeleton. The gizzard is equipped with chitinous teeth for the grinding of ingested food. A valve prevents passage of undigestible material from the gizzard into the stomach. A pair of elongated pouches (hepatic, or digestive, caecae) extend laterally from the stomach. Most digestion of food and absorption of nutrients occur in these hepatic caecae. Wastes pass through the rectum and out an anus, located ventrally at the base of the caudal spine.

b. Class Arachnida

Most arachnid species are carnivores. Because arachnids lack mandibles, preliminary tearing and grinding of food prior to ingestion must be accomplished by the chelicerae. Enzymes are also released into the prey from glands associated with the oral cavity, and the food is thus predigested externally. Digestive enzymes contributed by glands in the chelicerae or pedipalps may also participate in this process. After sufficient time has elapsed, the solubilized tissue is pumped into the foregut through a muscular pump, and moves from there into the stomach; with few exceptions, food is simply not ingested in particulate form among arachnids. Digestion and absorption take place primarily in the extensive tubular outgrowths of the stomach wall, the digestive diverticula. Undigested material and wastes pass through an intestine and exit from a posterior anus.

c. Classes Chilopoda and Diplopoda

Most species of chilopod (centipede) are predaceous carnivores on other invertebrates and on some smaller vertebrates, using their modified maxillipeds to hold and poison prey. The digestive tract is usually a straight tube. Diplopods (millipedes), on the other hand, are usually herbivores, feeding on both living and, especially, on dead or decaying plant material or on the juices of living plants; some carnivorous species also exist. A **peritrophic membrane** lines the midgut of millipedes, presumably to protect against abrasion. Food becomes enclosed by this membrane as it moves through the gut, and new peritrophic membrane is then secreted. As in the centipedes, the gut is essentially a linear tube.

d. Class Insecta

Insects feed on a variety of nutritive sources, including plant and animal tissues and fluids. The esophagus is highly muscularized and serves as a pump, moving food into a crop for storage and/or preliminary digestion. A **proventriculus** is generally present, functioning as a valve to regulate the passage of food and, in some species, to grind ingested food. Digestion and absorption occur in the midgut and its associated gastric caecae. Symbiotic bacteria and protozoans harbored in the gastric caecae of many species participate in the digestive process. The walls of the midgut are often lined by peritrophic membrane, as in the Diplopoda. This membrane is discarded and renewed periodically. Before wastes reach the anus, most of the water is resorbed from the fecal material by the rectal glands of the hindgut (see Fig. 16.17).

e. Class Crustacea

In the crustaceans, food is ground and strained in a muscular foregut consisting of a large **cardiac stomach** and a smaller **pyloric stomach** (Fig. 16.30). The food is ground by the chitinous, toothed ridges of a **gastric mill.** Stiff setae often prevent food particles from passing further until they have reached the proper consistency. Food passes on to the midgut, which is associated with an extensive array of digestive caecae in which the food is digested, absorbed, and stored. The digestive tubules often form a distinct organ, the **hepatopancreas,** or liver.

3. Excretion

a. Class Merostomata

Members of the Merostomata bear four pairs of excretory **coxal glands** adjacent to the gizzard. The glands empty into a common chamber, which then leads through a coiled tubule and into a bladder. Resorption of salts occurs in the bladder as needed, the final urine being discharged through pores at the base of the last pair of walking legs.

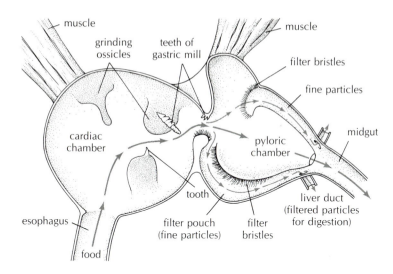

FIGURE 16.30. *The digestive system of a malacostracan. (After Hickman; after Yonge.)*

b. Class Arachnida

Coxal glands are also commonly encountered among arachnids. Here they are spherical sacs resembling annelid nephridia. Wastes are collected from the surrounding blood of the hemocoel, and discharged through pores on from one to several pairs of appendages. Recent evidence suggests that the coxal glands may also function in the release of pheromones.

Some arachnid species have Malpighian tubules instead of, or in addition to, the coxal glands. In some of these species, however, the Malpighian tubules seem to function in silk production rather than excretion. We have previously discussed the excretory role of Malpighian tubules with regard to the insects (p. 330). In addition to the coxal glands and Malpighian tubules, arachnids possess a number of strategically placed cells called

nephrocytes, which phagocytize waste particles. The major waste product of protein metabolism among arachnids is guanine, a purine biochemically related to uric acid.

c. Classes Chilopoda and Diplopoda

The major excretory structures in these groups are Malpighian tubules. Although some uric acid is produced, the major waste product of centipedes is ammonia.

d. Class Insecta

The major excretory structures are Malpighian tubules (Fig. 16.31a). Several other mechanisms also may be involved in the elimination of wastes in insects. In particular, there is evidence that some wastes are incorporated into the cuticle, to be shed at ecdysis.

It should be noted that the insects are the

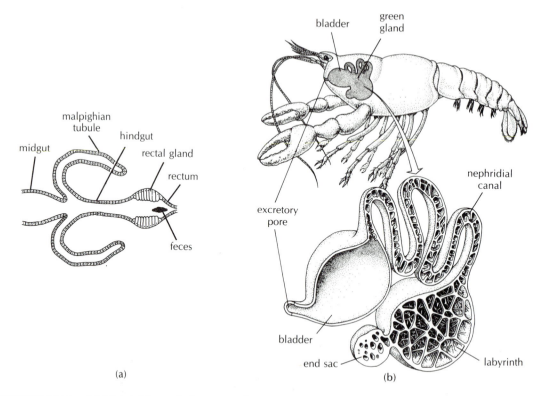

FIGURE 16.31. *(a) Malpighian tubule system from an insect. (b) Antennal gland of the crayfish. (From Purves and Orians, 1983.* Life: The Science of Biology. *Sinauer Associates/Willard Grant Press.)*

only arthropods to eliminate a significant fraction of their nitrogenous wastes as water-insoluble uric acid and related compounds. These compounds are eliminated from the body in nearly dry form, as most of the water in the urine is resorbed in transit through the rectum. This is an outstanding physiological adaptation for terrestrial existence.

e. Class Crustacea

Nitrogenous wastes are generally removed by diffusion across the gills—for those species that have gills. Most crustaceans release ammonia, although some urea and uric acid production also occurs. The so-called excretory organs may be more involved with resorption of salts than with discharge of nitrogenous wastes in freshwater species. In some species, these excretory organs are called **antennal glands** or **green glands** because of their location near the antennal segments and their color (Fig. 16.31). In other species, the organs are found near the maxillary segments and are termed **maxillary glands.** These excretory organs are structurally similar to the coxal glands of chelicerates. Fluid collects within the tubules from the surrounding blood of the hemocoel. The primary urine is modified substantially by selective reabsorption and secretion as it moves through the excretory system and rectum.

4. *Blood Pigments*

The bloods of many species lack respiratory pigments. This is particularly true of terrestrial species, in which gas exchange is accomplished primarily through tracheae. In other species, both hemocyanin (HCy) and hemoglobin (Hb) are found. Both pigments are sporadically distributed, even among the members of a given class.

TAXONOMIC SUMMARY

PHYLUM ARTHROPODA

  SUBPHYLUM TRILOBITOMORPHA
    CLASS TRILOBITA—THE TRILOBITES

  SUBPHYLUM CHELICERATA
    CLASS MEROSTOMATA—HORSESHOE CRABS
    CLASS ARACHNIDA—SPIDERS, MITES, TICKS,
      SCORPIONS

  SUBPHYLUM MANDIBULATA
    CLASS CHILOPODA—CENTIPEDES
    CLASS DIPLOPODA—MILLIPEDES
    CLASS INSECTA
      SUBCLASS APTERYGOTA—THE WINGLESS INSECTS
      SUBCLASS PTERYGOTA—THE WINGED INSECTS
    CLASS CRUSTACEA
      SUBCLASS BRANCHIOPODA—BRINE SHRIMP,
        FAIRY SHRIMP, WATER FLEAS
      SUBCLASS COPEPODA—COPEPODS
      SUBCLASS CIRRIPEDIA—THE BARNACLES
      SUBCLASS MALACOSTRACA
        ORDER ISOPODA—PILL BUGS, WOOD LICE
        ORDER AMPHIPODA—SAND FLEAS
        ORDER DECAPODA—CRABS, LOBSTERS,
          SHRIMP, HERMIT CRABS

TOPICS FOR FURTHER DISCUSSION AND INVESTIGATION

1. A small phylum (about 100 species) of invertebrates is closely related toArthropoda. The members of this phylum, the Pentastomida (Fig. 16.32) have become highly modified for a parasitic existence, so that their evolutionary history is obscure. All adult pentastomids are parasites on vertebrates, living primarily within the lungs and nasal passages of the host. Vertebrates generally serve as intermediate hosts for the larval stages as well. What characteristics do pentastomids have in common with annelids? What characteristics do pentastomids have in common with arthropods? What characteristics are different? With what arthropod group do they seem to be most closely allied? What characteristics do they share with parasitic members of other phyla that you have studied?

Riley, J., A.A. Banaja, and J.L. James, 1978. The phylogenetic relationships of the Pentastomida: the case for their inclusion within the Crustacea. *Int. J. Parasitol.*, 8: 245.

Self, J.T., 1969. Biological relationships of the Pentastomida: A bibliography on the Pentastomida. *Exp. Parasitol.*, 24: 63.

Self, J.T., and R. Kuntz, 1967. Host-parasite relations in some Pentastomida. *J. Parasitol.*, 53: 202.

Also: consult any recent parasitology text, e.g., T.C. Cheng, 1973. *General Parasitology.* N.Y.: Academic Press, pp. 776–785; or E.R. Noble and G.A. Noble, 1982. *Parasitology: the Biology of Animal Parasites*, 5th ed., Philadelphia: Lea & Febiger, pp. 389–391.

2. Arthropod affinities and origins are far from certain, even at the highest taxonomic levels. Arthropods are most certainly derived from annelid-like ancestors. However, embryological studies and studies of limb structure and function in various arthropod and polychaete species suggest that insects, millipedes, and centipedes may have had a very different evolutionary origin from that of other arthropod groups, such as the Crustacea and Arachnida. Indeed, a number of authorities include the insects, millipedes, and centipedes in a new taxonomic group, the Uniramia, which is given separate phylum status. What is the evidence in favor of an independent origin for these arthropods, and what difficulties does this scheme present in terms of the numbers of different arthropod characteristics that must be presumed to have evolved independently through convergent evolution?

Cisne, J.L., 1974. Trilobites and the evolution of arthropods. *Science*, 186: 13.

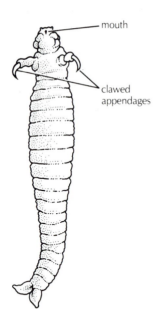

mouth

clawed appendages

FIGURE 16.32.   *A pentastomid, or "tongue worm." All species are parasitic, living within the respiratory system of vertebrates. The claws are used to cling to the tissues of the host. (After Noble and Noble.)*

Evans, H.E., 1959. Some comments on the evolution of the Arthropoda. *Evolution*, 13: 147.

Gupta, A.P., 1979. *Arthropod Phylogeny*. New York: Van Nostrand Reinhold.

Hessler, R.R., and W.A. Newman, 1975. A trilobitomorph origin for the Crustacea. *Fossils and Strata*, 4: 437.

Manton, S.M., 1958. Habits of life and evolution of body design in Arthropods. *J. Linn. Soc. (Zool.)*, 44: 58.

Manton, S.M., 1967. The polychaete *Spinther* and the origin of the Arthropoda. *J. Nat. Hist.*, 1: 1.

Manton, S.M., 1973. Arthropod phylogeny—a modern synthesis. *J. Zool., London*, 171: 111.

Tiegs, O.W., and S.M. Manton, 1958. The evolution of the Arthropoda. *Biol. Rev.*, 33: 255.

3. Living in the terrestrial environment poses many difficulties for invertebrates. Not least among these is that the air is dry and that temperatures fluctuate considerably, even on a day-to-day or hour-to-hour basis. Investigate some of the physiological and behavioral adaptations that enable terrestrial and semi-terrestrial arthropods to tolerate dehydrating conditions and marked shifts in environmental temperature.

References concerning temperature regulation:

Digby, P.S.B., 1955. Factors affecting the temperature excess of insects in sunshine. *J. Exp. Biol.*, 32: 279.

Dorsett, D.A., 1962. Preparation for flight by hawkmoths. *J. Exp. Biol.*, 39: 579.

Douglas, M.W., 1981. Thermoregulatory significance of thoracic lobes in the evolution of insect wings. *Science*, 211: 84.

Duman, J.G., 1977. Environmental effects on antifreeze levels in larvae of the darkling beetle, *Meracantha contracta*. *J. Exp. Zool.*, 201: 333.

Edney, E.B., 1953. The temperature of woodlice in the sun. *J. Exp. Biol.*, 30: 331.

Heinrich, B., 1972. Energetics of temperature regulation and foraging in a bumblebee, *Bombus terricola* Kirby. *J. Comp. Physiol.*, 77: 49.

Heinrich, B., 1974. Thermoregulation in endothermic insects. *Science*, 185: 747.

Lüscher, M., 1961. Air-conditioned termite nests. *Sci. Amer.*, 205: 138.

Wilkens, J.L. and M. Fingerman, 1965. Heat tolerance and temperature relationships of the fiddler crab, *Uca pugilator*, with reference to its body coloration. *Biol. Bull.*, 128: 133.

Willmer, P.G., 1982. Thermoregulatory mechanisms in Sarcophaga. *Oecologia (Berlin)*, 53: 382.

References concerning water balance:

Cohen, A.C., R.B. March and J.D. Pinto, 1981. Water relations of the desert blister beetle *Cysteodemus armatus* (Leconte) (Coleoptera: Meloidae). *Physiol. Zool.*, 54: 179.

Dresel, E.I.B., and V. Moyle, 1950. Nitrogenous excretion in amphipods and isopods. *J. Exp. Biol.* 27: 210.

Edney, E.B., 1966. Absorption of water vapour from unsaturated air by *Arenivaga* sp. (Polyphgidae, Dictyoptera). *Comp. Biochem. Physiol.*, 19: 387.

Gifford, C.A., 1962. Some observations on the general biology of the land crab, *Cardisoma guanhumi* (Latreille) in South Florida. *Biol. Bull.*, 123: 207.

Hamilton, W.J., III, and M.K. Seely, 1976. Fog basking by the Nambi desert beetle, *Onymacris unguicularis*. *Nature*, 262: 284.

Hurley, D.E., 1968. Transition from water to land in amphipod crustaceans. *Amer. Zool.*, 8: 327.

Lüscher, M., 1961. Air-conditioned termite nests. *Sci. Amer.*, 205: 138.

Noble-Nesbitt, J., 1970. Water uptake from subsaturated atmospheres: its site in insects. *Nature, London*, 225: 753.

Standing, J.D., and D.D. Beatty, 1978. Humidity behaviour and reception in the sphaeromatid isopod *Gnorimosphaeroma oregonensis* (Dana). *Canadian J. Zool.*, 56: 2004.

4. Invertebrates are basically poikilothermic. That is, in the absence of behavioral modification (including flight), body temperatures generally follow those of the air or water around them rather closely. This poses a particular problem for terrestrial and shallow-water invertebrates when temperatures fall below freezing. Investigate the mechanisms used by arthropods to prevent freezing of tissues during periods of cold weather.

Duman, J.G., K.L. Horwarth, A. Tomchaney, and J. L. Patterson, 1982. Antifreeze agents of terrestrial arthropods. *Comp. Biochem. Physiol.*, 73A: 545.

Horwarth, K.L., and J.G. Duman, 1982. Involvement of the circadian system in photoperiodic regulation of insect antifreeze proteins. *J. Exp. Zool.*, 219: 267.

van der Laak, S., 1982. Physiological adaptations to low temperature in freezing-tolerant *Phyllodecta laticollis* beetles. *Comp. Biochem. Physiol.*, 73A: 613.

Turnock, W.J., R.J. Lamb, and R.P. Bodnaryk, 1983. Effects of cold stress during pupal diapause on the survival and development of *Mamestra configurata* (Lepidoptera: Noctuidae). *Oecologia (Berlin)*, 56: 185.

Yingst, D. R., 1978. The freezing resistance of the arctic subtidal isopod, *Mesidotea entomon*. *J. Comp. Physiol.*, 125: 165.

5. Symbiotic relationships (including parasitism) are commonly encountered among the Crustacea, Insecta, and Arachnida. For one of these groups, discuss the morphological, behavioral, and/or physiological adaptations for the symbiotic life-style.

Christensen, A.M., and J.J. McDermott, 1958. Life-history and biology of the oyster crab, *Pinnotheres ostreum* Say. *Biol. Bull.*, 114: 146.

Day, J.H., 1935. The life-history of *Sacculina*. *Quart. J. Microsc. Sci.*, 77: 549.

Faxon, G.E.H., 1940. Notes on the life history of *Sacculina carcini* Thompson. *J. Mar. Biol. Assoc. U.K.*, 24: 253.

Gotto, R.V., 1979. The association of copepods with marine invertebrates. *Advances in Marine Biol.*, 16: 1.

Heatwole, H., and D.M. Davis, 1965. Ecology of three sympatric species of parasitic insects of the genus *Megarhyssa* (Hymenoptera: Ichneumonidae). *Ecology*, 46: 140.

Mitchell, R., 1968. Site selection by larval water mites parasitic on the damselfly *Cercion hieroglyphicum* Brauer. *Ecology*, 49: 40.

Moyse, J., 1983. *Isadascus bassindalei* gen. nov., sp. nov. (Ascothoracida: Crustacea) from northeast Atlantic with a note on the origin of barnacles. *J. Mar. Biol. Assoc. U.K.*, 63: 161.

Patterson, N.F., 1958. External features and life cycle of *Cucumaricola notabilis* nov. gen. et sp., a copepod parasite of the holothurian, *Cucumaria. Parasitology*, 48: 269.

Price, P.W., 1972. Parasitoids utilizing the same host: adaptive nature of differences in size and form. *Ecology*, 53: 190.

Salt, G., 1968. The resistance of insect parasitoids to the defense reactions of their hosts. *Biol. Rev.*, 43: 200.

A general treatment can be found in any recent parasitology text, such as Noble, E.R., and G.A. Noble, 1982. *Parasitology: the Biology of Animal Parasites*, 5th ed., Philadelphia: Lea & Febiger.

6. Investigate form and function in spider webs.

Denny, M., 1976. The physical properties of spiders' silk and their role in the design of orb-webs. *J. Exp. Biol.*, 65: 483.

Kenchington, W., 1983. The larval silk of *Hypera* spp. (Coleoptera: Curculionidae). A new example of the cross-protein conformation in an insect silk. *J. Insect Physiol.*, 29: 355.

Nentwig, W., 1982. Why do only certain insects escape from a spider's web? *Oecologia (Berlin)*, 53: 412.

Nentwig, W., 1983. The non-filter function of orb webs in spiders. *Oecologia*, 58: 418.

Palmer, J.M., F.A., Coyle, and F.W. Harrison, 1982. Structure and cytochemistry of the silk glands of the Mygalomorph spider *Antrodiaetus unicolor* (Araneae, Antrodiaetidae). *J. Morphol.*, 174: 269.

Popock, R.I., 1895. Some suggestions on the origin and evolution of web spinning in spiders. *Nature, London.*, 51: 417.

Rypstra, A.L., 1982. Building a better insect trap: an experimental investigation of prey capture in a variety of spider webs. *Oecologia (Berlin)*, 52: 31.

Witt, P.N., and R. Baum, 1960. Changes in orb webs of spiders during growth (*Araneus diadematus* Clerck and *Neoscona vertebrata* McCook). *Behaviour*, 16: 309.

Witt, P.N., and C.F., Reed, 1965. Spider-web building. *Science*, 149: 1190.

7. How do insects fly?

Alexander, R.M., 1979. *The Invertebrates*. New York: Cambridge Univ. Press, pp. 407–416.

Boettiger, E.G., and E. Furshpan, 1952. The mechanics of flight movements in Diptera. *Biol. Bull.* 102: 200.

Roeder, K.D. 1951. Movements of the thorax and potential changes in the thoracic muscles of insects during flight. *Biol. Bull.*, 100: 95.

Weis-Fogh, R., 1975. Unusual mechanisms for the generation of lift in flying animals. *Scientific Amer.*, 233: 80.

Wells, M., 1968. *Lower Animals*. New York: McGraw-Hill, pp. 101–115.

Consult any recent entomology textbook, such as Romoser, W.S., 1973. *The Science of Entomology*, New York: Macmillan, pp. 161–169, or Chapman, R.F., 1982. *The Insects—Structure and Function*, 3rd ed. Hodder and Stoughton, Ltd., London. pp. 216–243.

8. Hermit crabs are an active, marine group of decapod crustaceans that do not secrete a protective carapace. Instead, they protect their soft parts by taking up residence inside empty gastropod shells. What factors enter into the choice of shells by hermit crabs, and what selective pressures have probably molded these choices?

Bertness, M.D., 1982. Shell utilization, predation pressure, and thermal stress in Panamanian hermit crabs: an interoceanic comparison. *J. Exp. Mar. Biol. Ecol.*, 64: 159.

Conover, M.R., 1978. The importance of various shell characteristics to the shell-selection behavior of hermit crabs. *J. Exp. Mar. Biol. Ecol.*, 32: 131.

Mercando, N.A., and C.F. Lytle, 1980. Specificity in the association between *Hydractinia echinata* and sympatric species of hermit crabs. *Biol. Bull.*, 159: 337.

Mesce, K.A., 1982. Calcium-bearing objects elicit shell selection behavior in a hermit crab. *Science*, 215: 993.

Mitchell, K.A., 1975. An analysis of shell occupation by two sympatric species of hermit crab. I. Ecological factors, *Biol. Bull.*, 149: 205.

Scully, E.P., 1979. The effects of gastropod shell availability and habitat characteristics on shell utilization by the intertidal hermit crab *Pagurus longicarpus* Say. *J. Exp. Mar. Biol. Ecol.*, 37: 139.

Vance, R, 1972. Competition and mechanism of coexistence in three sympatric species of intertidal hermit crabs. *Ecology*, 53: 1062.

9. Through the process of natural selection, many arthropods have been able to develop remarkable abilities to disguise themselves, either to blend in with their surroundings or to imitate other species. Investigate the adaptive value of mimicry and camouflage among the Arthropoda.

Blest, A.D., 1957. The function of eyespot patterns in the Lepidoptera. *Behaviour*, 11: 209.

Blest, A.D., 1963. Longevity, palatability, and natural selection in five species of New World Saturuiid moth. *Nature, London*, 197: 1183.

Eisner, T., K. Hicks, M. Eisner, and D.S. Robson, 1978. "Wolf-in-sheep's-clothing" strategy of a predaceous insect larva. *Science*, 199: 790.

Kettlewell, H.B.D., 1956. Further selection experiments on industrial melanism in the Lepidoptera. *Heredity*, 10: 287.

Körner, H.K., 1982. Countershading by physiological colour change in the fish louse *Anilocra physodes* L. (Crustacea: Isopoda). *Oecologia (Berlin)*, 55: 248.

Platt, A., R. Coppinger, and L. Brower, 1971. Demonstration of the selective advantage of mimetic *Limentis* butterflies presented to caged avian predators. *Evolution*, 25: 692.

10. Much of the communication among arthropods is accomplished by means of chemicals. Investigate the mechanism and/or the adaptive significance of chemical communication among the Arthropoda.

Boeckh, J., H. Sass, and D.R.A. Wharton, 1970. Antennal receptors: reactions to female sex attractant in *Periplaneta americana*. *Science*, 168: 589.

Butler, C.G., D.J.C. Fletcher, and D. Walter, 1969. Nest-entrance marking with pheromones by the honey bee *Apis mellifera* L. and by a wasp, *Vespula vulgaris*. *Anim. Behav.*, 17: 142.

Dahl, E., H. Emanuelsson, and C. von Mecklenburg, 1970. Pheromone transport and reception in an amphipod. *Science*, 170: 739.

Derby, C.D., and J. Atema, 1980. Induced host odor attraction in the pea crab *Pinnotheres maculatus*. *Biol. Bull.*, 158: 26.

Derby, C.D., and J. Atema, 1982. Narrow-spectrum chemoreceptor cells in the walking legs of the lobster *Homarus americanus*: taste specialists. *J. Comp. Physiol.*, 146A: 181.

Eisner, T., and Y.C. Meinwald, 1965. Defensive secretion of a caterpillar (*Papilio*). *Science*, 150: 1733.

Fitzgerald, T.D., 1976. Trail marking by larvae of the eastern tent caterpillar. *Science*, 194: 961.

Myers, J., and L.P. Brower, 1969. A behavioural analysis of the courtship pheromone receptors of the queen butterfly, *Danaus gilippus berenice*. *J. Insect Physiol.*, 15: 2117.

Nault, L.R., M.E. Montgomery, and W.S. Bowers, 1976. Ant-aphid association: role of aphid alarm pheromone. *Science*, 192: 1349.

Price, P.W., 1970. Trail odors: recognition by insects parasitic on cocoons. *Science*, 170: 546.

Rust, M.K., T. Burk and W.J. Bell, 1976. Pheromone-stimulated locomotory and orientation responses in the American cockroach *Periplaneta americana*. *Anim. Behav.*, 24: 52.

Ryan, E.P., 1966. Pheromone: evidence in a decapod crustacean. *Science*, 151: 340.

Schneider, D., 1969. Insect olfaction: deciphering system for chemical messages. *Science*, 163: 1031.

Tobin, T.R., 1981. Pheromone orientation: role of internal control mechanisms. *Science*, 214: 1147.

11. It has recently become clear that most invertebrates possess some form of immune response; i.e., they have the ability to distinguish self from nonself at the cellular level. Investigate the ability of arthropods to make this distinction.

Brehélin, M., and J.A. Hoffmann, 1980. Phagocytosis of inert particles in *Locusta migratoria* and *Galleria mellonella*: a study of ultrastructure and clearance. *J. Insect Physiol.*, 26: 103.

Briggs, J.D., 1958. Humoral immunity in lepidopterous larvae. *J. Exp. Zool.*, 138: 155.

Edson, K.M., S.B. Vinson, D.B. Stoltz, and M.D. Summers, 1981. Virus in a parasitoid wasp: suppression of the cellular immune response in the parasitoid's host. *Science*, 211: 582.

Salt, G., 1963. Experimental studies in insect parasitism. XII. The reactions of six exopterygote insects to an alien parasite. *J. Insect Physiol.*, 9: 647.

Salt, G., 1968. The resistance of insect parasitoids to the defense reactions of their hosts. *Biol. Rev.*, 43: 200.

Sloan, B., C. Yocum, and L.W. Clem, 1975. Recognition of self from non-self in crustaceans. *Nature, London*, 258: 521.

Smith, V.J., and N.A. Ratcliff, 1980. Host defence reactions of the shore crab, *Carcinus maenas* (L.): clearance and distribution of injected test particles. *J. Mar. Biol. Assoc. U.K.*, 60: 89.

White, K.N., and N.A. Ratcliffe, 1982. The segregation and elimination of radio- and fluorescent-labelled marine bacteria from the haemolymph of the shore crab, *Carcinas maenas. J. Mar. Biol. Assoc. U.K.*, 62: 819.

12. Small planktonic crustaceans, such as copepods and cladocerans, are near the base of the food chain in aquatic ecosystems. That is, the size of a fish, or a whole population, depends, to a large extent, on the size of the herbivorous zooplankton population available to feed them. In turn, the size of the herbivorous zooplankton population depends on the amount of phytoplankton available, and the rate at which the phytoplankton can be captured, ingested, and converted into new biomass and offspring. Consequently, considerable effort has gone into the study of zooplankton feeding biology. What factors are involved in the collection of phytoplankton by crustacean zooplankton?

Anraku, M., and M. Omori, 1963. Preliminary survey of the relationship between the feeding habit and structure of the mouth parts of marine copepods. *Limnol. Oceanogr.*, 8: 116.

Frost, B.W., 1977. Feeding behavior of *Calanus pacificus* in mixtures of food particles. *Limnol. Oceanogr.*, 22: 472-492.

Gerritsen, J., and K.G. Porter, 1982. The role of surface chemistry in filter feeding by zooplankton. *Science*, 216: 1225.

Hamner, W.M., P.P. Hamner, S.W. Strand, and R.W. Gilmer, 1983. Behavior of antarctic krill. *Euphausia superba*: chemoreception, feeding, schooling, and molting. *Science*, 220: 433.

Huntley, M.E., K.-G., Barthel, and J.L. Star, 1983. Particle rejection by *Calanus pacificus*: discrimination between similarly sized particles. *Marine Biol.*, 74: 151.

Koehl, M.A., and J.R. Strickler, 1982. Copepod feeding currents: food capture at low Reynold's numbers. *Limnol. Oceanogr.*, 26: 1062.

Landry, M.R., 1980. Detection of prey by *Calanus pacificus*: implications of the first antenna. *Limnol. Oceanogr.*, 25: 545.

Paffenhöfer, G.-A., J.R. Strickler, and M. Alcaraz, 1982. Suspension-feeding by herbivorous calanoid copepods: a cinematographic study. *Marine Biol.*, 67: 193.

Poulet, S.A., and P. Marsot, 1978. Chemosensory grazing by marine calanoid copepods (Arthropoda: Crustacea). *Science*, 200: 1403.

Price, H.J., G.-A. Paffenhöfer, and J.R. Strickler, 1983. Modes of cell capture in calanoid copepods. *Limnol. Oceanogr.*, 28: 116.

Richman, S., D.R. Heinle, and R. Huff, 1977. Grazing by adult estuarine calanoid copepods of the Chesapeake Bay. *Marine Biol.*, 42: 69.

Richman, S., and J.N. Rogers, 1969. The feeding of *Calanus helgolandicus* on synchronously growing populations of the marine diatom *Ditylum brightwelli. Limnol. Oceanogr.*, 14: 701.

Yule, A.B., and D.J. Crisp. 1983. A study of feeding behavior in *Temora longicornis* (Müller) (Crustacea: Copepoda). *J. Exp. Marine Biol. Ecol.*, 71: 271.

13. What is the evidence indicating that color changes in arthropods are regulated by hormones?

Brown, F.A., Jr., 1935. Control of pigment migration within the chromatophores of *Palaemonetes vulgaris. J. Exp. Zool.*, 71: 1.

Brown, F.A. Jr., 1940. The crustacean sinus gland and chromatophore activation. *Physiol. Zool.*, 13: 343.

Brown, F.A., Jr., and H.E. Ederstrom, 1940. Dual control of certain black chromatophores of *Crago. J. Exp. Zool.*, 85: 53.

Fingerman, M., 1969. Cellular aspects of the control of physiological color changes in crustaceans. *Amer. Zool.*, 9: 443.

McWhinnie, M.A., and H.M. Sweeney, 1955. The demonstration of two chromatophorotropically active substances in the land isopod, *Trachelipus rathkei. Biol. Bull.*, 108: 160.

Pérez-González, M.D., 1957. Evidence for hormone-containing granules in sinus glands of the fiddler crab *Uca pugilator. Biol. Bull.*, 113: 426.

14. What are the roles of predation and competition in regulating the sizes of insect and arachnid populations?

Chew, F.S., 1981. Coexistence and local extinction in two pierid butterflies. *Amer. Nat.*, 118: 655.

El-Dessouki, S.A., 1970. Intraspecific competition between larvae of *Sitona* spp. (Coleoptera, Curculionidae). *Oecologia (Berlin)*, 6: 106.

Enders, F., 1974. Vertical stratification in orb-web spiders (Araneidae, Araneae) and a consideration of other methods of coexistence. *Ecology*, 55: 317.

Frank, J.H. 1967. The insect predators of the pupal stage of the winter moth, *Operophtera brumata* (L.) (Lepidoptera: Hydiomenidae). *J. Anim. Ecol.*, 36: 375.

MacKay, W.P., 1982. The effect of predation of western widow spiders (Araneae: Theridiidae) on harvester ants (Hymenoptera: Formicidae). *Oecologia (Berlin)*, 53: 406.

Moore, N.W., 1964. Intra- and interspecific competition among dragonflies (Odonata). *J. Anim. Ecol.*, 33: 49.

Wise, D.H., 1983. Competitive mechanisms in a food-limited species: relative importance of interference and exploitative interactions among labyrinth spiders (Araneae: Araneidae). *Oecologia (Berlin)*, 58: 1.

15. Based upon lectures and your reading in this text, what factors contribute to making arthropods the most abundant animals (with the possible exception of the nematodes) on Earth?

16. Based upon lectures and your readings in this text, what functions does the blood of insects serve, now that gas exchange is achieved by means of tracheae rather than by a circulatory system?

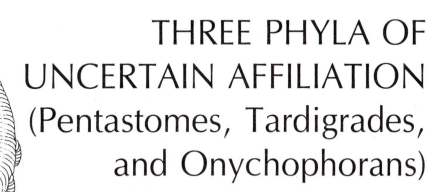

17

THREE PHYLA OF UNCERTAIN AFFILIATION
(Pentastomes, Tardigrades, and Onychophorans)

Phylum Penta•stomida
(G: FIVE MOUTHS)

Phylum Tardi•grada
(L: SLOW WALKER)

Phylum Onycho•phora
(G: CLAW BEARER)

INTRODUCTION

One would think that a group as stereotyped as the Arthropoda, and with as long a fossil record, would have a relatively uncontroversial phylogeny. The following features are considered to be characteristic of arthropods: external, jointed exoskeleton; tracheae; compound eyes; Malpighian tubules; mandibles; heart with ostia. Nevertheless, there is considerable controversy regarding the evolutionary history of these animals. In large part, the controversy centers on how many times the above characteristics have been independently evolved.

All of the classes listed in the preceding chapter are already represented in the earliest fossil records of 600 million years ago.

The difficulty in tracing the ancestry of the group lies in the fact that arthropods are most certainly derived from soft-bodied ancestral forms poorly represented in the fossil record. There seems to be unanimous agreement that arthropods are derived either from annelids, or at least from an annelid-like ancestor. In particular, both groups demonstrate a clearly segmented coelomic space during embryogenesis; absence of this characteristic in the arthropod adult is certainly a modification of the ancestral condition. The existence of animals possessing some characteristics of arthropods and some characteristics typical of other groups, or at least atypical for arthropods, tantalize one with phylogenetic implications. The animals in the three phyla we will discuss in this chapter all show this combination of arthropod-like characteristics and non-arthropod features.

GENERAL CHARACTERISTICS

The Pentastomida (tongue worms) have been referred to briefly in the preceding chapter (p. 356). As in some arthropods, the members of this phylum possess such characteristics as a chitinous exoskeleton that is periodically molted; striated musculature arranged metamerically; and larval stages often possessing three pairs of legs. However, the evolutionary history of the pentastomids is, and will remain, untraceable, since other aspects of external morphology—and of the sensory, digestive, excretory, and reproductive systems—have become highly modified as adaptations for the exclusively parasitic existence encountered among all members of the group (Fig. 17.1). All pentastomids are internal parasites of vertebrates and all of the characteristics of potential diagnostic utility have been lost.

Members of the phylum Tardigrada (the water bears) also have certain arthropod affinities and an uncertain place in the phylogenetic scheme of things. About 400 tardigrade species have been described, and all are quite small, ranging between about 50 and 500 μm in length. Most species live in surface films of water on terrestrial plants, especially on mosses and lichens. Some marine species have been described, many of them living in the spaces between sand grains (i.e., **interstitially**). Tardigrades possess a cuticle that is molted, but in contrast to the arthropod cuticle, that of the tardigrades is generally thought to be proteinaceous rather than chitinous. More studies are apparently needed to resolve this issue. Because the tardigrades are small and their cuticle is permeable to water and gases, gas exchange occurs across the general body surface; no specialized respiratory structures are found. Each individual tardigrade possesses four pairs of clawed appendages, with which the animal lumbers over the substrate in bearish fashion (Fig. 17.2). The appendages are not jointed.

The nervous system of tardigrades is organized in annelid/arthropod manner, with a paired ventral nerve cord (Fig. 17.3). Also, several glands are found that look suspiciously like Malpighian tubules, although their function in tardigrades has yet to be demonstrated. The mouthparts are a pair of stylets, which are used to pierce into plant cells.

Several characteristics of tardigrades are decidedly nonarthropod. Their embryological development is said to be entero-

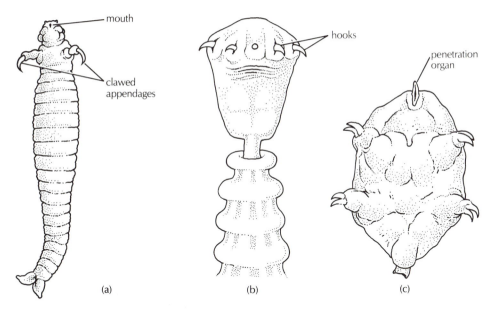

(a)   (b)   (c)

FIGURE 17.1.   *Representative pentastomids. The adults are
parasitic in the nasal passages of snakes and lizards, while
cockroaches, bats, raccoons, muskrats, and armadillos commonly
serve as intermediate hosts for the larval stages. (a)* Raillietiella
mabuiae, *entire individual. (From Baer, 1951,* Ecology of Animal
Parasites. University of Illinois Press.) *(b) Anterior end of*
Armillifer annulatus. *(From Baer.) (c) Intermediate larval stage
of* Porocephalus crotali. *(From Noble and Noble, 1982.*
Parasitology: The Biology of Animal Parasites, *5th ed. Lea & Febiger,
after Penn.)*

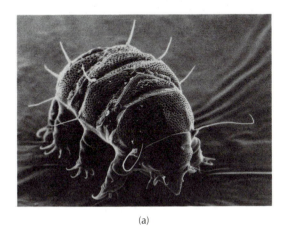

(a)   (b)

FIGURE 17.2.   *Scanning electron micrographs of (a)* Echiniscus
spiniger *and (b)* Macrobiotus hufelandi, *two tardigrades. (Courtesy
of D. R. Nelson.)*

365

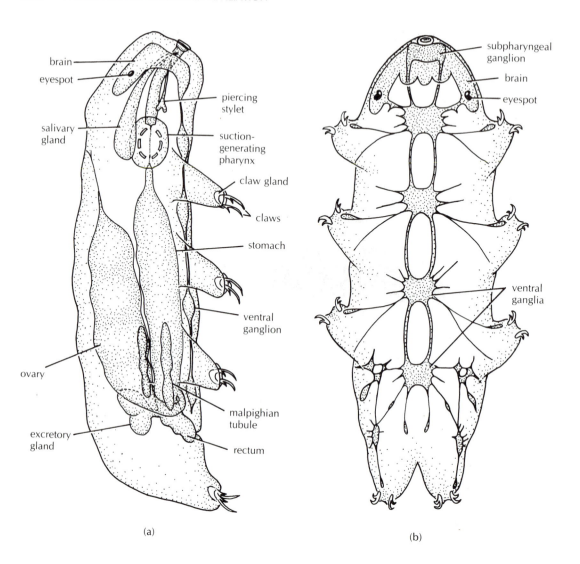

FIGURE 17.3. *(a) Internal anatomy of a typical tardigrade. Note that the nerve cord is ventrally located, as in annelids and arthropods. The claw glands secrete the claws. (b) Detail of the tardigrade nervous system. Note the large ganglia associated with each pair of appendages. (From Pennak, 1978. Fresh-Water Invertebrates of the United States, 2nd ed. John Wiley & Sons.)*

coelous, a typically deuterostome characteristic. Moreover, it is not clear that the body cavity is a hemocoel; it may well be a pseudocoel. In fact, the phylum is linked by some workers to the aschelminthes, based upon constancy of cell numbers in the tardigrade cuticle and the presence of a nonchitinous cuticle that is molted periodically. Moreover, like nematodes and rotifers, tardigrades exhibit **cryptobiosis,** a bizarre ability to dehydrate and reduce metabolic rate in order to withstand extreme environmental conditions of low-temperature and desiccation stress.[1]

The existence of the third phylum of uncertain affiliation, the phylum Onychophora, is very exciting, because the members of this group possess some characteristics that are clearly annelid in nature; some that are clearly arthropod in nature; and none that affiliate them with any other animal group. All members are free-living, and are clearly protostomous coelomates. More than 100 species have been described, with *Peripatus* being the best-known genus (Fig. 17.4a).

All onychophorans are terrestrial, and most are found in moist habitats in tropical environments and in southern temperate regions (e.g., New Zealand). Indeed, they appear to be restricted to such environments largely because they possess a thin, nonwaxy cuticle that is not effective in deterring evaporative loss of body water. Some species are carnivores, some are herbivores, some are omnivores, and all are active primarily at night. Individuals may attain a length of up to 15 centimeters. Onychophorans possess

---

1 See Topics for Discussion.

the following annelid-like characteristics: body wall musculature is smooth and composed of longitudinal, circular, and diagonal elements (Fig. 17.4b); a single pair of feeding appendages (jaws) is present; all appendages are unjointed; hydrostatic skeleton plays a role in locomotion (as described below); one pair of nephridia is found in most segments; light receptors are ocelli rather than compound eyes; outer body wall is deformable. The following arthropod-like characteristics are found: jaw musculature is striated; cuticle contains chitin; main body cavity is a hemocoel, not a true coelom; gas exchange is achieved by spiracles opening into a tracheal system (the spiracles cannot be closed, however—another factor preventing the onychophorans from invading drier habitats); mouth appendages are mandible-like; heart bears ostia; legs are extended by hemocoelic pressure rather than by direct muscular contraction (as also found for the legs of arachnids and merostomes and the maxillae of butterflies); excretory organs closely resemble the green glands of crustaceans; adhesive defense secretions are reminiscent of those produced by repugnatorial glands of the centipedes and millipedes(myriapods).The nervous system is of the annelid/arthropod type: segmented, with a pair of ventral nerve cords.

A few characteristics separate the onychophorans from both the arthropods and the annelids. The head appendages consist of a pair of antennae, a pair of jaws, and a pair of oral papillae. These are followed by a series of unjointed walking legs that are quite unlike the parapodia of polychaete annelids, both structurally and functionally. Blood pigment is lacking.

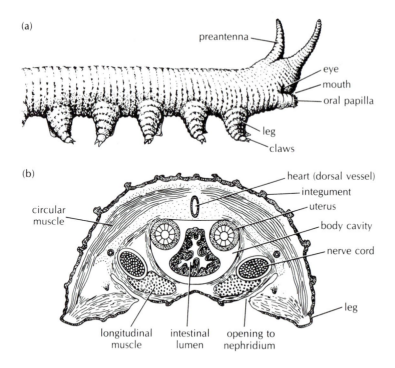

FIGURE 17.4.   *(a) The onychophoran* Peripatus, *in lateral view.*
(b) Same animal, seen in diagrammatic cross section . (Reprinted with
permission of Macmillan Publishing Company from The
Invertebrates: Function and Form, *2nd ed., by Irwin W. Sherman*
and Vilia G. Sherman. Copyright © 1976 by Irwin W. Sherman and
Vilia G. Sherman.)

Locomotion has been studied for a few species within this group, and has been found to be uniquely onychophoran. Propulsion is generated directly by the musculature of the limbs themselves, with the body remaining rigid, serving as an anchor point against which the limbs can operate. Of course, in the absence of a rigid skeleton, body rigidity is a function of the body wall musculature. The limbs, of which about 20 pairs are typically found, project ventrolaterally, elevating the body above the ground. This is quite different from polychaete parapodia, which project laterally, leaving the body surface in direct contact with the substrate.

As in polychaetes, waves of limb activity pass down the length of the body of onychophorans, and several waves are generally progressing concurrently (Fig. 17.5). In the preparatory stroke for any given onychophoran limb, the tip of the leg is raised above the substrate and extended forward. On the backstroke (power stroke), the tip of

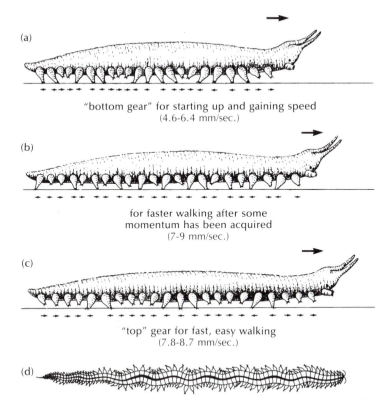

(a)

"bottom gear" for starting up and gaining speed
(4.6-6.4 mm/sec.)

(b)

for faster walking after some
momentum has been acquired
(7-9 mm/sec.)

(c)

"top" gear for fast, easy walking
(7.8-8.7 mm/sec.)

(d)

FIGURE 17.5. *(a–c) The three most common gaits observed during locomotion of the onychophoran* Peripatopsis sedgwicki. *Shifts in walking speeds are associated with changes in body length and width (not shown). Sufficient speed must be obtained using gaits (a) and (b) before the animal can switch into gait (c) for prolonged, rapid walking. (From Manton, 1950. J. Linn. Soc. (Zool.), 41:529.) (d) Locomotion of the polychaete (Annelida: Polychaeta)* Nereis diversicolor, *shown for comparison. The locomotory appendages (parapodia) project laterally. Waves of limb activity pass down the length of the body. Note that for each segment, the left parapodium and right parapodium are not in the same phase of the activity cycle. Waves of longitudinal muscle contraction aid in generating forward thrust. (From Gray, 1939. J. Exp. Biol., 16: 9.)*

the leg is applied to the substrate, and the leg is held straight and stiff. The tip of the limb remains stationary while the limb musculature contracts, sweeping the body past the point of contact between the limb and the substrate. As the body progresses past the stationary end of the leg, the leg shortens until it is essentially perpendicular to the substrate, and then elongates as the portion of the body wall above the limb continues to move forward. Because of these well-timed changes in the length of the limb, the body itself shows little undulation, changing position neither laterally nor vertically with respect to the ground. If you could not see the legs, the body would appear to be gliding. The jointed legs of arthropods also show such a shortening during movement, but the mechanism by which this is accomplished is necessarily quite different.

As so far described, the hydrostatic skeleton appears to have a rather passive role in the locomotion of onychophorans. However, a more active involvement has been demonstrated through analysis of movie footage. As the animal moves, the length and width of the body is continually changing; indeed, the dimensions of the body rarely remain constant for more than a few seconds. Changes in body lengths are correlated with changes in speed and gait (i.e., in the manner of walking) (Fig. 17.5). These alterations of body dimensions are brought about through the actions of the circular and longitudinal musculature, interacting through a constant-volume, internal hydrostatic skeleton (the hemocoel). Because there are no internal septa in onychophorans, contractions in one part of the body can bring about rapid distension, or rigidity, in any other part of the body. Changes in the speed of the animal are brought about through changes in body length, limb length, the distance to which the legs reach forward on each stroke, and the amount of time required between the initiation of the forward swing and the completion of the power stroke. Changes in speed appear to be achieved primarily through alterations in the gait.

Onychophoran locomotion is somewhat similar to that found within the Polychaeta, in that outfoldings of the body wall (projecting ventrally in the onychophoran and laterally in the polychaete) are used to generate thrust, and in that a hydrostatic skeleton is implicated. However, there are important differences. First, the onychophoran limb is stiffened not by internal rigid acicula, but rather by the intrinsic musculature of the limbs themselves and by hydrostatic forces generated by contraction of the body or limb musculature against the fluid-filled hemocoel. Moreover, the onychophoran body does not undulate; all thrust is generated by the limbs directly. In contrast, among the polychaetes, contractions of the longitudinal body wall musculature may be used to generate waves of body undulation (Fig. 17.5d). These undulations transmit additional thrust against the substrate through the stationary parapodia. Indeed, when polychaetes are really making headway, most of the progress is attributable to contractions of the body wall musculature rather than to the parapodial elements directly. An additional difference between the locomotory mechanics of the two groups is that the polychaete limb does not change length during the propulsive cycle as does the onychophoran limb. Locomotion in these two groups is clearly quite dissimilar. This fact has been used to support the contention that, although the Onychophora probably arose from annelid stock, onychophorans probably did not arise from a polychaete ancestor.

The evolutionary history of the Onychophora is unknown. Fossils of marine, onychophora-like animals from the mid-Cambrian period (approximately 600 million years ago) have been described, but it is impossible to tell whether these fossils are more like arthropods or more like annelids. Existence of these fossils suggests, at least, that the evolution of onychophoran-like limbs pre-dated movement of the group to land; limb development can thus be viewed as a pre-adaptation for a terrestrial existence in this group. But little else about the evolutionary history of onychophorans is even hinted at by examination of other animals, either living or extinct. Certainly there is no clear concensus about the degree of relationship of the Onychophora to the Arthropoda. Indeed, scientists who set the insects and myriapods (including the centipedes and millipedes) on a distinctly different evolutionary pathway from that of the Chelicerata and Crustacea, often place the Onychophora together with the above terrestrial arthropods (i.e., insects and myriapods) in a separate phylum, the Uniramia. The Uniramia are so-called in recognition of the fact that all members have only uniramous appendages.

Although the onychophorans are certainly related to both the arthropods and the annelids, for our purposes there would seem to be sufficient dissimilarities to warrant their placement in a separate phylum, serving as reminders that animals invariably refuse to be neatly pigeonholed into the categories of human-made classification schemes, and drawing our attention to what must be an intriguing phylogenetic linkage between the Annelida and the Arthropoda.

TOPIC FOR FURTHER DISCUSSION AND INVESTIGATION

Investigate the tolerance of low temperatures and desiccation by tardigrades.

Crowe, J. H., and A. F. Cooper, 1971. Cryptobiosis. *Sci. Amer.*, 225: 30.

Crowe, J. H., 1972. Evaporative water loss by the tardigrades under controlled relative humidities. *Biol. Bull.*, 142: 407.

Pigón, A., and B. Weglarska, 1955. Rate of metabolism in tardigrades during active life and anabiosis. *Nature*, 176: 121.

Pollock, L. W., 1975. Tardigrada, In: A. C. Giese and J. S. Pearse (Eds.), *Reproduction of Marine Invertebrates*, Vol. II. New York: Academic Press, pp. 43–54.

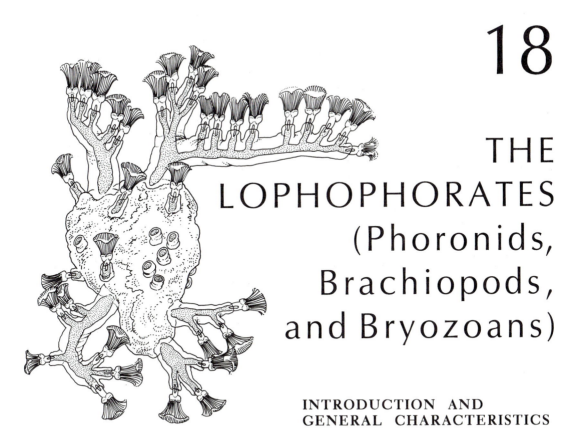

18

THE LOPHOPHORATES (Phoronids, Brachiopods, and Bryozoans)

INTRODUCTION AND GENERAL CHARACTERISTICS

The three phyla discussed in this chapter have uncertain phylogenetic relationships. Nevertheless, they have in common one major anatomical feature that may be homologous in the three groups—the **lophophore**. The lophophore is a circumoral (i.e., around the mouth) body region characterized by a circular or U-shaped ridge around the mouth. This ridge bears either one or two rows of ciliated, hollow tentacles. The space within the lophophore and its tentacles is always a coelomic cavity, and the anus always lies outside the circle of tentacles; both characteristics have come to be important parts of the definition of a lophophore. The lophophore of all species functions as a food collection organ and as a surface for gas exchange. A distinct head is never encountered among lophophorates. All members are sessile or sedentary suspension-feeders, employing the cilia of the lophophore to capture phytoplankton and small planktonic animals.

The coelomic cavity of the lophophore, called the **mesocoel,** is physically separated by a septum from another large coelomic cavity, the **metacoel** (Fig. 18.1a). In some groups, the septum is perforated or incomplete, so that the two coelomic cavities are interconnected. A third coelomic cavity, the **protocoel,** may be present anterior to the mesocoel. The protocoel is distinct but quite small in the adults of some lophophorate groups, but is particularly conspicuous during the larval stage of phoronids. The absence of a conspicuous protocoel in the re-

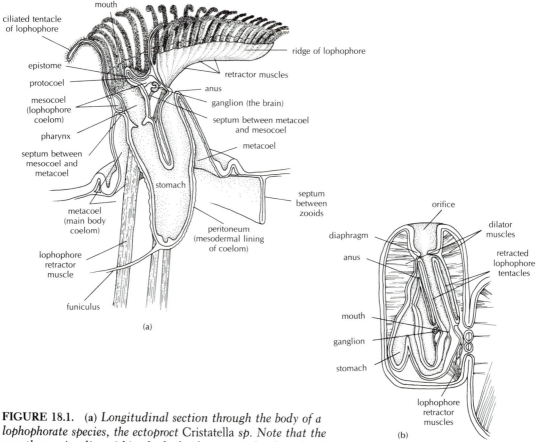

FIGURE 18.1. (a) *Longitudinal section through the body of a lophophorate species, the ectoproct* Cristatella *sp. Note that the mouth opening lies within the lophophore tentacles, but that the anus does not. (From Hyman; after Cori.) (b) Illustration of a polypide retracted into a cystid. Anatomical details are discussed further in the section on Phylum Bryozoa. (From Hyman; after Marcus.)*

maining lophophorate species appears to be related to the evolutionary reduction of the head.

One feature held in common among all lophophore-bearing animals is the possession of very simple gonads, arising from a portion of the mesodermal lining of the main (trunk) coelomic compartment, i.e. from the metacoel. Additionally, all species possess U-shaped digestive tracts, in which the anus (when present) terminates adjacent to the mouth (Fig. 18.1). Finally all species secrete some form of protective covering about the body.

Despite these similarities, there is no agreement about the likely origins of any of the three phyla, or on the most likely primitive lophophorate condition. The three groups differ markedly with respect to their circulatory and excretory systems; the nature of the protective coverings they secrete; and in important details of development. Indeed, the developmental patterns described for lophophorates are unlike those reported for any other coelomate group. Features characteristic of both protostome and deuterostome development occur within the lophophorate phyla, as do features uncharacteristic of either group. Cleavage is basically radial, as in deuterostomes, and cell fates are not fixed—at least in the species so far studied—until several cleavages have taken place. Such indeterminate cleavage is also a typical deuterostome characteristic. However, coelom formation is not distinctly schizocoelous (as it would be in a typical protostome) or enterocoelous(as it would be in a typical deuterostome). Both modes of coelom formation

have been observed among lophophorates. To further add to the intrigue, coelom formation in the Bryozoa occurs by neither of these mechanisms! The bryozoan coelom arises not during embryogenesis, but only following the completion of larval life. The mouth forms from the blastopore in at least one group (the Phoronida), a distinctly protostome characteristic, and typical protostome nephridia are found in larvae and adults of two of the groups (the Phoronida and the Brachiopoda). On the other hand, the tripartite coelom typically encountered among lophophorates is considered to be a feature unique to the deuterostomes. On balance, deuterostome characteristics are felt to outweigh the protostome characteristics, but the phylogenetic relationships of the three lophophorate phyla to other phyla, and to each other, are clearly uncertain. For this reason alone, lophophorates are a most intriguing group of animals.

Nearly all lophophorate species are marine. A few representatives of one of the three phyla (Bryozoa) are commonly encountered in fresh water. None of the species is terrestrial, in any sense of the word; the sedentary, suspension-feeding life-style that all lophophorates exhibit is not a likely preadaptation for life on land. Today, a total of approximately 4500 lophophorate species are known. Lophophorate diversity was far greater during the Paleozoic and Mesozoic Eras, approximately 135–500 million years ago. Seven to ten times as many species flourished in past millenia, but are with us now only as much-studied fossils.

PHYLUM PHORONIDA

The phoronids are the easiest of the lophophorate phyla to discuss, as all members conform closely to the same basic body plan. Members of the other two lophophorate phyla are not nearly so cooperative. Only about one dozen phoronid species are known, and all of these are marine.

Most phoronids live in permanent, chitinous tubes implanted in muddy or sandy sediments or attached to solid surfaces. A few species burrow into hard, calcareous substrates; even so, these species secrete a chitinous tube within the burrow. Adult phoronids do not move from place to place, although they can move within their tubes and, if artificially removed from these tubes, can burrow back into the sediment. A giant nerve fiber permits rapid withdrawal of the animal into the tube or burrow upon provocation.

Most of the phoronid body is nondescript (Fig. 18.2). Appendages are lacking, except for the anterior lophophore. A flap of tissue called the **epistome** covers the mouth (*epi* = G: around; *stoma* = G: mouth). The epistome is hollow, as it contains a remnant of the embryonic protocoel.

The lophophore is the only prominent external structure of phoronids. It consists of a conspicuous ring of tentacles, usually deeply indented to form a U-shape, and a less conspicuous, ciliated food groove (Fig. 18.3). Ciliary activity drives water into the ring of tentacles from the top of the lophophore, and outward through the narrow spaces between the tentacles. In this manner, suspended food particles can be captured by the

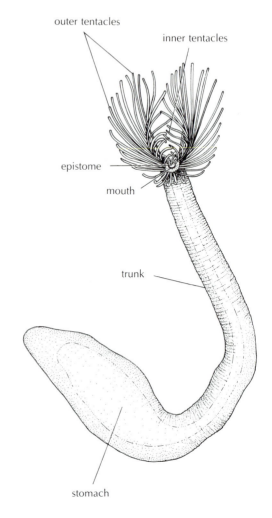

FIGURE 18.2. A *phoronid*, Phoronis architecta, *removed from its tube. (From Hyman; after Wilson.)*

tentacular cilia and mucus, transferred to the cilia of the food groove, and conducted to the mouth for ingestion.

Internally, phoronids possess a pair of metanephridia, with the ciliated nephrostomes collecting coelomic fluid and the nephridiopores discharging urine near the

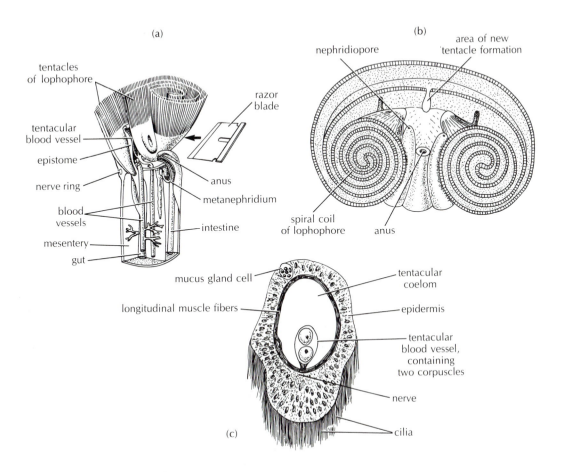

FIGURE 18.3. *Phoronid anatomy. (a) Lateral view of* Phoronis australis, *cut vertically to review internal anatomy. (b) Looking down on the lophophore, after the tentacles have been chopped off in the plane indicated by the razor blade in (a). The tentacles would be coming out of the page if they had not been chopped off. The lophophore is basically U-shaped, although the ends of the U are curled to form a spiral in this species. [(a, b) from Hyman; after Shipley and Benham.] (c) Cross section through a region of lophophore tentacle. Note the large coelomic space and the blood vessel running through the tentacle. (From Hyman; after Selys-Longchamps.)*

anus. A blood circulatory system with hemoglobin is present in all species. There is no distinct heart, but the major trunk blood vessel is contractile. Blood circulates largely through a series of interconnected, discrete vessels. Each tentacle of the lophophore is serviced by a single, small vessel, through which the blood both ebbs and flows.

Many biologists feel that the phoronids are closest to the primitive lophophorate condition, although most workers would not consider other lophophorates to have evolved from phoronids *per se*.

PHYLUM BRACHIOPODA

The external appearance of brachiopods, which include the so-called "lampshells," is quite unlike that of the phoronids. Brachiopods bear a superficial resemblance to the bivalved molluscs, in that the body is protected externally by a pair of convex, calcified shells (Fig. 18.4). Moreover, the shells are coated with a thin layer of organic periostracum, as in molluscs. Indeed, up until about 100 years ago, brachiopods were considered to be *bona fide* members of the Phylum Mollusca. Only about 300 brachiopod species currently exist, but more than 30,000 species are represented in the fossil record. This is a phylum that has clearly seen better times.[1]

The lophophore of brachiopods is similar to that of the phoronids except that it is drawn out into two arms (Fig. 18.5), increasing the effective surface area for food collection and gas exchange. Also, as in the

---

1 See Topics for Discussion, No. 2.

Phylum Brachio•poda
(G: ARM FOOT)

Phoronida, one or two pairs of metanephridia serve as excretory organs. A blood circulatory system is also present, but it contains no oxygen-binding pigments. A blood pigment is found in the coelomic fluid, but it is hemerythrin rather than hemoglobin. Hemerythrin, although it does contain iron, is structurally and functionally quite different from hemoglobin. The distribution of hemerythrin among animals has long been a puzzle to zoologists; it is found only in brachiopods, sipunculids, priapulids, and a few species of polychaetes, groups with no direct phylogenetic connection. Hemerythrin is not encountered among any other lophophore-bearing animals. Among brachiopods, blood is circulated by the action of a well-developed heart and from one to several contractile vessels associated with the main dorsal blood vessel. Subsequent to leaving this main vessel, the blood moves somewhat haphazardly through a system of interconnected blood sinuses.

Most brachiopods live permanently attached to a solid substrate or firmly implanted in sediment. Attachment is generally achieved by means of a stalk, called the **pedicle** (L: "little foot"), which protrudes posteriorly through a notch or hole in the ventral shell valve (Figs. 18.4, 18.5). The stalk is quite long and flexible in some species, perhaps serving to keep the body up above the substrate and in a zone of greater water flow.[2] This would benefit the brachiopod both in terms of increased gas exchange and increased rate of food capture. The pedicle is often muscular and hollow, hous-

---

2 See Topics for Discussion, No. 3.

FIGURE 18.4. *Photographs of an inarticulate brachiopod,* Lingula *sp. (right), and an articulate brachiopod, the lampshell* Terebratella *sp. (left).* Lingula *sp. has a long, muscular pedicle, which anchors the animal in a burrow in the sediment. In contrast, lampshells generally have a short, nonmuscular pedicle that attaches to rocks. The pedicle of* Terebratella *sp. is seen protruding from an opening in the ventral shell valve in one of the specimens. In a few articulate and inarticulate brachiopod species, the pedicle is completely lost; the left valves cement directly to solid substrate. (Photo by J. Pechenik and L. Eyster.)*

ing an extension of the main coelomic cavity (the metacoel) through which coelomic fluid can circulate.

The shells of brachiopods are composed of a protein matrix plus either calcium carbonate or calcium phosphate, and are secreted by two lobes of tissue referred to as the **mantle tissue**. This mantle tissue is by no means homologous with the shell-secreting molluscan tissue of the same name. Like the "mantle" of barnacles, the terminology simply reflects past ideas about the relationships

379

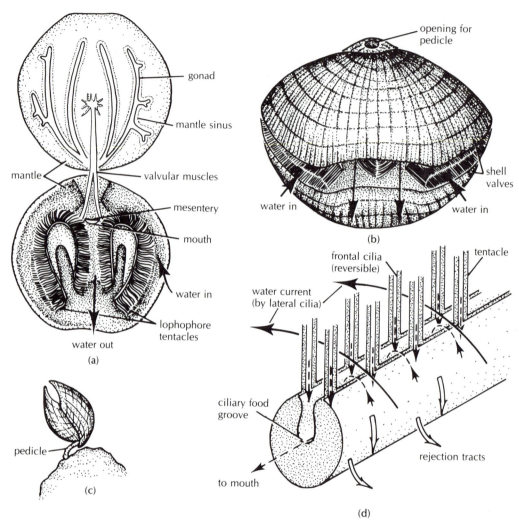

FIGURE 18.5. *(a) Brachiopod with valves opened to show orientation of lophophore. (From Beck and Braithwaite, 1968.* Invertebrate Zoology Laboratory Workbook, *3rd ed. Burgess Publishing Company, Minneapolis, Minnesota.) (b) Path of water flow through the lophophore of the brachiopod (*Terebratella *sp.) in the process of feeding. Note the opening through which the pedicle protrudes in life. (After Hyman; after Blochmann.) (c) The brachiopod* Hemithyris psittacea *in its normal feeding posture. (After Hyman.) (d) Detail of one part of a lophophore arm. Dotted arrows show paths of captured food particles in transit to the mouth, or being rejected (open arrows). Bold arrows indicate water flow through adjacent tentacles. (Modified after Russell-Hunter.)*

between animal groups, ideas that have changed substantially over the past hundred years or so. (Recall that both barnacles and brachiopods were once classified as molluscs.) The body of the brachiopod is oriented so that the shell valves are ventral and dorsal. This is quite unlike the situation encountered among the bivalved molluscs, in which the shell valves are to the left and right sides of the body.

In some brachiopod species, comprising the class Inarticulata, the shell valves are held together entirely by adductor muscles. In members of the only other brachiopod class, the Articulata, the shell valves are hinged, as in molluscan bivalves. That is, the margins on one side of the shells possess a series of interlocking teeth and sockets, locking the shells together and preventing substantial sliding of one valve relative to the other. In both classes, the shell valves are brought together by contracting the adductor muscles, as in bivalve molluscs (Fig. 18.6a). However, there is no springy bivalve-style hinge ligament to separate the valves when the adductors relax. Instead, separation of the valves from each other is an active process and depends upon the contraction of an opposing set of muscles, the **diductor muscles.** The shell thus acts as a complete skeletal system; not only do the shell valves protect the soft body parts, but they also serve as the vehicles through which the two muscle groups, the adductors and the diductors, antagonize each other.

Members of the two brachiopod classes also differ with respect to the chemical composition of their shells and the morphology of their digestive tracts (Table 18.1). The shells of articulate brachiopods are all strengthened with calcium carbonate. In contrast, those of inarticulate species usually contain calcium phosphate. The digestive tract of the inarticulates is always U-shaped, with a mouth and separate terminal anus. The digestive tract of articulate brachiopods, on the other hand, terminates blindly (Fig. 18.6a). Thus, the articulate brachiopods, sophisticated in so many other respects, make do without an anus. Lastly, the lophophore of some articulate brachiopod species contains a calcified internal support. Such a rigid internal support is never encountered among the Inarticulata.

PHYLUM BRYOZOA (= ECTOPROCTA) (= POLYZOA)

All three names for this phylum are widely used in the literature. The name "bryozoa" (moss animal) refers to the finely branched appearance of many common species. "Ectoprocta" emphasizes the fact that the anus lies outside the ring of tentacles, as in all lophophorates (*ecto* = G: outside; *proct* = G: anus). Finally, "polyzoa" (i.e., multiple animals) refers to the fact that all species are colonial. I will use the name Bryozoa, as it seems to be the most conspicuous term used in recent writings about this phylum.

All bryozoans secrete a house around the body. The contents of the house—i.e., the

Phylum Bryo•zoa
(G: MOSS ANIMALS)

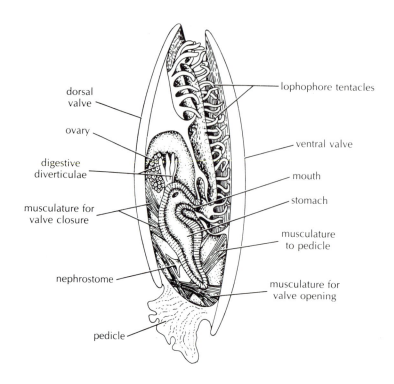

dorsal valve

ovary

digestive diverticulae

musculature for valve closure

nephrostome

pedicle

lophophore tentacles

ventral valve

mouth

stomach

musculature to pedicle

musculature for valve opening

FIGURE 18.6. *Diagrammatic illustration of the internal anatomy of an articulate brachiopod. Note the blind-ending intestine and the complex musculature operating the shell valves. Also note the conspicuous nephrostome. (After Harmer and Shipley.)*

TABLE 18.1. Comparison of the Two Major Brachiopod Classes

CHARACTERISTIC	ARTICULATA	INARTICULATA
shell	always calcium carbonate	usually calcium phosphate
	tooth and socket hinge	no articulating hinges
digestive tract	ends blindly	open at mouth and anus
rigid internal support	present in some species	never present

lophophore, gut, nerve ganglia, and most of the musculature—are referred to as the **polypide** (Fig. 18.7a). The house itself, plus the body wall that secretes it, is called the **cystid,** while the secreted, nonliving part of the house is termed the **zooecium** (*zoo* = G: animal; *oecus* = G: house). This rather confusing terminology evolved from the mistaken idea that the body wall of the house and the contents of the house were two separate individuals. That misconception was corrected quite some time ago, but the

terminology had been so widely used by that time that it has persisted. The body wall is actually attached to the zooecium, so that the bryozoan is essentially glued to its house. The entire individual (cystid and polypide) is termed a **zooid.**

Perhaps it is just as well to draw particular attention to the bryozoan body wall by giving it a special name, since it has a unique developmental potential; the entire zooid can be generated, or regenerated, from the body wall of the cystid. Such generation and

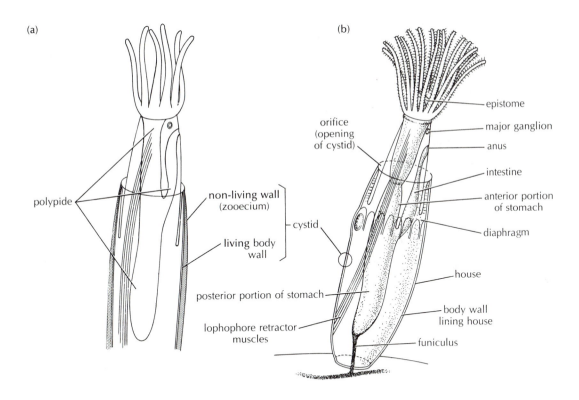

FIGURE 18.7. *(a) Diagrammatic illustration demonstrating terminology of the polypide, cystid, and zooecium. (b) The freshwater bryozoan (Phylactolaemata) Fredericella, showing details of the internal anatomy. (From Hyman; after Allman.)*

regeneration is, in fact, an integral part of the life cycle of all bryozoan species.

During the life of an individual zooid, the entire polypide periodically degenerates into a dark-pigmented, spherical mass known as a **brown body**. In most species, a new polypide is subsequently produced from the cystid. In some species, the brown body remains conspicuously present in the coelomic space of the new polypide. In other species, the brown body becomes engulfed by the digestive system of the regenerating polypide, soon to be discharged unceremoniously through the anus. Bryozoan zooids may go through four or more such cycles of degeneration and rebirth during their lives. In some instances, brown-body formation may provide a mechanism for bypassing unfavorable environmental conditions. Brown-body formation has also been thought to be a mechanism for the elimination of insoluble waste products, by disposing of the entire polypide—a rather remarkable method of garbage disposal, to say the least. Nephridia are lacking among the Bryozoa.

The septum dividing the metacoel from the mesocoel is very incomplete in bryozoans, so that the coelomic fluid of the body cavity is continuous with that of the lophophore and tentacle cavities. The lophophore of bryozoans is strikingly similar to that of phoronids, except that it can be retracted within the zooecium for protection and protruded at will for feeding and gas exchange. The lophophore of bryozoans is protruded through an **orifice** in the zooecium (Fig. 18.7b). In many species, a muscular **diaphragm** is positioned just beneath this orifice. Protrusion of the lophophore is accomplished indirectly, by increasing the hydrostatic pressure within the main coelomic cavity and dilating the diaphragm. The means of achieving the temporary increase in pressure within the coelomic cavity differ among bryozoan species, and will be discussed below.

Bryozoans show a great variety of external morphology, far greater than that encountered within other lophophorate groups. Moreover, all bryozoans are colonial; i.e., a single individual reproduces itself asexually to form a contiguous grouping of genetically identical individuals, up to about two million zooids in a single colony (Fig. 18.8a). All of these individuals are extremely small, usually less than one mm each.

Bryozoan colonies show a considerable variety of species-specific geometrical patterns. Colonies may be erect and branching or flat and encrusting. Great diversity of form is encountered within each of these two basic patterns. Many people have erect bryozoans in their homes without knowing it. Colonies of these marine species are dried, dyed, and sold commercially as "air ferns." No wonder they never need to be watered!

In many bryozoans, thick mesenchymal cords form tissue connections between the individual members of a colony. These cables are called **funicular cords.** A single cable is termed a **funiculus** (Figs. 18.1, 18.7). The funicular system of a single zooid may extend across the coelomic cavity from the stomach to a pore in the body wall, providing a mechanism through which direct transfer of nutrients among adjacent zooids may be possible. The funicular system may be homologous with the circulatory systems of other metazoans, including other lophophorates.

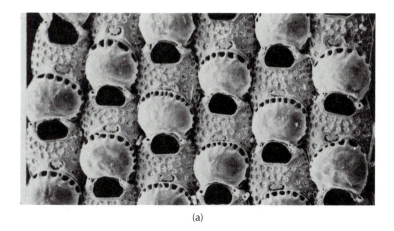

(a)

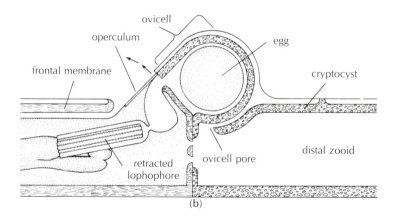

(b)

FIGURE 18.8. *(a) Scanning electron micrograph of the marine bryozoan* Fenestrulina malusii. *Each zooid is associated with an ovicell, in which a single embryo is brooded after fertilization, as indicated in (b). The arrows adjacent to the operculum indicate the direction of opercular movement. (Courtesy of C. Nielsen, from Nielsen, 1981. Ophelia, 20: 91.)*

The majority of bryozoans (about 4000 species) are marine, but about 50 species are restricted to fresh water. These 50 species constitute the only nonmarine lophophorates. Approximately 15,000 additional marine bryozoan species are known only from the fossil record.

Bryozoans are divided into two major

classes and one minor class, based largely on differences in the morphology of the lophophore, mechanism of lophophore protrusion, chemical composition of the external body covering, presence or absence of an epistome, presence or absence of body-wall musculature, and habitat.

CLASS PHYLACTOLAEMATA

All members of this class (about 50 species) are found in fresh water, and most (but not all) freshwater bryozoans are members of this class. The zooids of a given colony are morphologically identical; i.e., the colonies are **monomorphic**. Some species secrete a chitinous outer covering, but others produce thick gelatinous surroundings (Fig. 18.9g). In some species, the diameter of a single colony can exceed 60 centimeters. Colonies develop on a variety of submerged, solid surfaces, including shells, rocks, and the leaves and branches of freshwater vegetation. Most species are permanently affixed to these substrates and are incapable of locomotion. However, the colonies of a very few species are capable of slowly moving from place to place, using the single muscular "foot" shared by all of the zooids in the colony.

The lophophore is U-shaped in all phylactolaemate species, as in phoronids. A flap of tissue, termed the **epistome** as in phoronids, hangs over the mouth. A protocoel is evident within the epistome of some species. As in the phoronids, the body wall contains both circular and longitudinal muscles. A pronounced, hollow funiculus extends from the stomach into the coelomic cavity shared by all of the zooids (Fig. 18.7).

Protraction of the lophophore is brought about by the contraction of muscles in the deformable body wall.[3] Because the coelomic fluid is essentially incompressible, this contraction increases the pressure within the metacoel and, since the septum between the two main body cavities is incomplete, squeezes coelomic fluid from the metacoel into the mesocoel of the lophophore. Contraction of the body wall musculature thus inflates the tentacles of the lophophore and increases the hydrostatic pressure within the metacoel as well. Once the orifice diaphragm is opened by the contraction of specialized dilator muscles, the lophophore is forced out from the cystid by the elevated hydrostatic pressure within the metacoel. Retractor muscles extending from the body wall to the lophophore are responsible for bringing the lophophore back within the zooecium.

Perhaps the most intriguing feature unique to the phylactolaemates is the formation of **statoblasts**. These structures, produced seasonally, can withstand considerable desiccation and thermal stress. The statoblasts consist of a cell mass enclosed within a bivalved protective capsule of species-specific morphology. They are produced along the funiculus of each zooid (Fig. 18.10), often in great numbers, and are usually released from the zooecium upon the degeneration of the polyp in the late fall. When environmental conditions improve the following spring, the two valves of the statoblast separate along a preformed suture line, and a polypide soon emerges. Colony

---

3 See Topics for Discussion, No. 4.

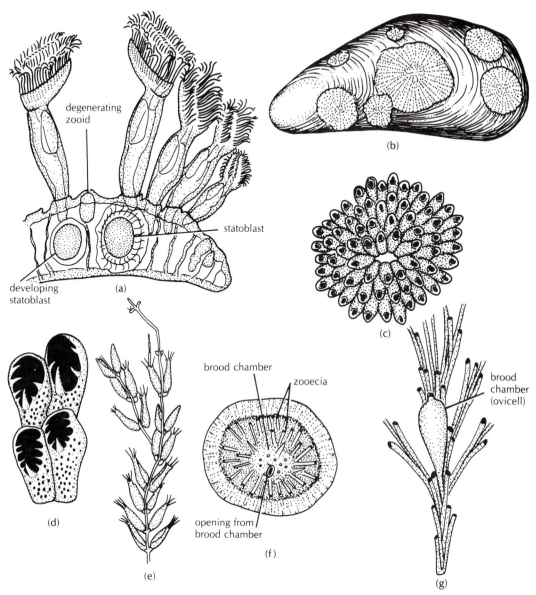

FIGURE 18.9. *Bryozoan diversity.*
Phylactolaemata: (a) Cristatella colony. This
species is found commonly encrusting lily pads
in freshwater ponds. Note the gelatinous,
creeping sole. The degenerating zooid is forming
a brown body. (After Hyman.) Gymnolaemata:
(b) Mussel shell valve encrusted with colonies of
Cryptosula, *shown in detail in frame (c). (From*
Hyman; after Rogick.) (d) Electra pilosa. Note
the conspicuous spines guarding the frontal

opening. (From Smith, 1964. Keys to Marine
Invertebrates of the Woods Hole Region.
Marine Biological Laboratory, Woods Hole; after
Rogick and Croasdale.) (e) Bowerbankia gracilis,
with tentacles withdrawn. (From Smith; after
Rogick and Croasdale.) (f) Lichenopora colony.
Note the communal brood chamber. (From
Hyman; after Hincks.) Stenolaemata: (g) Crisia
ramosa, *showing tubular zooecium. (After*
Harmer and Shipley.)

387

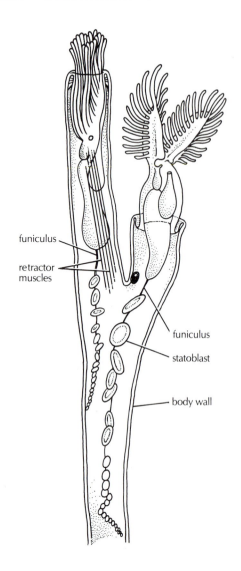

funiculus

retractor
muscles

funiculus

statoblast

body wall

FIGURE 18.10. *Statoblast formation in the freshwater bryozoan Plumatella repens. Note the association of the statoblasts with the funiculus and the lack of boundaries between adjacent zooids. (After Harmer and Shipley.)*

formation ensues by standard means, through asexual budding of additional zooids.

The discerning reader may see the relationship between statoblast formation by phylactolaemate bryozoans and the formation of gemmules and resting eggs by freshwater sponges and tardigrades, respectively. Clearly, formation of resting (i.e., **diapause**) stages is a common adaptation to the vagaries of freshwater existence.

CLASS GYMNOLAEMATA

Most extant bryozoan species are assigned to the class Gymnolaemata (Fig. 18.9a–f). The gymnolaemates are primarily marine. They are especially fascinating animals, even by bryozoan standards, displaying a broad range of morphological and functional diversity. Bryozoans are heavily preyed upon by turbellarians, polychaetes, insect larvae, crustaceans, arachnids (mites), gastropods, asteroids, and fish. To a great extent, the story of gymnolaemate evolution is one of increasing the degree to which the polypide is protected from these predators. In general, this is achieved by strengthening the zooecia. This is not so simply accomplished among the Bryozoa. Recall that lophophore protrusion depends upon a change of shape of the animal, causing an elevation in coelomic pressure, and that the body wall is joined to the zooecium. The trick, then, is to strengthen the zooecium without losing the ability to generate elevated pressures within the metacoel.

Gymnolaemates differ from the phylactolaemates in a number of major respects. In phylactolaemates, the walls of the cystid are very incomplete, so that morphological boundaries between adjacent zooids are lacking (Figs. 18.9g, 18.10). Although each zooid protracts its lophophore through its own orifice, the polypides of a colony share a common metacoel. In contrast, the zooids of gymnolaemates are morphologically distinct individuals, although small **pore plates** allow the exchange of coelomic fluid between neighboring zooids (Fig. 18.11). The two classes also differ with respect to lophophore morphology. Whereas the lophophore of phylactolaemate bryozoans has the deeply invaginated U-shape seen in phoronids, that of the gymnolaemates is circular in appearance.

In a few species, the feeding zooids are borne on stolons, i.e., on tubular extensions of the body wall. The stolons may be upright (**erect**) or flat against the substrate. In most gymnolaemate species, however, stolons are absent. Instead, zooids are contiguous, with the zooecium of one zooid supported by the zooecium of adjacent zooids (Fig. 18.8a), thereby strengthening the entire colony.

The colonies of most marine bryozoan species are composed of a large array of small, boxlike or elliptical houses, typically encrusting on any submerged, solid surface. Zooids are arranged in species-specific patterns, determined by the pattern of asexual budding. The zooecia themselves are characterized by species-specific morphology and are therefore of great taxonomic value.

There are two orders within the Gymnolaemata: the Ctenostomata and the Cheilostomata. The ctenostomes are characterized by a flexible, chitinous zooecium. Protraction of the lophophore is achieved in much the same way as in the phylactolaemates. Muscle contractions draw the zooecium inward, and the resulting increase in hydrostatic pressure of the coelomic fluid forces the lophophore out through the orifice.

Most gymnolaemate bryozoans are cheilostomes. Cheilostomes show varying degrees of calcification of the zooecia. In many species, the orifice is sealed by a hinged, calcareous **operculum** when the polypide is fully withdrawn within the zooecium (Fig. 18.8b, 18.11). Calcium carbonate is liberally deposited between the chitinous cuticle of the zooecium and the epidermis of the cystid of most species. The **frontal membrane** often remains uncalcified. Muscles attach from the calcified body wall to this flexible surface (Fig 18.11a, top); protraction of the lophophore is achieved by the contraction of this musculature and the resulting increase in coelomic pressure. The body wall musculature is dramatically reduced or, more frequently, completely absent. Although calcification in these species certainly strengthens the colony structurally and probably provides some protection from predators, the zooid remains vulnerable through the uncalcified frontal membrane.

Often, the frontal membrane is partially shielded by spines projecting from the frontal margins of the zooecium, or, more rarely, from the frontal membrane itself. Nevertheless, the frontal membrane remains the Achilles' heel of the zooid.

In some cheilostome species, the frontal

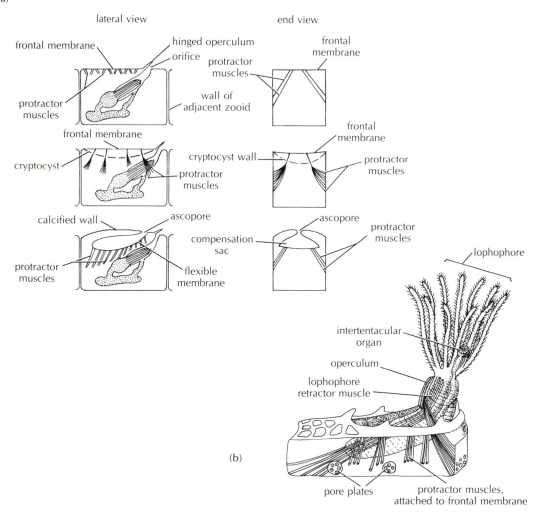

(a)

lateral view   end view

frontal membrane   hinged operculum   frontal membrane

orifice

protractor muscles

protractor muscles   wall of adjacent zooid

frontal membrane

cryptocyst   cryptocyst wall   frontal membrane

protractor muscles   protractor muscles

calcified wall   ascopore   ascopore

compensation sac   protractor muscles

protractor muscles   flexible membrane

lophophore

intertentacular organ

operculum

lophophore retractor muscle

(b)

pore plates   protractor muscles, attached to frontal membrane

FIGURE 18.11. *Functional diversity among the gymnolaemates, showing different degrees of protection of the frontal membrane. (After Clark; after Harmer.) (b) Electra pilosa, a species with a well-exposed frontal membrane, shown with lophophore protruded from the zooecium. (After Clark; after Marcus.)*

membrane remains uncalcified and flexible, but a calcareous shelf, called the **cryptocyst,** is secreted beneath it (Fig. 18.11a, middle). Muscles pass to the frontal membrane through small holes in the cryptocyst. Thus, the lophophore can still be protracted by muscles pulling downward on the frontal membrane, but the polypide can be protected within the zooecium by the calcified cryptocyst. The frontal membrane itself, however, remains vulnerable to attack.

A different means of protecting the polypide has evolved in yet a third group of cheilostomes. In these species, the entire zooecium calcifies, including the frontal membrane. All of the soft tissues are thus completely protected within the zooecium, but the frontal membrane can no longer serve as the vehicle through which internal hydrostatic pressure can be elevated. Instead, these species possess a new, uncalcified membrane separating the metacoel from the calcified frontal surface. A space exists between the frontal wall and this new membrane, forming a sac—specifically, a **compensation sac** (Fig. 18.11a, bottom). The compensation sac opens to the outside through a single pore, the **ascopore,** in the frontal surface (*ascus* = G: sac). Muscles extend from the calcified sides of the zooecium to the undersurface of the compensation sac. Contraction of these muscles pulls the compensation sac tissue inward, elevating the pressure within the coelomic cavity and forcing the lophophore out through the orifice of the zooecium. A vacuum does not form within the compensation sac during this process because sea water is drawn into the compensation sac through the ascopore.

The zooids described so far are the feeding and reproductive members of the colony, the **autozooids.** All gymnolaemate colonies contain some nonfeeding members as well. These **heterozooids** take a variety of forms. The stolons and holdfasts of stoloniferous species are composed of a series of short heterozooids. Strangely enough, the pore plates permitting the exchange of coelomic fluid between adjacent zooids are also highly specialized heterozooids in some species. Other heterozooids, found only among the cheilostomes, are specialized for protection and cleaning of the colony. One such heterozooid is little more than a highly modified operculum, drawn out into a long moveable bristle. The continual sweeping movements made by these **vibracula** (Fig. 18.12d) presumably discourage invertebrate larvae from attaching to the colony and keep the colony surface free of debris. Other heterozooids form immobile, protective spines.

Probably the most intriguing of the heterozooids are the **avicularia.** Avicularia, in their least modified form, have the external appearance of a normal autozooid (Fig. 18.12c). Internally, the avicularia consist of little more than well-developed muscle fibers extending across a capacious coelom to the vicinity of the ventral, hinged operculum. The polypide is much reduced, or completely absent. The operculum can be closed suddenly and with considerable force, discouraging, mutilating, or even killing potential predators. The avicularia of some species are highly modified for their tasks. In the most extreme form, the avicularia consist primarily of a modified opercular system, now referred to as a **mandible** (Fig. 18.12a, b). These avicularia may be mounted on a stalk, permitting some rotation. Some of these very specialized avicularia closely resemble a bird's head. In fact, it is this resemblance that gave rise to the name "avicularia" (*aves* = L: bird, thus "avicularia" = "little bird").

CLASS STENOLAEMATA

A small number of bryozoans are sufficiently dissimilar from the phylactolaemates and gymnolaemates to warrant placement into a third class, the Stenolaemata. The members of this class are all marine, and the living

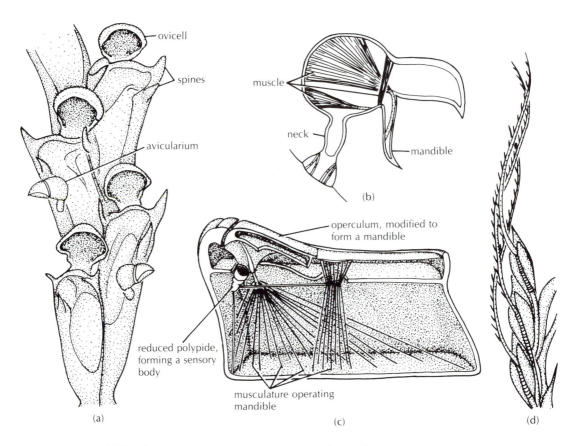

FIGURE 18.12. (*a*)Bugula *sp.*, *an erect, marine bryozoan with ovicells and highly modified avicularia. (From Hyman; after Rogick and Croasdale.) (b) Detail of the avicularia shown in (a). (From Laverak and Dando, 1979. Essential Invertebrate Zoology, 2nd ed. John Wiley & Sons; after Ryland.) (c) Relatively unmodified avicularium of* Flustra foliacea, *showing more clearly the relationship between the avicularium and a normal zooid. (From Silen, in Woollacott and Zimmer (Eds.), 1977. The Biology of Bryozoa, Academic Press.) (d) Caberea ellisi, showing three members of a colony, each in possession of a single long vibraculum. (From Smith.)*

In the figure, the labels read: ovicell, spines, avicularium (panel a); muscle, neck, mandible (panel b); operculum, modified to form a mandible, reduced polypide, forming a sensory body, musculature operating mandible (panel c). Panels are labeled (a), (b), (c), (d).

species belong to a single order, the Cyclostomata. Cyclostome zooids are always tubular and erect, and the zooecia are completely calcified (Fig. 18.9f). Not surprisingly, the cystid is nonmuscular. A thin, cylindrical membrane divides the main coelomic space into two compartments, one of which contains the polypide. Protraction of the lophophore is accomplished by the sideways, muscular displacement of this membrane. As in all bryozoans, the lophophore is drawn back into the zooecium by powerful lophophore retractor muscles.

OTHER FEATURES OF LOPHOPHORATE BIOLOGY

1. *Reproduction*

Asexual reproduction is most characteristic of the Bryozoa, in keeping with their habit of colony formation. Asexual reproduction by budding or fission is encountered in only one or two phoronid species, and reproduction is exclusively sexual in brachiopods. Brachiopods are also the exception when it comes to sexuality, most species being dioecious. In contrast, most bryozoans and phoronids are hermaphrodites. Gametes are formed by simple gonads within the metacoel. Discharge of gametes in phoronids and brachiopods is usually through the nephridia. Bryozoans lack nephridia; instead, sperm are released through the tentacles of the lophophore, and eggs are released through a special pore located between two of the tentacles. Copulation does not occur among lophophorates.

Phylum Phoronida

In phoronid reproduction, sperm emitted through the nephridiopores of one individual are harvested from the sea water by another, neighboring individual. A characteristic, ciliated feeding larval stage called an **actinotroch** is produced following fertilization and cleavage (Fig. 18.13a). Metamorphosis to adult form involves a rapid and dramatic "turning inside out" of the larval body.

Phylum Brachiopoda

Among brachiopods, sperm and eggs are discharged through the nephridiopores and fertilization occurs in the surrounding sea water. Free-swimming, ciliated larvae are produced (Fig. 18.13b).

Phylum Bryozoa

Among the bryozoans, sperm are released into the coelomic cavity and exit from openings in the tentacles of the lophophore. Neighboring individuals collect the sperm from the surrounding sea water. Once the eggs have been fertilized, a period of brood protection generally follows. Some species brood their embryos within the metacoel, but a variety of other brooding sites are found within the phylum. Most cheilostomes possess specialized brooding chambers, called **ovicells,** at one end of the zooid (Fig. 18.8). As the embryos develop within the various brood chambers, the polypide of the parent usually degenerates. This presumably reflects a transformation of parental tissues into nutrition for the offspring. The polypide of the parent may later be regenerated from the cystid.

One or, more rarely, several nonfeeding, ciliated **coronate larvae** eventually emerge from each ovicell (Fig. 18.13c). After a limited period of dispersal away from the parent colony, attachment to a substrate is accomplished through the eversion of a sticky **adhesive sac,** and metamorphosis to adult form ensues (Fig. 18.14). During metamorphosis, all larval tissues move to the interior of the animal and are destroyed through a combination of phagocytosis and autolysis. Only the outer body wall remains intact, and this becomes the cystid. The first zooid of a colony is termed an **ancestrula.** The rest of the colony is subsequently generated via asexual budding.

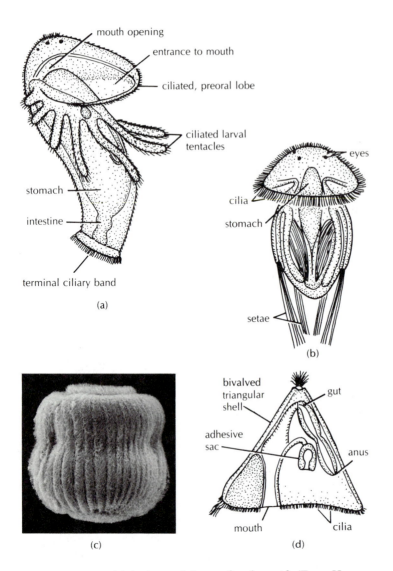

mouth opening

entrance to mouth

ciliated, preoral lobe

ciliated larval
tentacles

stomach

intestine

terminal ciliary band

(a)

eyes

cilia

stomach

setae

(b)

(c)

bivalved
triangular
shell

gut

adhesive
sac

anus

mouth

cilia

(d)

FIGURE 18.13. *(a) Actinotroch larva of a phoronid. (From Hyman; after Wilson.) (b) Larva of the brachiopod* Argyrotheca *sp. (From Hyman; after Kowalevsky.) (c) Scanning electron micrograph of the coronate larva of the bryozoan* Bugula neritina. *(Courtesy of R.M. Woollacott and C.G. Reed.) (d) Cyphonautes larva of a bryozoan. (After Hardy.)*

The fertilized eggs of a few, nonbrooding species develop into feeding larval stages called **cyphonautes larvae** (Fig. 18.13d). The cyphonautes larva is enclosed within a pair of triangular, chitinous valves, and is ciliated on the marginal surfaces not covered by these "shells." The shell valves are lateral, as in bivalved molluscs. The cyphonautes is

394

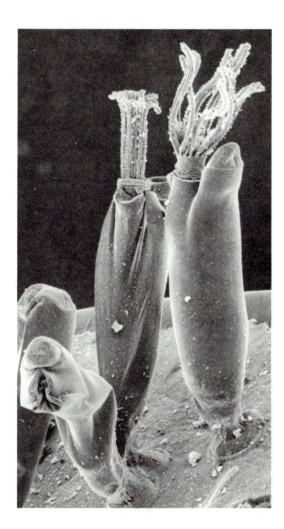

FIGURE 18.14. *Four ancestrulae of* Bugula neritina, *newly metamorphosed from four coronate larvae. (Courtesy of R.M. Woollacott and C.G. Reed.)*

equipped with a complete digestive tract, enabling it to feed and remain in the plankton for long periods of time, perhaps as long as several months. Attachment to a substrate is eventually accomplished by the eversion of

an adhesive sac, and metamorphosis to the ancestrula quickly follows.

2. Digestion

The digestive tract of all phoronids and bryozoans is U-shaped, with a mouth and separate anus. The anus always lies outside the circle of lophophore tentacles, by definition. Digestion occurs within the relatively large stomach, and has both extracellular and intracellular components. Brachiopods may (Inarticulata) or may not (Articulata) possess a U-shaped, one-way digestive system. In either case, the stomach connects to a large **digestive gland,** within which food material is broken down. Digestion is believed to be primarily intracellular.

3. Nervous System

The nervous system takes the form of a ring in all lophophorate species. This ring is found at the base of the lophophore in phoronids; encircling the esophagus in brachiopods; and adjacent to the pharynx in bryozoans. The ring may or may not be distinctively ganglionated. From the nerve ring, nerve fibers innervate the tentacles and musculature. An epidermal nerve network is commonly found in the body wall. Discrete sense organs are rare among lophophorates. Although a balance organ (**statocyst**) has been described in one species of inarticulate brachiopod, the sensory apparatus of other lophophorate species consists of scattered mechanoreceptor and chemoreceptor cells and/or setae.

TAXONOMIC SUMMARY

PHYLUM PHORONIDA

PHYLUM BRACHIOPODA—THE LAMPSHELLS
    CLASS INARTICULATA—THE INARTICULATE
          BRACHIOPODS
    CLASS ARTICULATA—THE ARTICULATE BRACHIOPODS

PHYLUM BRYOZOA (= ECTOPROCTA)(= POLYZOA)—
          THE MOSS ANIMALS
    CLASS PHYLACTOLAEMATA
    CLASS GYMNOLAEMATA
      ORDER CTENOSTOMATA
      ORDER CHEILOSTOMATA
    CLASS STENOLAEMATA
      ORDER CYCLOSTOMATA

TOPICS FOR FURTHER DISCUSSION AND INVESTIGATION

1. The Entoprocta is a phylum of colonial, suspension-feeding animals restricted to marine environments. Like the lophophorates, entoprocts collect food particles using an anterior organ bearing numerous ciliated tentacles (Fig. 18.15). However, since the anus lies within this ring of tentacles (*ento* = G: within; *procta* = G: anus), the organ is not, by definition, a lophophore. Discuss the anatomical similarities and differences between bryozoans (ectoprocts) and entoprocts, and the evidence for and against a close evolutionary relationship between these two groups.

Hyman, L.H., 1951. *The Invertebrates*, Vol. III. Acanthocephala, Aschelminthes and Entoprocta. New York: McGraw-Hill.

Mariscal, R.N., 1965. The adult and larval morphology and life history of the entoproct *Barentsia gracilis* (M. Sars, 1835). *J. Morphol.*, 116: 311.

Nielsen, C., 1977. The relationships of Entoprocta, Ectoprocta and Phoronida. *Amer. Zool.*, 17: 149.

Nielsen, C., 1977. Phylogenetic considerations: the protostomian relationships. In: Woollacott, R.M., and R.L. Zimmer (Eds.), *Biology of Bryozoans*. New York: Academic Press, pp. 519–534.

2. Discuss the potential role of bivalve molluscs in bringing about the dramatic evolutionary decline of brachiopods.

Gould, S.J., and C.B. Calloway, 1980. Clams and brachiopods—ships that pass in the night. *Paleobiology*, 6: 383.

Stanley, S.M., 1968. Post-Paleozoic adaptive radiation of infaunal bivalve molluscs; a consequence of mantle fusion and siphon formation. *J. Paleontol.*, 42: 214.

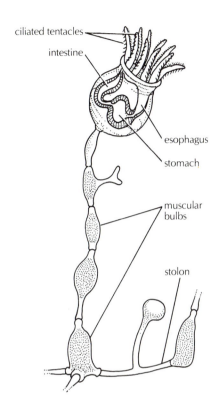

ciliated tentacles

intestine

esophagus

stomach

muscular
bulbs

stolon

FIGURE 18.15. *An entoproct*, Arthropodaria
sp. (From Hyman; after Nasonov.)

3. Investigate the role of water currents in the feeding biology of lophophorates.

Eshleman, W.P., and J.L. Wilkins, 1979. Brachiopod orientation to current direction and substrate position (*Terebratalia transversa*). *Canadian J. Zool.*, 57: 2079.

Gilmour, T., 1979. Ciliation and function of food-collecting and waste-rejecting organs of lophophorates. *Canadian J. Zool.*, 56: 2142.

LaBarbera, M., 1977. Brachiopod orientation to water movement. I. Theory, laboratory behavior, and field orientations. *Paleobiology*, 3: 270.

Strathmann, R.R., 1973. Function of lateral cilia in suspension-feeding of lophophorates (Brachiopoda, Phoronida, Ectoprocta). *Marine Biol.*, 23: 129.

4. Based upon your readings in this book, compare and contrast the operation of the bryozoan lophophore with the operation of the nemertine proboscis.

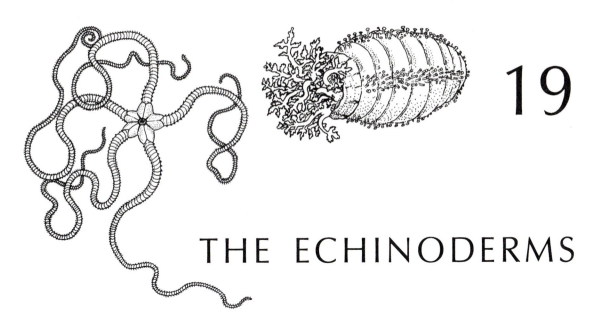

19

THE ECHINODERMS

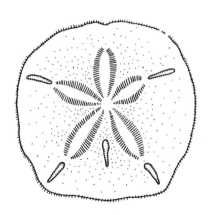

Phylum Echino•dermata
(G: SPINE SKIN)

INTRODUCTION AND GENERAL CHARACTERISTICS

This phylum includes the sea lilies, feather stars, brittle stars, sea stars, sand dollars, sea urchins, sea biscuits, and sea cucumbers.

There are approximately 6000 echinoderm species living today, all of which are marine or estuarine. An additional 13,000 or so species, distributed among approximately 20 classes, are known from the fossil record. Most adult echinoderms show a basic five-point radial symmetry, although most members of this phylum pass through bilaterally symmetric larval stages during development. As radially symmetric animals, cephalization is lacking. Thus, echinoderms do not generally have anterior and posterior ends. Instead, body surfaces are designated as being either **oral** (bearing the mouth) or **aboral** (not bearing the mouth).

Most echinoderms possess a well-developed internal skeleton composed

399

largely (up to 95%) of calcium carbonate, with smaller amounts of magnesium carbonate (up to 15%), even lesser amounts of other salts and trace metals, and a small amount of organic material. The components of the echinoderm skeleton are individually manufactured within specialized cells. This is in sharp contrast to the method of shell production in molluscs and other invertebrate groups, in which minerals are deposited into an extracellular protein matrix.

The major unifying characteristic of the phylum Echinodermata is the presence of what is known as the **water vascular system** (often abbreviated as the **WVS**). The WVS consists of a series of fluid-filled canals derived primarily from one of three pairs of coelomic compartments that form during development. These canals lead to thin-walled tubular structures called **podia**, or **tube feet** (*podium* = G: a foot). (See Fig. 19.1a.) The podia are best visualized as tubular extensions of the WVS that penetrate the echinoderm body wall and skeleton in particular regions. The areas in which podia are found are known as **ambulacral zones**, or, in some groups, **ambulacral grooves** (*ambulacr* = L: walk). The system of internal WVS canals is generally linked to the outside sea water through a sieve-plate called the **madreporite**, which leads down a **stone canal** (so named because it is reinforced with spicules or plates of calcium carbonate), and then to a **ring canal,** which forms a ring around the esophagus in all echinoderm species (Fig. 19.1a). Accessory fluid-storage structures called **Polian vesicles** and **Tiedemann's bodies** are often associated with the ring canal. Five (or some multiple thereof) **radial canals** radiate

symmetrically from the ring canal. Pairs of bulb-shaped **ampullae** usually connect to these radial canals. Each ampulla services a single tube foot. Both ampullae and tube feet are supported by a system of calcareous ossicles, the **ambulacral ossicles.**

Tube feet often lack circular muscles (Fig. 19.1b). In most species, fluid is pumped into the tube foot by the contraction of the ampulla, extending the foot hydraulically. A one-way valve at the juncture of the ampulla and radial canal ensures that fluid flows from ampulla to tube foot when the ampulla contracts, rather than to the radial canal. Retraction of the tube foot occurs when longitudinal muscles in each tube foot contract.

A single echinoderm may possess more than 2000 tube feet.[1] Locomotion often requires the coordinated protraction and retraction of all these feet. The inner surface of each tube foot is well ciliated, so that the fluid in the WVS circulates. The thin-walled podia can thus function effectively as respiratory structures as well as locomotory structures, and the fluid of the WVS, together with that of the other coelomic compartments, functions as the primary circulatory medium. Tube feet may also be the primary sites of excretion (by simple diffusion), and, in some groups, function in chemoreception and food collection as well.

Specialized excretory organs are never found among echinoderms. A true heart is also absent.

Associated with the WVS of echinoderms is a peculiar system of tissues and organs

---

1 See Topics for Discussion, No. 5.

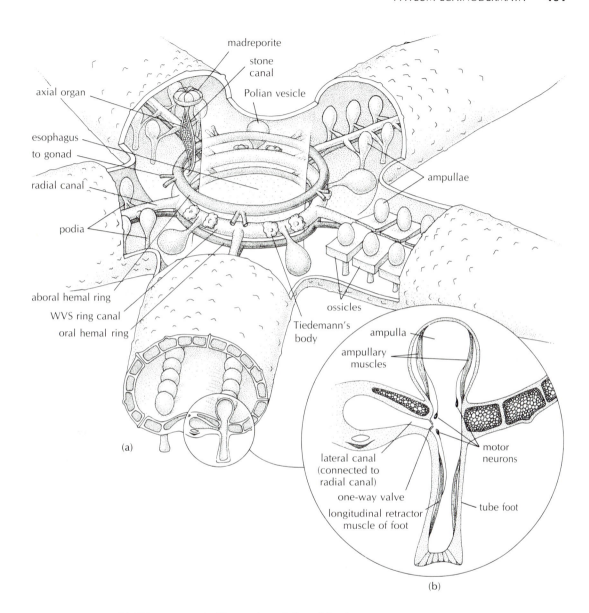

FIGURE 19.1. *(a) Diagrammatic illustration of the echinoderm water vascular system. The hemal system is also shown. Modified after Fretter and Graham, and others.) (b) Detail of individual tube foot. (Modified after Fretter and Graham; after Smith.)*

known as the **hemal system.** A major component of this hemal system is a spongy **axial organ** that lies adjacent to the stone canal of the WVS (Fig. 19.1a). The axial organ is housed in its own coelomic compartment, the **axial sinus.** The axial organ connects to two **hemal rings,** one oral and the other aboral. From the aboral hemal ring, strands of tissue, each contained within a coelomic, **perihemal canal,** extend outward to the gonads. Another series of strands radiate from the oral hemal ring to the tube feet.

The functional significance of the hemal system has long been uncertain. There do not appear to be any direct connections between the hemal system and the WVS, or between the hemal and digestive systems. Recent studies with asteroids (sea stars) and holothurians (sea cucumbers) suggest that the hemal system functions in the transport of nutrients from the coelomic fluid to the gonads. This conclusion is based in part on examination of the sites of uptake of ^{14}C-labeled food material; radioactivity appeared sequentially in the digestive system, hemal system, and, finally, in the gonads. In addition, concentrations of carbohydrates, lipids, proteins, and amino acids in hemal fluid may be ten times higher than in other echinoderm fluids. How nutrients move from the digestive system to the hemal system remains uncertain.

The axial organ may also be involved in producing intriguing cells called **coelomocytes,** which are found in nearly all echinoderm tissues and body fluids. These coelomocytes are involved in recognizing and ingesting foreign material; synthesizing pigments and collagen (for connective tissue);

transporting oxygen (some coelomocytes contain hemoglobin) and nutritive material; and digesting food particles. They also play a role in wound repair.[2]

CLASS CRINOIDEA

The class Crinoidea is the oldest of the echinoderm classes extant, and its members show many characteristics that, based upon studies of fossils, appear to be primitive. (It should be remembered that "primitive" characteristics are those that appear to show the least change from the presumed ancestral condition; the word does not imply a relative lack of complexity.) The Crinoidea is comprised of the stalked crinoids (the sea lilies, of which about 80 species exist), and the non-stalked, motile comatulid crinoids (the feather stars), of which about 550 species are known.

The sea lilies, although a small group at present, were quite numerous 300–500 million years ago.[3] Most fossil crinoids were, as the sea lilies remain today, permanently attached to the substrate by a **stalk** (Fig. 19.2). The stalk is flexible, being composed of a series of calcareous discs (**columnals**) stacked one on top of the other. The feeding and reproducing part of the animal is situ-

---

2 See Topics for Discussion, No. 2.
3 See Topics for Discussion, No. 7.

Class Crinoidea
(G: LILY-LIKE)

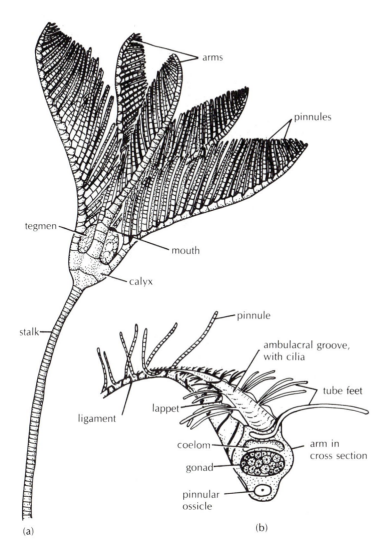

FIGURE 19.2. *(a) Diagrammatic illustration of a stalked crinoid. (Modified from Hyman, 1955. The Invertebrates, Vol. IV. McGraw-Hill. Reproduced with permission.) (b) Detail of a crinoid arm. (From Hyman; after Carpenter.)*

ated at the top of the stalk. The digestive system is tubular, consisting of a mouth, intestine, and terminal anus, and is confined entirely to a calyx/tegmen complex. The **ca**-lyx is a cup-shaped structure containing the complete digestive system. The calyx is covered by a membrane (the **tegmen**) that bears the mouth. The mouth, and therefore the oral surface in general, is always directed away from the base of the calyx. Both the calyx and tegmen generally bear numerous protective calcareous plates on the outer surface.

From 5 to more than 200 arms (usually in some multiple of 5) extend outward from the calyx. These arms, like the stalk, consist of a series of jointed calcareous ossicles, so that the arms can bend. The arms bear tube feet (podia).

Two rows of tubular **pinnules** extend outward from each arm, one row on each side of the ambulacral groove, which runs the length of the arm. Thin elongated podia, grouped in triplets, also flank the ambulacral groove of each arm. These podia are studded with mucus-secreting glands. Food collection is accomplished by extending the arms, pinnules, and tube feet into the surrounding water current.[4] Movement of the arms and pinnules is accomplished by contractions of extensor muscles and ligaments. As food particles in the water current contact the podia, the particles are entangled in mucus and are flicked into the ambulacral grooves. The food particles are then transported to the mouth by the ambulacral cilia. Note that the tube feet of crinoids have no locomotory function. Their use in crinoids is limited to food collection, gas exchange, and, probably, elimination of nitrogenous wastes by diffusion.

Crinoid tube feet are never associated with ampullae, distinguishing the crinoids from most other echinoderms. Protraction of the podia in crinoids is accomplished directly, by the contraction of muscles in the radial canals. Crinoids also lack a madreporite, although stone canals opening into the coelom are numerous. The WVS opens to the outside through a large number of ciliated tubes penetrating the tegmen.

Feather stars, also referred to as **comatulids** after the name of the single order in which all are placed (Order Comatulida), resemble the sea lilies from the calyx upward. However, in place of a long attached stalk, a series of jointed, flexible appendages called **cirri** occur near the base of the body (Fig. 19.3). The cirri are used to grasp solid substrates during periods of resting and feeding, which, for a comatulid, is nearly all the time. Comatulids are often observed perching atop sponges, corals, and other structures, living and nonliving; in this way, the feather star can use local water currents to best advantage for feeding. Food collection is as described for stalked crinoids.

Locomotion of feather stars is accomplished either by walking on the tips of the arms or by swimming above the substrate. This short-distance swimming is achieved by the forceful downward movement of the arms. A number of arms distributed uniformly around the calyx beat downward in unison; another group of arms then beat downward while the first group of arms makes its recovery stroke. Swimming thus involves a highly coordinated series of arm movements. The ability to move gives the comatulids a means of escaping or avoiding predators that is not available to the stalked crinoids.

---

4 See Topics for Discussion, No. 3.

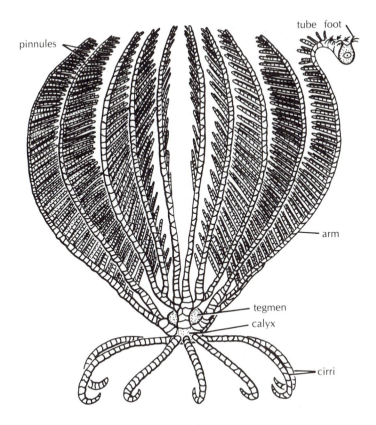

pinnules

tube foot

arm

tegmen

calyx

cirri

FIGURE 19.3. *Diagrammatic illustration of a comatulid crinoid. (Modified from Hyman; after Clark.)*

CLASS OPHIUROIDEA

The approximately 2000 species in this class are motile, as are all living echinoderms other than the stalked crinoids (sea lilies). As with

> ### Class *Ophiuroidea*
> (G: SNAKE-LIKE)

the crinoids, arms extend from a central body, are built of jointed, **vertebral ossicles (vertebrae)**, and are quite flexible (Fig. 19.4). This class is named in recognition of the snake-like movements made by these arms during locomotion. Even so, tube feet often play a role in ophiuroid locomotion, particularly in very young individuals; in species that remain small as adults; and in burrowing species.

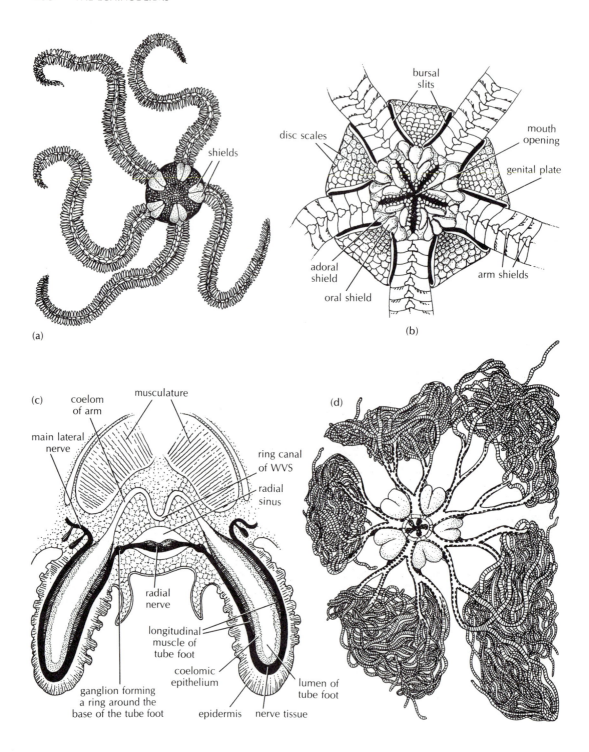

(a)

shields

(b)

bursal
slits

disc scales

mouth
opening

genital plate

adoral
shield

arm shields

oral shield

(c)

coelom
of arm

musculature

main lateral
nerve

ring canal
of WVS

radial
sinus

radial
nerve

longitudinal
muscle of
tube foot

coelomic
epithelium

ganglion forming
a ring around the
base of the tube foot

lumen of
tube foot

epidermis

nerve tissue

(d)

The ophiuroids generally possess five long arms, radiating symmetrically from a small central disc (generally only a few centimeters in diameter). In some species (the basket stars) each arm branches several to many times. The discs of basket stars may reach 10 cm in diameter, and the outstretched arms may be about one meter across! The common name for a typical ophiuroid is "brittle star," reflecting the tendency of the arms to detach from the central disc when provoked.[5]

The oral surface of the body disc is often covered with a thin layer of minute, calcareous scales, while the aboral surface of the body usually bears a number of protective calcareous plates, or **shields** (Fig. 19.4a). The arms are similarly encased in a series of plates, as illustrated. The bulk of each arm, however, consists of a series of thick, articulating, calcareous discs (the vertebral ossicles). The tube feet (Fig. 19.4c) penetrate these "vertebrae" to the outside through a series of minute holes.

As in the crinoids, the ophiuroid digestive system is generally confined to the central disc. Unlike most other members of the Echinodermata, however, the ophiuroids possess only a single opening to the digestive

---

5 See Topics for Discussion, No. 1.

system: a mouth is present, but an anus is lacking.

One final similarity between the crinoids and ophiuroids should be noted. The podia of ophiuroids are generally not operated by ampullae. In most other respects, the WVS follows the typical echinoderm pattern described earlier, except that a single animal may possess numerous madreporites, all of which open on the oral surface.

One feature found among most ophiuroids that distinguishes them from all other echinoderms is the occurrence of slit-shaped infoldings on the oral surface along the arm margins, adjacent to the arm shields. These ten invaginations are known as **bursae** (Fig. 19.4b), and they project well into the coelomic space in the central disc. Sea water is constantly circulated through the bursae, presumably for gas exchange and, perhaps, for the elimination of wastes. This circulation of external fluid is accomplished by cilia and, in some species, by muscular contractions. The bursae may also function in reproduction; in many species, the bursae serve as brood chambers within which the embryos develop.

Many ophiuroids are deposit feeders, ingesting sediment and assimilating the organic fraction. They also capture small animals in the sediment and ingest them

FIGURE 19.4. *(a) An ophiuroid,* Ophiothrix fragilis, *seen from the aboral surface. (After Kingsley.) (b) Oral view of the disc of* Ophiomusium sp., *showing the location of the mouth opening and bursal slits. (From Hyman, 1955.* The Invertebrates, Vol. IV. McGraw-Hill. *Reproduced with permission.) (c) Cross section through an arm of* Ophiothrix sp. *Note the absence of ampullae. (From Hyman; after Cuenot.) (d) The basket star,* Gorgonocephalus sp. *(After Pimentel.)*

individually. Some other brittle star species are suspension feeders, filtering food particles from the water, while still other species function as carnivores or scavengers.

CLASS ASTEROIDEA

Approximately 1600 species of sea stars (often misleadingly termed starfish) have been described from the living fauna, making this

Class Asteroidea
(G: STAR-LIKE)

the second largest class (behind the Ophiuroidea) in the Echinodermata. The asteroids and the ophiuroids are superficially similar, in that the members of both groups possess arms and a basically star-shaped body (Fig. 19.5a). However, the arms of the sea stars are not distinct from the central body disc, and they do not generally play an active, direct role in locomotion. Moreover, there are important morphological and functional differences in the digestive system and WVS. In addition, adult asteroids tend to be larger than adult ophiuroids. Few sea stars are smaller than several centimeters in diameter. Most species are about 15–25 cm in

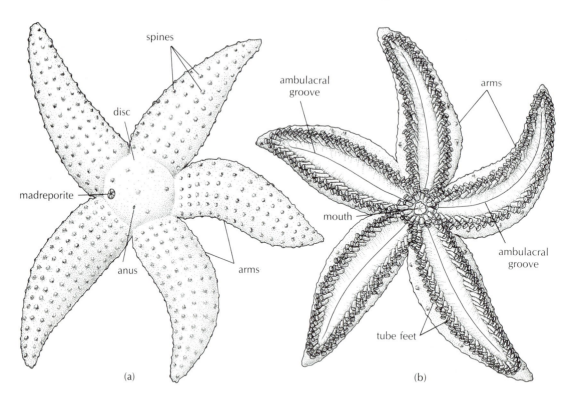

FIGURE 19.5. (a) A sea star, Asterias vulgaris, in aboral view. (After Hyman.) (b) Oral view of Asterias vulgaris, showing the tube feet in ambulacral grooves. (After Sherman and Sherman.)

diameter, and some are considerably larger.

Locomotion in sea stars is slow, and is accomplished by the highly coordinated activities of all the tube feet that radiate out along the oral surface of each arm. The WVS associated with the tube foot is essentially as described at the beginning of this chapter, with the madreporite opening on the aboral surface. The tube feet lie in distinct **ambulacral grooves** on the oral surface (Fig. 19.5b). Such grooves are not encountered among ophiuroids (or in any other group of echinoderms, other than the crinoids). Each tube foot is individually operated by an ampulla and generally terminates in a small suction cup at the distal end (Fig. 19.6). A given

tube foot is extended through contraction of the associated ampulla, and is swung forward or backward by contraction of the longitudinal musculature on one side or the other of the podium. The **ambulacral ossicles** apparently support the podia during these movements. Once the terminal portion of the podium contacts the substratum, slight contractions of the longitudinal muscles pull the central portion of the distal end of the podium upward, creating suction. Suction-cup locomotion seems best adapted for movement over firm substrates. The tube feet of asteroid species that move over or burrow into soft substrates do not terminate in suction cups.

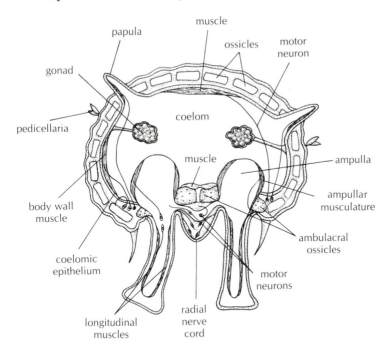

FIGURE 19.6. *Cross section through the arm of an asteroid. Note that the podia protrude through minute pores in the ambulacral ossicles. Other anatomical features, including the papulae and pedicellaria, are discussed later in the text (From Hyman; after J.E. Smith.)*

The sea star mouth is directed downward, opens into a very short esophagus, and passes upward into a lower stomach (the **cardiac stomach**) (Fig. 19.7). This "lower" stomach is confined to the central body disc, and is chiefly responsible for the digestion of food. Above the lower stomach is an upper, or **pyloric stomach**, branches of which radiate out into each arm as **pyloric caeca**. The great surface area of the pyloric caeca, achieved through outfolding of the tissue, is in keeping with the primary functions of these organs: secretion of digestive enzymes and absorption of digested nutrients. The pyloric caeca are also primary storage sites for assimilated food. The anus lies on the aboral surface, nearly in line with the mouth.

Asteroids are typically predators on large invertebrates and small vertebrates, including gastropods, polychaetes, bivalves, and fishes.[6] During feeding on large prey, the cardiac stomach of some species is actually protruded out of the body disc through the mouth, and is placed in contact with the soft tissues of the prey. The cardiac stomach can be protruded through spaces as narrow as 1–2 mm (e.g., the gap between the shell valves of a bivalve). Digestion in such species is frequently external, the resulting nutrient broth being transferred to the pyloric stomach by means of the ciliated channels of the cardiac stomach. No other echinoderms feed

_____

6 See Topics for Discussion, No. 3.

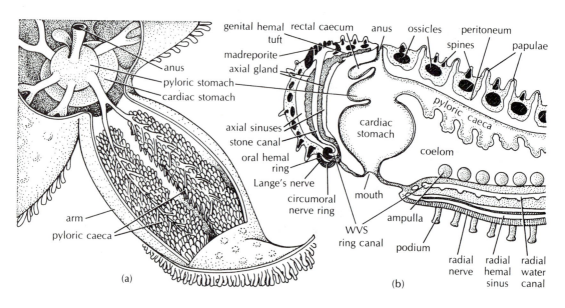

FIGURE 19.7.   *(a) Diagrammatic illustration of the digestive system of a sea star. (b) Diagrammatic illustration of a sea star digestive system, viewed laterally. (After Chadwick.)*

in this manner. If the prey is sufficiently small, the stomach will be retracted while holding the victim, and digestion will proceed internally. Alternatively, the small prey item may be ingested directly at the mouth, without the stomach having to be protruded at all. Some suspension-feeding asteroid species have also been described.

The calcareous skeleton of the sea star takes the form of discrete rods, crosses, and plates, which are embedded in connective tissue (Fig. 19.8). Ossicles other than those supporting the ambulacral grooves often bear outwardly directed spines that can be moved from side to side by muscles connecting to the underlying ossicles. The

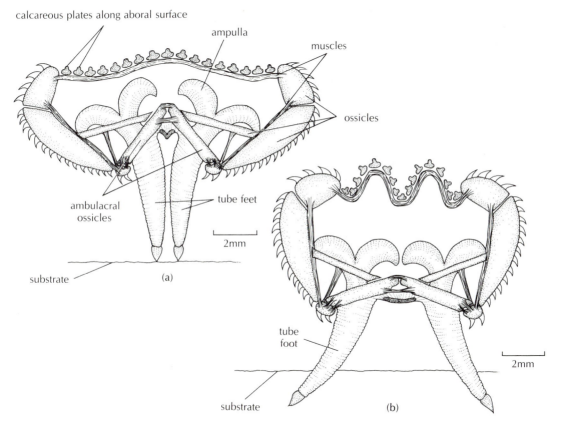

FIGURE 19.8. *Transverse section through the arm of the asteroid* Astropecten irregularis, *showing the arrangement of calcareous ossicles. In frame (a), the animal is walking over a substrate. In frame (b), the animal is burrowing. Note the changes in orientation of the ossicles, made possible by the elastic connective tissue that joins adjacent plates. (Modified from D. Heddle, 1967. Symp. Zool. Soc. London, 20: 125.)*

spines and outer surface of the skeletal ossicles are covered by a fairly thick cuticle, which is secreted by the underlying, ciliated epidermis.

Asteroids possess two other types of appendage in addition to spines and podia. Thin, noncalcified outfoldings of the outer body wall serve a respiratory function. These structures, called **papulae**, are found protruding from the spaces between ossicles and are connected directly with the main coelomic cavity. The second type of appendage is much more dynamic. These appendages, called **pedicellariae**, consist of two (sometimes three) calcium carbonate ossicles (**valves**), whose ends can be moved together or apart by muscles (Fig. 19.9). The two jaws are supported by a nonmovable, basal ossicle. The pedicellariae generally function in the removal of unwanted organisms and debris that contact the surface of the animal,

and they have also been shown to function in the capture of living prey (including fish!) in several asteroid species. Pedicellariae are found in only one other echinoderm class, the Echinoidea.

CLASS ECHINOIDEA

The last two classes of the Echinodermata remaining to be discussed consist of species that lack arms entirely. The Echinoidea include the sea urchins, heart urchins, and sand dollars, somewhat less than 1000 species in total. The class is perhaps best represented

> **Class Echinoidea**
> (G: SPINE-LIKE)

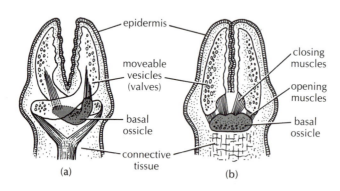

FIGURE 19.9. *Two types of pedicellariae found on the body of Asterias sp. In type (a), the moveable ossicles (valves) cross, as in the blades of a scissors. In type (b), the valves are straight and nearly parallel, and operate in the manner of forceps. In both types of pedicellariae, the basal ossicle supports the valves, but does not itself move. (After Hyman.)*

by the sea urchins, which possess large numbers of long, rigid calcium carbonate spines. The Greek word *"echinus"* means, literally, a hedgehog. The spines serve for protection and, in some species, are actively involved in locomotion. Most sea urchins are free-living, roaming individuals, but a number of species bore into rock.

The spines of echinoids attach to the underlying skeleton via ball-and-socket joints, and can be declined rapidly in various directions by contraction of specialized muscle fibers that connect between the ball (**tubercle**) and the spine (Fig. 19.10). The spines are often thin and sharp, but are thick and blunt in a few species. Toxins may be extruded

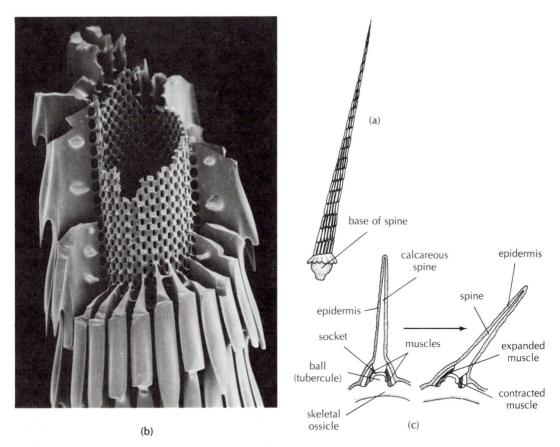

(a)

base of spine

calcareous spine

epidermis

epidermis

spine

socket

muscles

ball (tubercule)

expanded muscle

contracted muscle

skeletal ossicle

(b)

(c)

FIGURE 19.10. *(a) A spine of the tropical sea urchin* Diadema antillarus. *(After Hyman.) (b) A spine of* Diadema setosum, *as seen with the scanning electron microscope. The spine has been broken to reveal the complexity of the internal construction. (Courtesy of K. Märkel and Springer-Verlag.) (c) Movement of a sea urchin spine. (After Hyman.)*

through the spines, or from glands associated with the spines in some species, as many an unwary, warm-water vacationer has discovered.

The ossicles comprising the skeleton of most echinoids are flat and fused together, so that movement of ossicles relative to one another cannot occur. In most species, the skeleton thus forms a solid, inflexible **test,** a feature setting the typical member of this class apart from the members of all other classes in the Echinodermata. Tube feet are widely distributed on the body, protruding through five double rows of pores in the **ambulacral plates** of the test (Fig. 19.11a). The areas of the body containing tube feet (i.e., the ambulacral plates) are distributed symmetrically about the body in strips extending orally/aborally. These regions are separated from each other by distinct **interambulacral** areas, which are strips devoid of tube feet. The tube feet are highly modified and generally bear suction-cup ends; the podia function in locomotion as in the Asteroidea.

Pedicellariae are also prominent echinoid appendages. However, they are generally borne on stalks, and, unlike asteroid pedicellariae, are often equipped with calcareous support rods and bear three opposing jaws rather than two (Fig. 19.11c). Certain types of pedicellariae, found in most urchin species, discharge poisons for defense. Some echinoid species have thin-walled outfoldings surrounding the mouth. These small appendages, called **gills,** presumably function in gas exchange.

Echinoids may be either regular or irregular (Fig. 19.12). **Regular urchins** have an almost perfect, spherical symmetry. All sea urchins fall into this group. Other echinoids are classified as **irregular** and display varying degrees of bilateral symmetry. This bilateral symmetry may be associated with a life-style of burrowing through sand, mud, or gravel. In association with the burrowing habit, the tube feet of irregular urchins tend to lack terminal suckers. In all irregular echinoids, ambulacral areas (and thus the tube feet) are restricted to the oral and aboral surfaces, rather than extending in an unbroken line from the oral to the aboral surfaces. The ambulacral areas on the aboral surface form a conspicuous five-pointed pattern, resembling the petals of a flower. The heart urchins have distinct anterior and posterior ends; the mouth is located anteriorly and the anus posteriorly. The spines of heart urchins are much more numerous than they are in regular urchins, but they are also much shorter. Sand dollars also bear very short spines, most likely as an adaptation for burrowing. Unlike the heart urchins and regular urchins, in which the aboral surface is convex, the test of most sand dollars is greatly flattened to form a very thin disc.

The digestive system of an echinoid is quite unlike that of an asteroid. In fact, it differs significantly from the digestive system of all other echinoderm species. A complex system of ossicles and muscles, called **Aristotle's lantern,** surrounds the esophagus in all regular echinoids and in some irregular species as well (Fig. 19.13). The teeth of Aristotle's lantern can be protruded from the mouth and moved in various directions. They are used to scrape food, especially al-

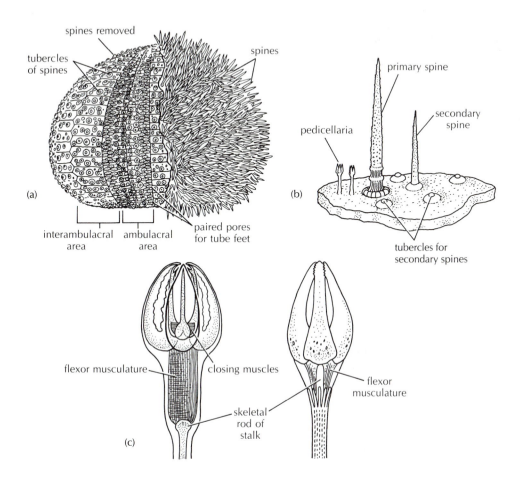

FIGURE 19.11. (a) A sea urchin, Echinus esculentus, with the spines removed from a portion of the test to reveal the ambulacral and interambulacral zones. Small pores, through which tube feet protrude in the living urchin, are seen in the ambulacral plates. Tubercules, upon which the spines pivot, are conspicuous on the interambulacral plates. The sea urchin test is comprised of five double rows of ambulacral plates and five double rows of interambulacral plates, for a total of twenty rows of plates. (b) Schematic illustration of an urchin test with appendages. (After Jackson.) (c) Pedicellariae from the echinoids Strongylocentrotus droebochiensis (left) and Eucidaris sp. (right), showing the stalks and three-part jaws. (From Hyman; after Mortensen.)

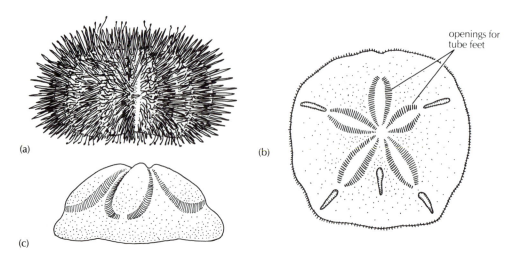

FIGURE 19.12. *(a) A spiny urchin, a regular echinoid. (b) A sand dollar, an irregular echinoid. (c) A sea biscuit, another irregular echinoid. Regular echinoids (the "sea urchins") are characterized by pentamerous symmetry, a globular test, and long spines. Irregular echinoids (including the "heart urchins" and "sand dollars") have a flattened test, relatively short spines, and tend toward bilateral symmetry; the periproct with anus is displaced posteriorly and the mouth may be displaced anteriorly. The ambulacra of irregular species often resemble flower petals in outline, as in the sand dollar shown. (All after Pimentel.)*

gae, from solid substrates. Echinoids lacking Aristotle's lantern generally feed on small organic debris, which is often collected by means of modified tube feet, spines, and/or external ciliary tracts.[7] The stomach of echinoids is not protrusible. Indeed, there is no true stomach in this group, the esophagus

leading instead into a very long, convoluted intestine (Fig. 19.14), where the food is both digested and absorbed (**assimilated**). The anus is located aborally, and is surrounded by a series of plates comprising the **periproct** (G: around the anus).

Assimilated food passes into the coelomic fluid. The coelomic space of echinoids is immense, particularly in the regular urchins (Fig. 19.14). Coelomic fluid is the principal agent of transport of both food and wastes, i.e., the coelomic fluid is the principal circu-

---

7 See Topics for Discussion, No. 3.

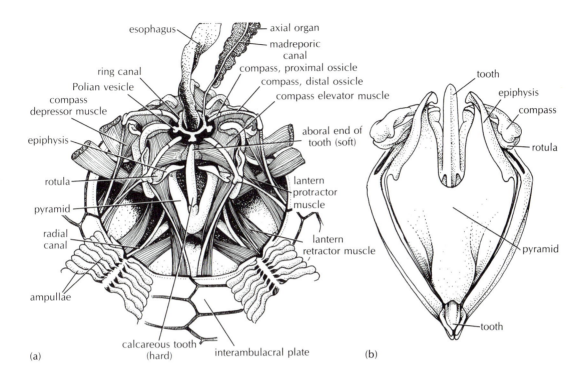

FIGURE 19.13. *(a) Aristotle's lantern and its associated complex musculature, from the sea urchin* Arbacia punctulata. *(b) Detail of the lantern ossicles, with the musculature omitted for clarity. Lantern morphology differs considerably among echinoids and is an important tool for species identification. The lantern consists of five bulky pyramids, which support the five teeth; a series of bars (epiphyses) running along the aboral ends of the pyramids; a series of five thin compass ossicles aborally; and a series of five similar pieces, the rotulas, lying below the compass. Protractor muscles push the teeth outward; retractor muscles move the teeth apart and draw them back into the test. The teeth are very hard at the tips but soft at the aboral end; new tooth material is formed continually at the aboral end, compensating for tooth wear distally. (After Brown.)*

latory fluid. The inner surface of the mesodermal lining of the coelomic cavity is ciliated, maintaining a constant movement of the circulatory medium.

The WVS of echinoids follows the archetypic pattern closely, with a single madreporite opening aborally, as in the asteroids (Fig. 19.14).

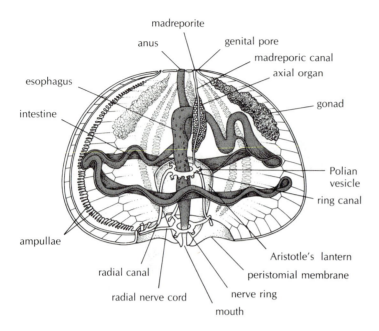

FIGURE 19.14. *A sea urchin,* Arbacia punctulata, *seen laterally in diagrammatic section. Note the large coelomic space contained within the test. (After Brown; after Petrunkevitch.)*

CLASS HOLOTHUROIDEA

This group of animals contains over 900 species, making it approximately equal in size to the Echinoidea. The members of these two groups resemble each other in that arms are lacking. In certain respects, the typical holothurian may be regarded as a flexible echinoid, whose morphological modifications reflect adaptation for a different life-style. To transform an echinoid into a typical holothurian, one must first remove all spines and pedicellariae, and discard Aristotle's lantern. (This is a thought experiment, not something that echinoderms actually do!) Next,

the ossicles of the test must be separated from each other and greatly reduced in size. This mutual detachment and reduction in size of the ossicles renders the body wall stretchable, since it is now largely composed of connective tissue. From this point to a finished holothurian is largely a matter of stretching the imaginary animal, increasing the distance between the oral and aboral surfaces (Fig. 19.15).

Thus, holothurians are typically soft-bodied, bilaterally symmetric, vermiform creatures with distinct anterior and posterior ends and with podia generally confined to distinct ambulacral strips (as in the Echi-

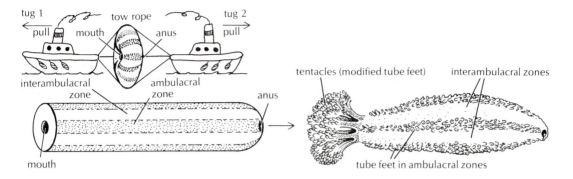

FIGURE 19.15. *Relationship between the basic body plans of the echinoids and the holothurians. The holothurian shown at the end of the sequence is* Cucumaria frondosa. *(After Hyman.)*

noidea) (Fig. 19.15). The calcareous ossicles, so conspicuous in other echinoderm classes, are microscopic in holothurians and embedded in the body wall; these ossicles are often exquisitely shaped (Fig. 19.16). The outer body wall is often warty and dark-colored. Some species actually do resemble very closely the fruit from which they derive their common name of sea cucumber. Adults range from several centimeters to over one meter in length. Strangely enough, cephalization is not pronounced among holothurians, despite the presence of a distinct anterior end.

The oral tube feet of holothurians are modified as large, feathery tentacles at the anterior end; these tentacles can be protracted from the mouth and used to capture food (Figs. 19.15, 19.17). Each tentacle may be operated by a single, large ampulla. In some species, the tentacles are coated with sticky mucus to trap food particles from suspension, but most sea cucumber species are deposit-feeders, ingesting sediment and extracting the organic component. It has been

estimated that some holothurians may pass nearly 60 pounds of substrate through their digestive systems per year!

The ambulacral tube feet of surface-living sea cucumber species generally bear suckers, and are used in locomotion as in the echinoids and asteroids.

The holothurian digestive system closely resembles that of the echinoids, except that it is now greatly elongated (Fig. 19.17). The WVS follows the typical echinoderm pattern, with the ring canal forming a ring about the esophagus as usual. The ring canal is supported, however, by a calcareous ring, which may have an evolutionary origin in common with the Aristotle's lantern of echinoids. The madreporite generally lies free in the coelomic cavity (Fig. 19.17), so that there appears to be no direct connection between the WVS and the outside. The coelomic space of holothurians is very large, as in the echinoids and asteroids, and the coelomic fluid is the primary circulatory medium.

In contrast to the situation in members of any other echinoderm class, the body wall of

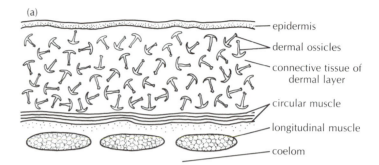

- epidermis
- dermal ossicles
- connective tissue of dermal layer
- circular muscle
- longitudinal muscle
- coelom

FIGURE 19.16. *(a) Schematic cross section of the body wall of a typical holothurian, showing the arrangement of the musculature and location of the ossicles. Separation and size reduction of the ossicles permits holothurians to undergo major changes in shape; the body wall, musculature, and coelom form a functional hydrostatic skeleton. (b) Scanning electron micrograph of the microscopic ossicles removed from the body walls of several holothurian species. Ossicle morphology plays a major role in species identification.* (A, B) Eostichopus regalis *(Cuvier), dorsal and lateral views;* (C) Euapta lappa *(Müller);* (D) Holothuria (Cystipus) occidentalis *Ludwig;* (E) Holothuria (Cystipus) pseudofossor *Deichmann;* (F) Holothuria (Semperothuria) surinamensis *Ludwig. These ossicles range from 60–400 μm in longest dimension. (Courtesy of John E. Miller, Harbor Branch Foundation.)*

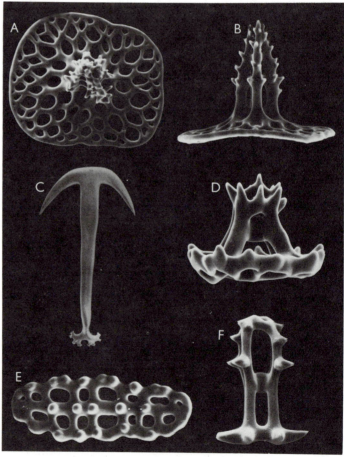

(b)

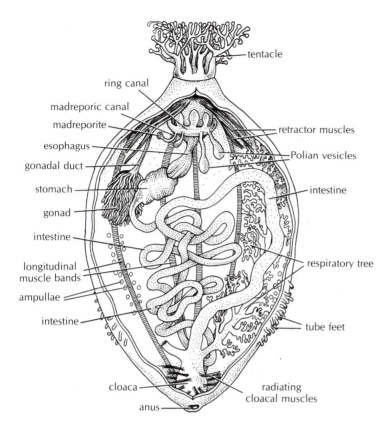

tentacle

ring canal

madreporic canal

madreporite

esophagus

gonadal duct

stomach

gonad

intestine

longitudinal muscle bands

ampullae

intestine

retractor muscles

Polian vesicles

intestine

respiratory tree

tube feet

cloaca

anus

radiating cloacal muscles

FIGURE 19.17.   *Internal anatomy of the holothurian* Thyone briareus. *Note the extensive respiratory trees (only one shown), large coelomic space, and long intestine. (After Hyman; after Coe.)*

most holothurians contains well-developed layers of both circular and longitudinal musculature (Fig. 19.16a), and is considered by some to be a culinary delicacy (as are the gonads of echinoids!).

Clearly, the holothurian has the characteristics necessary for the operation of a hydrostatic skeleton: a large, fluid-filled, constant-volume body cavity; a deformable body wall; and an appropriate musculature. It should not be surprising to learn, then, that many holothurian species are burrowers in sand and mud (Fig. 19.18). The tube feet of many of these species are much reduced, and some of the more specialized burrowing species lack tube feet entirely. Locomotion

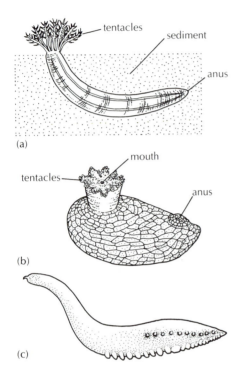

(a)

(b)

(c)

FIGURE 19.18. *Holothurian diversity. (a) A burrowing holothurian,* Leptosynapta inhaerens, *in sand. This species is only a few centimeters long; has a very thin, flexible body wall; lacks respiratory trees; and, except for the tentacles, lacks podia. (From Fretter and Graham, 1976. A Functional Anatomy of Invertebrates.* Academic Press.) *(b) A highly modified holothurian,* Psolus fabricii. *The ventral surface is flattened, forming a "creeping sole." Podia are restricted to the flattened surface; the dorsal surface is covered with protective, calcareous scales. (From Fretter and Graham.) (c) A deep-sea holothurian,* Psychropotes longicauda, *with the tube feet modified as legs. Holothurians comprise one of the dominant elements of the macrofauna of the deep sea. The specimen illustrated was obtained from a depth of perhaps 800 meters; the body may reach nearly 0.5 meter in length. (From Marshall, 1979. Deep-Sea Biology.* Garland STPM Press.)

in these burrowing species is accomplished in part by using the tentacles to push substrate away, but is primarily the result of waves of contraction of the circular and longitudinal muscles, earthworm style.

Two more characteristics of sea cucumber biology must be mentioned. Holothurians are the only group of echinoderms to possess truly specialized respiratory structures. These are called **respiratory trees.** The coelom of most holothurians holds a pair of these highly branched muscular structures. The respiratory trees connect to the cloaca, which pumps water into the trees. Water is expelled through the cloaca by contraction of the respiratory tree tubules themselves.

I have saved the best for last. Many holothurian species respond to a variety of physical and environmental factors by eviscerating. In some species, this is limited to the expulsion of incredibly sticky and/or toxic nondigestive organs called **Cuverian tubules,** which are attached to the respiratory trees and apparently used only for the purpose of discouraging potential predators. In many other species, true **evisceration** occurs, in which the entire digestive system may be expelled, along with the respiratory trees and gonads. All lost body parts are eventually reformed, reflecting the substantial regenerative capabilities possessed by most echinoderms.[8]

OTHER FEATURES OF ECHINODERM BIOLOGY

1. Reproduction and Development

Asexual reproduction is commonly encountered among asteroids and ophiuroids. The central disc separates into two pieces, and each piece proceeds to reform the missing arms and organs. At least one asteroid species reproduces asexually from pieces of the arms alone. However, reproduction in most echinoderms is exclusively sexual. The sexes are usually separate. All echinoderms except holothurians and crinoids bear multiple gonads. In the asteroids, at least one pair of gonads extends into each arm. Each bursa houses from one to many gonads in ophiuroids, and echinoids usually have five gonads. The gonads of crinoids are borne on the arms or pinnules. Holothurians are unique among the Echinodermata in possessing a single gonad. In most echinoderm species, gametes are liberated into the surrounding sea water, so that fertilization is external. A distinctive, ciliated larval stage is characteristic of each class, as illustrated in Fig. 19.19. A delicate, internal skeleton supports the larval arms in echinoids and ophiuroids. Metamorphosis to the adult morph is often dramatic in echinoderms, in terms of both speed and the complexity and magnitude of morphological reorganization.[9]

2. Nervous System

There is no centralized brain in any echinoderm, nor are distinct ganglia generally found. Instead, the nervous system is composed of three diffuse nerve networks. An **ectoneural** system receives sensory input from the epidermis. This system, highly developed in all but the crinoids, consists of a ring around the esophagus with five associated nerves radiating outward. In species with arms, these radial nerve cords extend down each arm to the tube feet, ampullae (where present), and pedicellariae (where present). A second, **hyponeural** system, is exclusively concerned with motor function. It, too, consists of a circumoral nerve ring with five associated radial nerves, but it lies deeper within the tissues of the animal. The hyponeural system is well developed only in ophiuroids, and, to a lesser extent, in asteroids. In crinoids, the major nerve network is an **entoneural** system (Fig. 19.20), associated with the aboral end of the animal. From a central mass in the calyx/tegman complex, nerves radiate down the stalk of the crinoid to the cirri, and up into each arm. The entoneural system is inconspicuous or entirely absent in the other echinoderm classes.

---

8 See Topics for Discussion, No. 1.

9 See Topics for Discussion, No. 6.

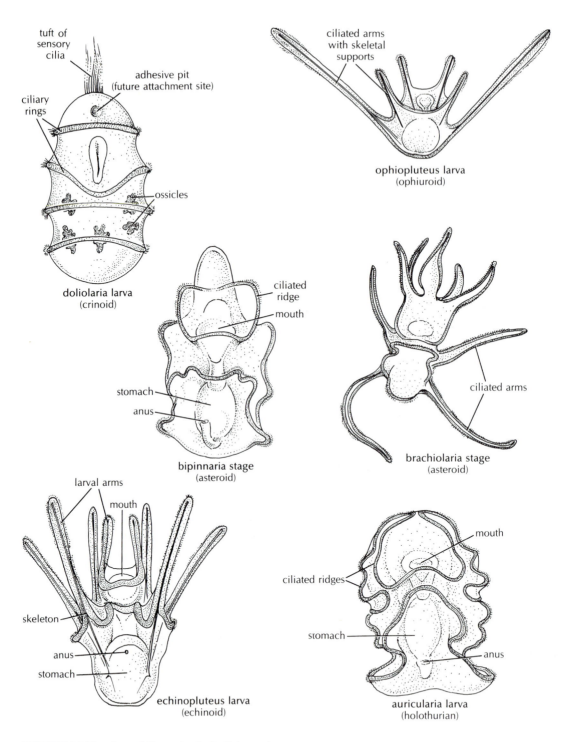

FIGURE 19.19. *Larval forms typical of the various extant echinoderm classes. (Modified from various sources.)*

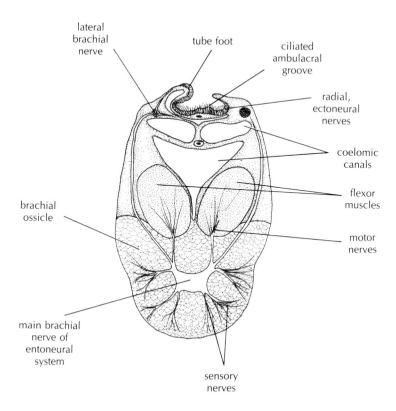

lateral brachial nerve

tube foot

ciliated ambulacral groove

radial, ectoneural nerves

coelomic canals

flexor muscles

brachial ossicle

motor nerves

main brachial nerve of entoneural system

sensory nerves

FIGURE 19.20. *Cross section through the arm of a crinoid, showing the arrangement of the nervous system. (From Hyman; after Hamann.)*

T A X O N O M I C   S U M M A R Y

PHYLUM ECHINODERMATA
    CLASS CRINOIDEA—THE SEA LILIES AND FEATHER STARS
    CLASS OPHIUROIDEA—THE BRITTLE STARS
    CLASS ASTEROIDEA—THE SEA STARS
    CLASS ECHINOIDEA—THE SEA URCHINS, HEART URCHINS,
            AND SAND DOLLARS
    CLASS HOLOTHUROIDEA—THE SEA CUCUMBERS

TOPICS FOR FURTHER DISCUSSION AND INVESTIGATION

1. Investigate the factors controlling the loss and regeneration of body parts among echinoderms.

Anderson, J.M., 1965. Studies on visceral regeneration in sea stars. III. Regeneration of the cardiac stomach in *Asterias forbesi* (Desor). *Biol. Bull.*, 129: 454.

Mladenov, P.V., R.H. Emson, L.V. Colpit, and I.C. Wilkie, 1983. Asexual reproduction in the West Indian brittle star *Ophiocomella ophiactoides* (H.L. Clark) (Echinodermata: Ophiuroidea). *J. Exp. Marine Biol. Ecol.*, 72: 1.

Smith, G.N., Jr., 1971. Regeneration in the sea cucumber *Leptosynapta*. I. The process of regeneration. *J. Exp. Zool.* 177: 319.

Smith, G.N., Jr., 1971. Regeneration in the sea cucumber *Leptosynapta*. II. The regenerative capacity. *J. Exp. Zool.*, 177: 331.

Smith, G.N., and M.J. Greenberg, 1973. Chemical control of the evisceration process in *Thyone briareus*. *Biol. Bull.*, 144: 421.

Wilkie, I.C., 1978. Arm autotomy in brittle stars (Echinodermata: Ophiuroidea). *J. Zool., London*, 186: 311.

Zeleny, C., 1903. A study of the rate of regeneration of the arms in the brittle star *Ophioglypha lacertosa*. *Biol. Bull.*, 6: 12.

2. What is the role of coelomocytes in mediating the immune response of echinoderms?

Bang, F.B., 1982. Disease processes in seastars: a Metchnikovian challenge. *Biol. Bull.*, 162: 135.

Hilgard, H.R., and J.H. Phillips, 1968. Sea urchin response to foreign substances. *Science*, 161: 1243.

Kaneshiro, E.S., and R.D. Karp, 1980. The ultrastructure of coelomocytes of the sea star *Dermasterias imbricata*. *Biol. Bull.*, 159: 295.

Yui, M.A., and C.J. Bayne, 1983. Echinoderm immunology: bacterial clearance by the sea urchin *Strongylocentrotus purpuratus*. *Biol. Bull.*, 165: 473.

3. Discuss the role of tube feet, spines, and cilia in the collection of food by echinoderms.

Burnett, A.L., 1960. The mechanism employed by the starfish *Asterias forbesi* to gain access to the interior of the bivalve *Venus mercenaria*. *Ecology*, 41: 583.

Chia, F.S., 1969. Some observations on the locomotion and feeding of the sand dollar, *Dendraster excentricus* (Eschscholtz). *J. Exp. Marine Biol. Ecol.*, 3: 162.

Ellers, O., and M. Telford, 1984. Collection of food by oral surface podia in the sand dollar, *Echinarachnius parma* (Lamarck). *Biol. Bull.*, 166: 574.

Fankboner, P.V., 1978. Suspension-feeding mechanisms of the armoured sea cucumber *Psolus chitinoides* Clark. *J. Exp. Marine Biol. Ecol.*, 31: 11.

Ghiold, J., 1983. The role of external appendages in the distribution and life habits of the sand dollar *Echinarachnius parma* (Echinodermata: Echinoidea). *J. Zool., London*, 200: 405.

Goodbody, I., 1960. The feeding mechanism in the sand dollar *Mellita sexiesperforata* (Leske). *Biol. Bull.*, 119: 80.

Hendler, G., 1982. Slow flicks show star tricks: elapsed-time analysis of basketstar (*Astrophyton muricatum*) feeding behavior. *Bull. Marine Sci.*, 32: 909.

Lasker, R., and A.C. Giese, 1954. Nutrition of the sea urchin, *Strongylocentrotus purpuratus*. *Biol. Bull.*, 106: 328.

LaTouche, R.W., 1978. The feeding behavior of the feather star *Antedon bifida* (Echinodermata: Crinoidea). *J. Marine Biol. Ass. U.K.*, 58: 877.

Lavoie, M.E., 1956. How sea stars open bivalves. *Biol. Bull.*, 111: 114.

Macurda, D.B., and D.L. Meyer, 1974. Feeding posture of modern stalked crinoids. *Nature, London*, 247: 394.

Mauzey, K.P., C. Birkeland, and P.K. Dayton, 1968. Feeding behavior of asteroids and escape responses of their prey in the Puget Sound region. *Ecology*, 49: 603.

Meyer, D.L., 1979. Length and spacing of the tube feet in crinoids (Echinodermata) and their role in suspension-feeding. *Marine Biol.*, 51: 361.

O'Neill, P.L., 1978. Hydrodynamic analysis of feeding in sand dollars. *Oecologia*, 34: 157.

Telford, M., A.S. Harold, and R. Mooi, 1983. Feeding structures, behavior, and microhabitat of *Echinocyamus pusillus* (Echinoidea: Clypeasteroidea). *Biol. Bull.*, 165: 745.

4. Sea water contains fairly high concentrations (up to 3×10^{-3} g carbon/liter) of dissolved organic matter (DOM), especially in shallow coastal waters. To what extent are echinoderms capable of meeting their nutritional needs through the uptake of DOM directly from sea water?

Ferguson, J.C., 1980. The non-dependency of a starfish on epidermal uptake of dissolved organic matter. *Comp. Biochem. Physiol.*, 66A: 461.

Fontaine, A.R., and F.S. Chia, 1968. Echinoderms: an autoradiographic study of assimilation of dissolved organic molecules. *Science*, 161: 1153.

Stephens, G.C., M.J. Volk, S.H. Wright, and P.S. Backlund, 1978. Transepidermal accumulation of naturally occurring amino acids in the sand dollar, *Dendraster excentricus*. *Biol. Bull.*, 154: 335.

5. How are individual echinoderm tube feet operated and integrated into the total behavior of the animal?

Binyon, J., 1964. On the mode of functioning of the water vascular system of *Asterias rubens* L. *J. Marine Biol. Assoc. U. K.*, 44: 577.

Kerkut, G.A., 1953. The forces exerted by the tube feet of the starfish during locomotion. *J. Exp. Biol.*, 30: 575.

Lavoie, M.E., 1956. How sea stars open bivalves. *Biol. Bull.*, 111: 114.

Polls, I., and J. Gonor, 1975. Behavioral aspects of righting in two asteroids from the Pacific coast of North America. *Biol. Bull.*, 148: 68.

Prusch, R.D., and F. Whoriskey, 1976. Maintenance of fluid volume in the starfish water vascular system. *Nature, London*, 262: 577.

Smith, J.E., 1937. The structure and function of the tube feet in certain echinoderms. *J. Marine Biol. Assoc. U.K.*, 22: 345.

Smith, J.E., 1947. The mechanics and innervation of the starfish tube foot-ampulla system. *Phil. Trans. Roy. Soc. B*, 232: 279.

6. What major morphological changes take place during the metamorphosis of echinoderms from larval to adult form?

Cameron, R.A., and R.T. Hinegardner, 1978. Early events in sea urchin metamorphosis, description and analysis. *J. Morphol.*, 157: 21.

Hardy, A., 1965. Pelagic larval forms. In: *The Open Sea: Its Natural History*, Boston: Houghton Mifflin, pp. 178–198.

Hendler, G., 1978. Development of *Amphioplus abditus* (Verrill) (Echinodermata: Ophiuroidea). II. Description and discussion of ophiuroid skeletal ontogeny and homologies. *Biol. Bull.*, 154: 79.

Mladenov, P.V.M., and F.S. Chia, 1983. Development, settling behaviour, metamorphosis and pentacrinoid feeding and growth of the feather star *Florometra serratissima. Marine Biol.*, 73: 309.

7. The echinoderms have an extensive fossil record, in keeping with their ancient origin, marine habitat, and solid skeleton. In fact, a number of classes are known only from the fossil record, there having been no living representatives of these classes for tens of thousands of years. Most of these extinct echinoderms were sessile animals, permanently attached to a substrate. What are the morphological similarities and differences between these extinct species and the present-day sea lilies (Class Crinoidea)?

Hyman, L.H., 1955. *The Invertebrates*, Vol. IV. New York: McGraw-Hill.

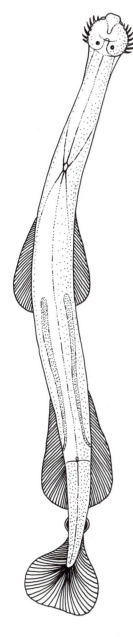

20

THE CHAETOGNATHS

INTRODUCTION AND GENERAL CHARACTERISTICS

Chaetognaths are all free-living, marine carnivores. The average chaetognath is only a few centimeters long, and even the largest individuals are no more than 10 cm long. Apparently the surface area/volume ratio is sufficiently large that gas exchange and excretion requirements can be met by diffusion across the general body surface; chaetognaths bear no specialized respiratory or excretory organs. A blood circulatory system is also lacking. The inner lining of the adult body cavity is ciliated; gas exchange and nutrient transport must be achieved by circulation of the coelomic fluid.

Typically, the nearly transparent chaetognath floats in the water, motionless, until something edible comes its way. The chaetognath then darts forward and grasps its prey with two rows of long, curved, stiff spines adjacent to the mouth (Fig. 20.1). A row of short teeth on either side of the mouth

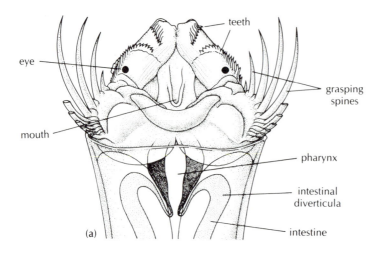

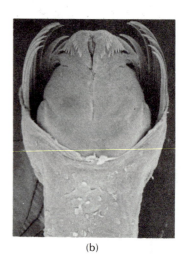

(b)

FIGURE 20.1. *(a) The anterior end of* Sagitta elegans, *in ventral view. (From Hyman; after Ritter-Zahony.) (b) Scanning electron micrograph of* S. setosa, *anterior end. (Courtesy of Q. Bone and Cambridge University Press. From Bone et al., 1983. J. Marine Biol. Assoc. U.K., 63: 929.*

aids in holding prey during ingestion. The spines themselves may be sensitive to vibration, enabling chaetognaths to detect the presence and location of the swimming fish larvae and copepods that constitute the primary prey items. Bristles and tufts of bristles distributed elsewhere on the body apparently serve as additional mechanoreceptors. Chaetognaths also bear a pair of eyes, each of which is composed of five pigment-cup ocelli. The ocelli are oriented in several different directions within a single eye, so that the chaetognath has a wide visual field; several ocelli point downward, so that the chaetognath actually sees through its own, transparent body. The ocelli are probably not image-forming, but may enable the animal to detect motion.[1]

---
1 See Topics for Discussion, No. 1.

The chaetognath body is arrow-like, and is covered by a thin cuticle secreted by the underlying epidermis. The common name for a member of this phylum is "arrow worm." The body is substantially longer than it is wide, and lacks appendages. The body does bear one or two pairs of **lateral fins** and a terminal **caudal fin** (Fig. 20.2). The lateral fins probably serve to increase the body surface area, reducing the rate at which the animal sinks while motionless. The caudal fin must also aid flotation, and generates forward thrust for swimming.

The body cavity of adult chaetognaths is compartmentalized. One septum isolates the head compartment from that of the trunk. A second septum divides the trunk compartment from that of the tail.

Most chaetognath species are planktonic, spending their entire lives being passively

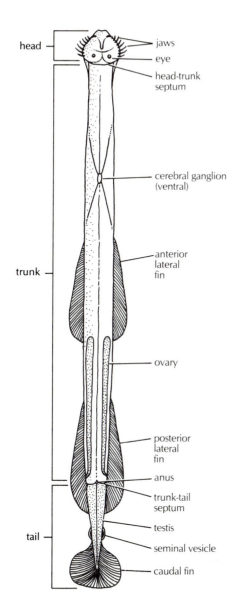

head — jaws
eye
head-trunk
septum

cerebral ganglion
(ventral)

anterior
lateral
fin

trunk —

ovary

posterior
lateral
fin

anus

trunk-tail
septum

testis

tail —

seminal vesicle

caudal fin

FIGURE 20.2. Sagitta elegans, *seen in ventral view. (From Hyman; after Ritter-Zahony.)*

transported by ocean currents. Although, as planktonic animals, arrow worms cannot swim against any substantial water current, many species can and do undertake exten-

sive vertical migrations of hundreds of meters daily (Fig. 20.3). Many other planktonic animals, including pteropods (Gastropoda), copepods (Crustacea), and a variety of invertebrate larval stages display similar patterns of behavior. Generally, the animals migrate downward during the day and upward at night, although the timing and extent of the migrations often differ significantly among species and among different developmental stages of a given species. No single rationale for such **diurnal vertical migrations** has been conclusively demonstrated in laboratory or field experiments, although several reasonable adaptive benefits have been proposed: avoiding visual predators; increasing the energetic efficiency of feeding and digestion; reducing competitive interactions both within and among species; improving physiological well-being by experiencing a varied temperature and salinity regime; and achieving dispersal by taking advantage of vertical differences in speeds and directions of water currents.[2]

One genus of chaetognaths (*Spadella*) is entirely benthic. These arrow worms use specialized **adhesive papillae** to form temporary attachments to solid substrates, usually rocks and/or macroalgae. In the typical feeding posture, the attachment is made posteriorly, with the rest of the body held elevated above the substrate. When the need arises, the attachment can be broken and the animal can dart off to a new location.

Only about 50–60 species of chaetognaths are known (Fig. 20.4). This phylum is nevertheless an important one. Local concentrations as high as several hundred chaetognaths per cubic meter of sea water have

2 See Topics for Discussion, No. 2.

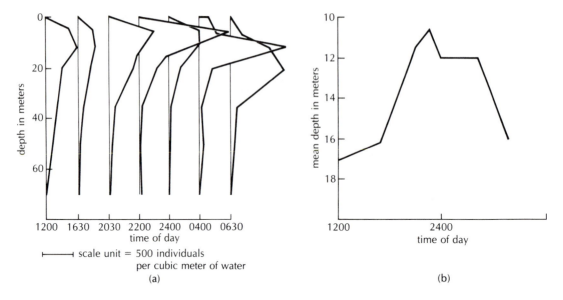

FIGURE 20.3. *Diurnal vertical migration by the chaetognath* Sagitta elegans. *(a) Changes in numbers of individuals per unit volume of sea water at different depths over time. Note that a pronounced population peak begins to form in the early evening, and that the average depth of this peak shifts toward early morning. (b) The average depth for the population over time is shown, using the data illustrated in frame (a) above. The population clearly migrates upward toward evening and downward in the early morning. (From Pearre, 1973. Ecology, 54: 300.)*

been reported. As significant predators of fish embryos, fish larvae, and copepods, chaetognaths must be an important component of oceanic food chains and, in particular, must play a major role in determining the size of commercially exploitable fish populations.

Chaetognaths are also intriguing for their enigmatic place in the phylogenetic scheme of things. Their relationship to other animal groups is most unclear. Studies of chaetognath embryology establish the arrow worms as deuterostomous coelomates. Cleavage is basically radial and indeterminant (i.e., cell fates are not irrevocably fixed following the first cell division). The site of the blastopore gives rise to the anus; as in other deuterostomes, the mouth arises elsewhere (recall that "deuterostome" = G: "second-mouth"). Finally, the embryonic coelom arises from an archenteron, although, in detail, the method of coelom for-

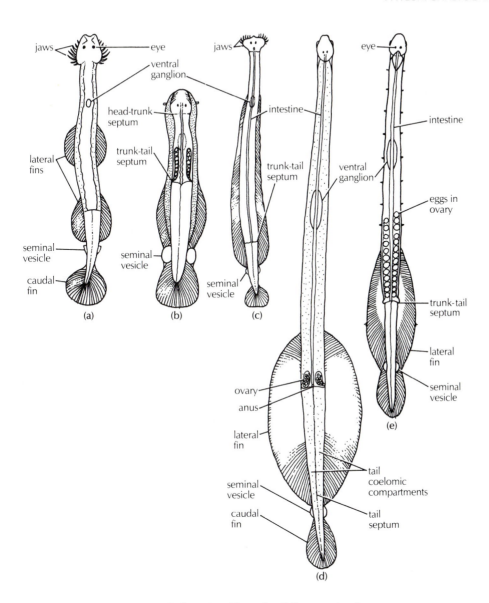

FIGURE 20.4. *Chaetognath diversity. Note the differences in the number and extent of the lateral fins and in the external appearance of the seminal vesicles. (a)* Sagitta macrocephala. *(b)* Sagitta cephaloptera. *(c)* Eukrohnia fowleri. *(d)* Krohnitta subtilis. *(e)* Krohnitta pacifica. [(a)–(c) modified from various sources. (d, e) from Hyman; after Tokioka.]*

mation by chaetognaths differs significantly from the basic deuterostome plan. As discussed previously (Chapter 2), the standard deuterostome coelom arises from a symmetrical outpouching of the archenteron. In chaetognaths, on the other hand, the coelom is formed by invaginations of the archenteron. Nevertheless, coelom formation is clearly enterocoelous, rather than schizocoelous, in nature.

In many other respects, however, chaetognaths show very weak affinities with other deuterostomes. Unique among deuterostomes, there is no ciliated larval stage in the life history; the morphology of the young chaetognath closely resembles that of the adult. The body cavity and musculature of adult chaetognaths are also unlike those of other deuterostomes. In fact, the only conspicuous similarity with other deuterostomes is the compartmentalization of the body cavity. Even here, we find deviation from the basic deuterostome plan. The body cavity of deuterostomes is typically divided during embryogenesis into three pairs of coelomic compartments; that of chaetognaths is divided into only two pairs of compartments. These trunk and tail compartments are clearly illustrated in Fig. 20.5.

Chaetognath development is unusual in several respects. In particular, the embryonic coelom, which is formed by enterocoely, is apparently obliterated as development proceeds. The adult body cavity forms later and by an entirely different mechanism. Nevertheless, the adult body cavity is lined with mesodermally derived peritoneum and is thus a true coelom, by definition.

The musculature of chaetognaths bears some resemblance to that encountered within the pseudocoelomate phylum Nematoda. No circular muscles are found among the members of either phylum. The musculature of the chaetognath body wall is exclusively longitudinal and, like that of the nematodes, is arranged in discrete bundles. Alternating contraction of the ventral and dorsal musculature of the chaetognath trunk and tail provides the thrust for locomotion, the two sets of musculature antagonizing each other through the hydrostatic skeleton of the fluid-filled body cavity. Unlike nematodes, however, chaetognaths do not generate sinusoidal waves. Instead, muscle contractions are sporadic, a single set of rapid muscle contractions being followed by a period of gliding. Unlike pseudocoelomates, chaetognaths are not eutelic (i.e., growth of chaetognaths occurs primarily by an increase in cell numbers, not by an increase in individual cell size), and adults possess substantial regenerative powers.

Chaetognaths have a long fossil record. Representatives are encountered among the earliest fossils of some 500 million years ago, so that we can learn little about the evolutionary origins of this group from the fossil record. Chaetognath embryology clearly marks this phylum as deuterostome. However, like the lophophorates, chaetognaths must have diverged from the major line of deuterostome evolution very early in the evolutionary history of metazoans. The present resemblance between adult chaetognaths and adult pseudocoelomates appears to reflect convergent evolution rather than any direct phylogenetic relationship.

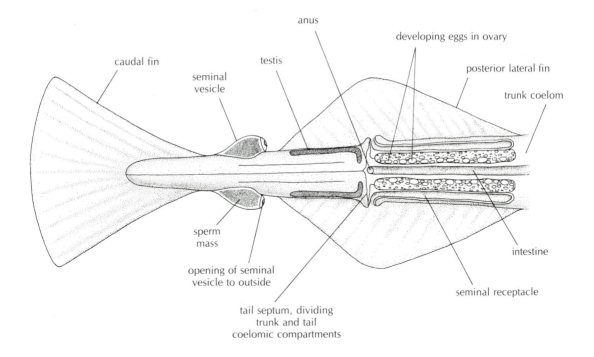

FIGURE 20.5. *Reproductive system of* Sagitta elegans. *Chaetognaths are simultaneous hermaphrodites, possessing both male and female gonads. The seminal vesicles are the sites of packing of sperm into spermatophores. The seminal receptacles are sites for storage of sperm received from another individual. (From Brown, 1950. Selected Invertebrate Types. John Wiley & Sons.)*

OTHER FEATURES OF CHAETOGNATH BIOLOGY

1. Reproduction

All chaetognaths are simultaneous hermaphrodites (Fig. 20.5), with the male gonads generally maturing earlier than the female gonads. Self-fertilization may occur at least occasionally in some species. There is no copulation in cross-fertilization; instead, sperm transfer is indirect, being mediated through an often mutual exchange of sperm-filled containers called **spermatophores.** The spermatophores are manufactured within a conspicuous pair of seminal vesicles and attached to the outside of the body of the recipient. Once sperm escape from the spermatophores, they must swim along the body to the opening of the female reproductive tract. Fertilization is internal. The fertilized eggs are released and develop into miniatures of the adult. There is no morphologically distinct larval stage in the life history.

435

2. Digestion

The chaetognath gut is linear and open at each end (see Fig. 20.4). Digestion is extracellular, and occurs within the intestine. A distinct stomach is lacking.

3. Nervous System

The nervous system of chaetognaths is fairly complex (Fig. 20.6). The pharynx is encircled by a ring of nervous tissue bearing several ganglia, including a large, cerebral (or "ventral") ganglion. Sensory and motor nerves extend from this and associated ganglia to the various light, tactile, and chemosensory systems and to the musculature of the trunk, tail, spines, and digestive tract.

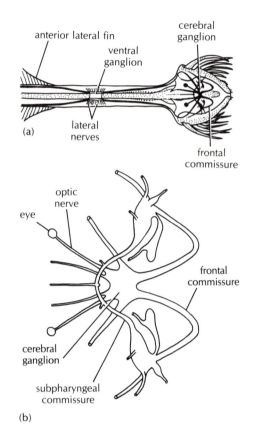

FIGURE 20.6. (a) Nervous system of Sagitta elegans, anterior. (After Brown; after monographs by Burfield and Kuhl.) (b) Detail of nervous system. (From Hyman; after Ritter-Zahony.)

TOPICS FOR FURTHER DISCUSSION AND INVESTIGATION

1. Discuss the evidence suggesting that chaetognaths can orient themselves to light and to vibrations.

Feigenbaum, D., and M.R. Reeve, 1977. Prey detection in the Chaetognatha: response to a vibrating probe and experimental determination of attack distance in large aquaria. *Limnol. Oceanogr.*, 22: 1052.

Goto, T., and M. Yoshida, 1983. The role of the eye and CNS components in phototaxis of the arrow worm, *Sagitta crassa* Tokioka. *Biol. Bull.*, 164: 82.

Newbury, T.K., 1972. Vibration perception by chaetognaths. *Nature, London*, 236: 459.

2. What environmental cues appear to regulate the cycle of vertical migration of chaetognaths? Discuss the likely adaptive benefits and costs associated with this migratory behavior.

Pearre, S., Jr., 1973. Vertical migration and feeding in *Sagitta elegans* Verrill. *Ecology*, 54: 300.

21

THE HEMICHORDATES

Phylum Hemi•chordata
(G: ONE-HALF STRING)

INTRODUCTION AND
GENERAL CHARACTERISTICS

The hemichordates are a small group of marine worms with an apparent and intriguing relationship with two major phyla: the Echinodermata and the Chordata. As typical deuterostomes, cleavage is radial, coelom formation is enterocoelic, and the coelom forms as three distinct compartments (protocoel, mesocoel, and metacoel). Although a notochord is lacking, excluding the hemichordates from membership in the Phylum Chordata, hemichordates do exhibit two other chordate characteristics: pharyngeal gill slits and, in some species, a dorsal, hollow nerve cord. Most hemichordate species are contained within a single class, the Enteropneusta.

CLASS ENTEROPNEUSTA

The enteropneusts are common inhabitants of shallow water, most species forming mucus-lined burrows in sandy or muddy sediment. The body is long and narrow, and is divided into three distinct regions corresponding to the tripartite compartmentation of the coelom (Fig. 21.1). The anterior-most section, called the **proboscis,** houses a single coelomic chamber, the **protocoel** (*proto* = G: first; *coel* = G: cavity). This anterior section of the body is generally conical in shape, and has given rise to the common name for enteropneusts, "acorn worms." The proboscis is highly muscular and has major responsibility for burrowing and food collection. Indeed, the proboscis is the only truly active part of the body; enteropneusts are sedentary animals, rarely moving from place to place as adults. Locomotion is largely restricted to movements within burrows (Fig. 21.2).

The proboscis is followed by a narrow **collar** region, containing a pair of coelomic chambers derived from the embryonic mesocoel (*meso* = G: middle). The mouth of the animal opens on the ventral anterior surface of the collar. The bulk of the body, termed the **trunk,** contains a pair of coelomic compartments derived from the embryonic metacoel. External ciliation of the trunk may aid

Class Entero•pneusta
(G: GUT BREATHING)

in locomotion within burrows. The total length of the body is typically about 8–45 cm, although the members of one South American enteropneust species attain a length of up to 2.5 meters!

Many hemichordate species are **deposit-feeders,** ingesting sediment, extracting the organic constituents, and producing a coil (**casting**) of mucus-bound, organically deprived sediment from the anus (Fig. 21.3). Other species are **suspension-feeders;** planktonic organisms and detritus adhere to mucus on the proboscis, and are then conveyed along ciliated tracts to the mouth.

The digestive tract of enteropneusts is tubular, consisting of a mouth (located behind the proboscis, on the collar—see Figs. 21.1, 21.4) leading into an esophagus (which compacts ingested particles into a mucus-bound rope); a pharynx; an intestine (the main site of digestion and absorption); and a terminal anus. Food is moved through the gut primarily by the action of ciliated cells lining the inner wall of the digestive tract.

An anterior extension of the pharynx forms a **buccal tube** within the collar of the animal. For some time, this tube was thought to be a notochord and the hemichordates were thus included within the Phylum Chordata. The two structures are no longer considered to be **homologous;** that is, the buccal tube of the enteropneusts and the notochord of the chordates are presumed to have had independent origins.

The pharynx opens to the outside through a series of lateral, paired gill slits (more than one hundred in some species). Cilia lining these gill slits beat in coordinated fashion so

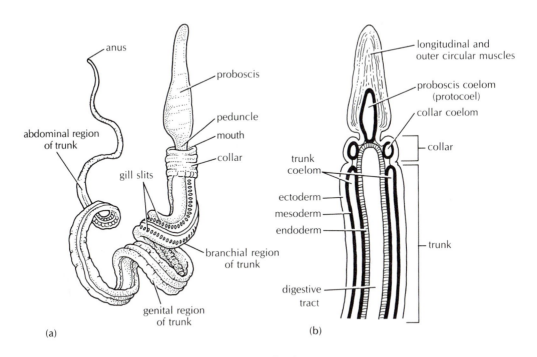

FIGURE 21.1. *(a) The acorn worm,* Saccoglossus kowalevskii, *removed from its burrow to show the basic body structure. (b) Schematic illustration of the coelomic compartments in an enteropneust. (Modified after various sources.)*

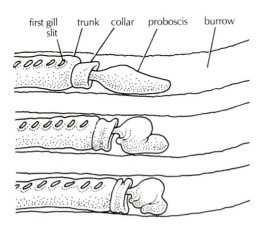

FIGURE 21.2. *Use of the proboscis in locomotion within the burrow by an acorn worm,* Saccoglossus horstii. *(From Clark, 1964. Dynamics in Metazoan Evolution.* Oxford University Press; *after Burdon-Jones.)*

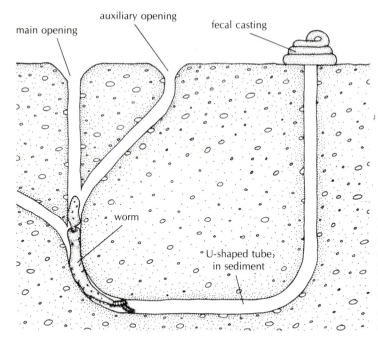

main opening

auxiliary opening

fecal casting

worm

U-shaped tube; in sediment

FIGURE 21.3. *The U-shaped burrow of a typical enteropneust. Note the coil of fecal material that has been deposited at the posterior end of the burrow. (Modified from Hyman; after Stiasny and after Burdon-Jones.)*

that water is drawn in at the mouth and discharged through the slits (Fig. 21.4). This flow of water is believed to serve for gas exchange, and the meaning of the name "enteropneust" then becomes apparent; the animals essentially breathe through a portion of their gut.

Acorn worms possess a true blood circulatory system. The blood, which lacks pigment, is circulated through dorsal and ventral blood vessels and associated blood sinuses by pulsations of the muscular blood vessels themselves; a distinct heart is never found. Evaginations of the blood sinuses are pronounced in the pharynx, so that the gill slits are highly vascularized, in keeping with their likely function in respiration.

The nervous system of acorn worms is echinoderm-like in that much of it is in the

form of an epidermal nerve network. In some regions of the enteropneust body, however, the network is consolidated to form distinct longitudinal nerve cords. The nerve cord in the collar region of the body actually lies below the epidermis and, in some species, is hollow. In such cases, this hollow nerve cord may well be homologous with the dorsal hollow nerve cord of vertebrates and other chordates.

Acorn worms are **dioecious** (sexes are separate), with the gonads housed in the trunk. The eggs of enteropneusts are fertilized externally in the sea water. A free-living larval form is found among a number of enteropneust species (Fig. 21.5). This planktonic, **tornaria** larva is equipped with a series of sinuous ciliated bands reminiscent of those encountered among the echinoderms.

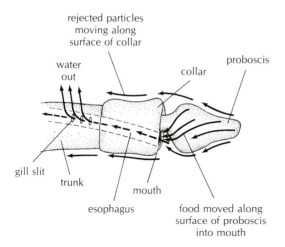

FIGURE 21.4. *Diagrammatic illustration of the respiratory and feeding pathways of a typical enteropneust. Fine food particles are trapped in mucus and transferred by cilia along the proboscis to the mouth, which is located in the collar; rejected particles are moved posteriorly over the surface of the collar. Ciliary activity draws water in at the mouth, through the pharynx, and out at the gill slits. The general body surface also plays a major role in gas exchange. (After Russell-Hunter.)*

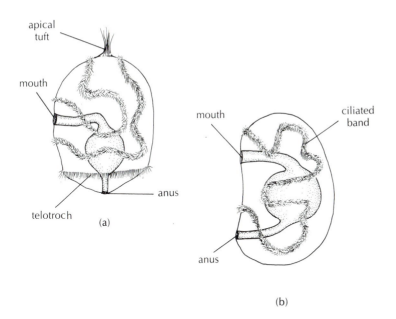

FIGURE 21.5. *(a) The tornaria larva of* Balanoglossus *sp. The tornaria bears a conspicuous terminal ring of cilia (the telotroch), which is lacking in the auricularia larva of echinoderms. (b) In other respects, ciliation patterns for larvae of the two groups are similar. (From Hardy, 1965.* The Open Sea: Its Natural History. *Houghton Mifflin.)*

CLASS PTEROBRANCHIA

A small number of hemichordataes are markedly dissimilar from the enteropneusts and are placed in a separate class, the Pterobranchia (*ptero* = G: a feather or wing; *branch* = G: a gill). Pterobranchs have been collected primarily by dredging in fairly deep water, especially in the antarctic; they are rarely encountered by the average invertebrate zoologist.

Unlike the enteropneusts, pterobranchs possess anterior tentacles and a U-shaped gut, and most species occupy rigid tubes that they secrete. A number of pterobranch species are colonial (Fig. 21.6).

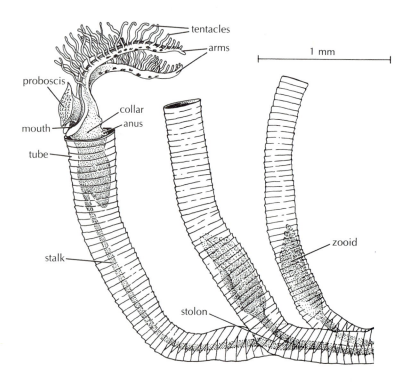

FIGURE 21.6. *A colony of the pterobranch* Rhabdopleura *sp.* (*Modified from Hyman and from Dawydoff; after Schepotieff and after Delage and Hérouard.*)

PHYLUM HEMICHORDATA
CLASS ENTEROPNEUSTA—THE ACORN
WORMS
CLASS PTEROBRANCHIA

◀ **T A X O N O M I C   S U M M A R Y**

**T O P I C   F O R   F U R T H E R   D I S C U S S I O N
A N D   I N V E S T I G A T I O N**

What morphological features of pterobranchs ally these animals with the enteropneusts?

Barrington, E.J.W., 1965. *The Biology of the Hemichordata and Protochordata.* Edinburgh: Oliver & Boyd.

Hyman, L.H. 1959. *The Invertebrates, Vol. V. Smaller Coelomate Groups.* New York: McGraw-Hill, pp. 155–191.

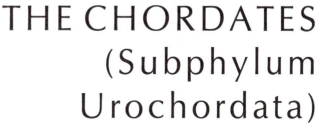

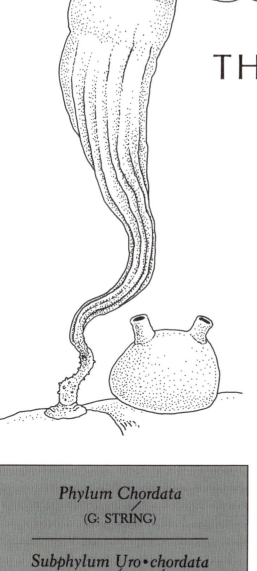

THE CHORDATES
(Subphylum
Urochordata)

Phylum Chordata

(G: STRING)

———

Subphylum Uro•chordata

(G: TAIL STRING)

INTRODUCTION AND
GENERAL CHARACTERISTICS

Yes indeed, the phylum Chordata includes some 2500 species of invertebrates, in addition to the 48,000 species contained in the subphylum Vertebrata. All urochordates are marine. In common with vertebrates, urochordates generally show, at some point in their life histories, the following characteristics: pharyngeal gill slits; hollow, dorsal nerve cord; and a notochord. The **notochord** (G: back-string) consists of a linear series of cells, each of which contains a large, fluid-filled vacuole. In vertebrates, the notochord is eventually replaced by cartilaginous tissue or by calcareous vertebrae during **ontogeny** (i.e., development).

The Urochordata is one of the few major taxonomic groups to contain no parasitic

445

species; all urochordates feed by straining small particles (especially phytoplankton) from the surrounding water. The method of generating the water current from which these food particles are obtained differs dramatically among the different urochordate classes. Members of this subphylum are commonly referred to as "tunicates," for reasons that become apparent in the next paragraph.

CLASS ASCIDIACEA

The class Ascidiacea contains at least 90% of all described urochordate species; its members are found throughout the world's oceans, in both shallow and deep water. Most adult ascidians, commonly known as "sea squirts," live attached to solid substrates. Some species live anchored in soft sediments. With few exceptions, ascidian adults are **sessile** (i.e., incapable of locomotion). The body is bag-like, and covered by a secretion of the epidermal cells (Fig. 22.1). This secreted, protective **test,** also called the **tunic,** is composed largely of protein and a polysaccharide (tunicin) that closely resembles cellulose. Amoeboid cells, blood cells and, in some species, blood vessels, are found within the tunic. The coloration of the tunic varies from nearly transparent in some species to dramatically pigmented in other species. Reds, browns, greens, and yellows are commonly encountered among the Ascidiacea.

> **Class Ascidia•cea**
> (G: A LITTLE BAG)

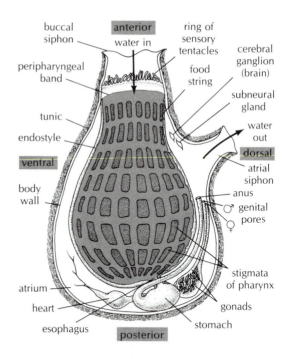

FIGURE 22.1. *Diagrammatic illustration of the anatomy of a typical ascidian. A thin sheet of mucus, produced by the endostyle, coats the basket-like pharynx. All incoming water must pass through this mucus coating before reaching the atrium, so that food particles are strained out. Cilia move the sheet dorsally, where it compacts to form a discrete mucus-food string, ready for ingestion and digestion. The water current is generated by the action of ciliated cells along the margins of the stigmata. (Modified after Bullough and other sources.)*

The digestive tract is especially noteworthy in that the pharynx is of unusually great diameter and is perforated with numerous slits called **stigmata;** stigmata morphology differs substantially among species, and is therefore an important taxonomic character. The perforated pharynx thus forms a large **pharyngeal** (or **branchial**) **basket** (Fig. 22.1). The outer margin of the pharynx is lined by a double ring of ciliated cells (the **peripharyngeal band**—*peri* = G: around), and the general surface and the stigmata of the pharynx are also ciliated; these cilia create a flow of water through the ascidian for feeding and gas exchange. Water is drawn into the ascidian through a **buccal** (**oral**) siphon. The buccal siphon is studded with sensory receptors on the outer surface and sensory tentacles on the inner surface. Incoming water is also apparently sampled by a **subneural gland** located beneath the brain; the precise functional significance of the subneural gland is still uncertain.

An elongated gland called the **endostyle** extends along the ventral surface of the pharynx (Fig. 22.1). Mucus secreted by the endostyle is drawn across the pharynx in a thin sheet, so that water passing through the stigmata must pass through the mucous sheet as well. Food particles as small as approximately 1 μm can thus be filtered from the inflowing water. Pharyngeal cilia continually move the particle-laden sheet of mucus dorsally (away from the endostyle) to form a mucus-food string. The string is moved posteriorly (away from the siphons) and enters the esophagus and then the stomach for digestion. Digestion is extracellular, with absorption of nutrients taking place in the intestine. Food is moved through the gut by cilia lining the digestive organs, and solid wastes are discharged through the anus.

The water that has been relieved of its particles (and oxygen) during passage through the stigmata enters a chamber, termed the **atrium,** enclosed by the tunic (Fig. 22.1). Water leaves the atrium through an **atrial siphon,** taking with it feces, excretory products, respired carbon dioxide, and any gametes that may have been discharged from the gonad. The buccal and atrial siphons can be closed off by contracting sphincter muscles ringing the incurrent and excurrent openings.

The body wall of the ascidian contains both circular and longitudinal musculature. The animal is thus capable of making major shape changes, although it cannot, in general, move from place to place.

Although many ascidian species are **solitary** (i.e., individuals are physically separated from each other), a large number of species are **colonial.** As in the colonial members of most other groups (e. g., the Hydrozoa and Bryozoa), individuals are produced by asexual budding and remain connected to the "parent." Commonly, the individual ascidians of a colony have separate buccal siphons and separate mouths, but share a single atrial siphon—the epitome of communal living (Fig. 22.2).

A small tubular heart is located adjacent to the stomach. The heart is notable primarily because it reverses the direction of its pumping many times each hour; the opening to the heart through which circulatory fluid enters will be the opening through which fluid exits several minutes later, and what was a vein becomes an artery and vice versa.[1] The circulatory fluid itself is peculiar in that it contains **amoebocyte cells,** which accumulate and store excretory wastes, and **morula cells,** which, in a number of species, accumulate vanadium and sulfuric acid and deposit these substances in the tunic.[2] The circulatory

---

1 See Topics for Discussion, No. 1.
2 See Topics for Discussion, No. 2.

FIGURE 22.2. *(a) A colonial ascidian,*
Botryllus violaceus. *As many as one dozen
individuals may share a single excurrent
opening. (After Milne-Edwards.)*
*(b) Diagrammatic section through a colony,
showing the anatomy of the individual
ascidians. Note that the circulatory system
extends into the tunic; this is also true in
solitary species. (Modified from various sources;
after Delage and Hérouard.)*

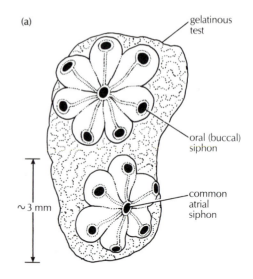

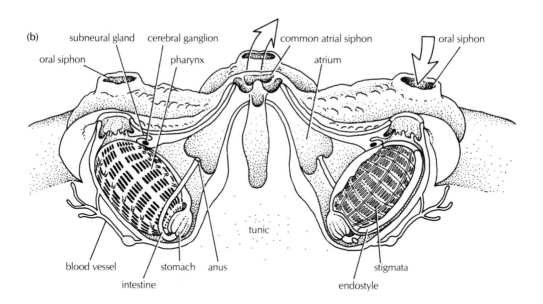

fluid of ascidians contains no oxygen-carrying respiratory pigments, so that the oxygen-carrying capacity of the blood is determined entirely by the solubility of oxygen in the fluid.

The only chordate characteristic encountered among adult ascidians is the presence of **pharyngeal gill slits** (stigmata). To see your chordate reflection in the ascidians, you must look to the larval stage. The ascidian larva is known as a **tadpole** because of its superficial resemblance to the developmental stage of frogs (Fig. 22.3). The heart and digestive system of the tadpole are restricted to the head region. The larvae are not capable of feeding, so that the digestive tract becomes functional only following metamorphosis to the adult form. The tail of the tadpole contains a conspicuous dorsal hollow nerve cord, a stiff notochord, and longitudinal muscles extending the length of the tail. The notochord can be bent but can be neither elongated nor shortened. The longitudinal muscles on one side of the tail can

thus antagonize the longitudinal muscles on the other side of the tail, enabling the tadpole to swim by flexing the tail from side to side. In the absence of the notochord, contractions of the longitudinal musculature in the tail would merely cause the tail to buckle or shorten. Swimming of tadpole larvae is utterly dependent upon the skeletal function subserved by the notochord.[3]

Attachment to a substrate by means of anterior suckers and a subsequent dramatic

---

3 See Topics for Discussion, No. 6.

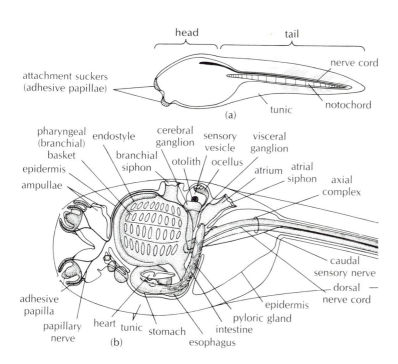

FIGURE 22.3. *(a) Diagrammatic illustration of an ascidian tadpole larva, with the digestive system omitted for clarity. (b) Internal anatomy of an ascidian tadpole. (From Cloney, 1982.* Amer. Zool., *22: 817.)*

metamorphosis to adulthood generally follow a free-swimming larval life of less than one day. In particular, metamorphosis includes the resorption of the notochord and tail of the larva. Muscular activity thus plays a significant, but very short-lived, role in the life history of ascidians.

CLASS LARVACEA

The members of this class appear to be derived from ascidian ancestors through the process of **neoteny**, in which the rates at which **somatic** (i.e., non-gonadal) body structures differentiate is slowed down relative to the rate of differentiaton of reproductive structures. Thus, the animal becomes sexually mature while retaining the larval morphology. The former adult morphology is thereby deleted from the life cycle, and the new adult morph gradually becomes increasingly better adapted to its new lifestyle through natural selection.

The larvacean heart, respiratory system, digestive system, and reproductive system are confined to the head (Fig. 22.4a), as in the ascidian tadpole larva. Now, however, the tadpole morph is a feeding individual. The mechanism of food collection differs considerably from that encountered among the Ascidiacea. The larvacean tail, like that of the ascidian tadpole, contains a notochord. The waving of the tail generates thrust for locomotion, and creates a feeding current as well. Unlike the ascidian tadpole, the larvacean secretes a most interesting gelatinous house around itself (Fig. 22.4b). This house plays a role in both locomotion and food collection; the undulation of the larvacean tail

drives water through the house. Water exits through a narrow-diameter opening that is partially occluded by a fine mesh filter. The orientation of this opening determines the direction of movement of the house and animal resulting from the expulsion of fluid. Small particles are filtered out by the mesh of mucus extending across the opening as water leaves the house. The particle-laden mucous net is then drawn into the mouth by the action of the oral cilia. Very coarse particles, which might damage the delicate house, are screened out from the incoming water current by a coarser mesh positioned across the incurrent opening. The larvacean abandons its house periodically as the meshes become clogged and as the house becomes littered with feces, and secretes a new house in the space of an hour or so.

Members of this class generally reach only a few millimeters in length, but a few species are said to reach lengths of up to 100 centimeters. Larvaceans are encountered in most areas of the ocean, at all latitudes.

CLASS THALIACEA

The members of this class are mostly free-living, planktonic individuals, but they achieve their mobility through modification of the adult ascidian body plan rather than through exploitation of the tadpole morph.

Thaliaceans, including the three orders of animals known commonly as pyrosomes, salps, and doliolids, are planktonic and nearly transparent. The buccal and atrial siphons are at opposite ends of the body, and bands of circular muscle are generally highly developed. These two features account for

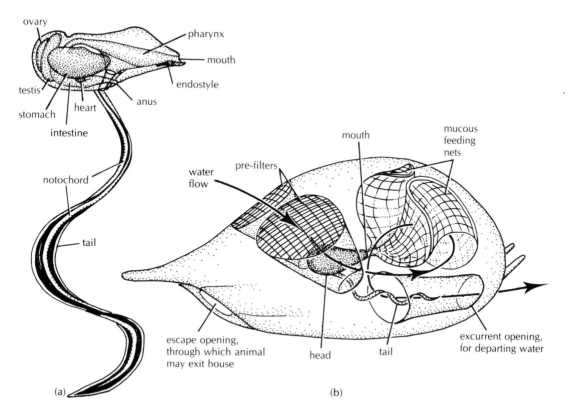

FIGURE 22.4. *(a) Schematic illustration of a larvacean, removed from its gelatinous house. (After Pimentel.) (b) The larvacean* Oikopleura *sp., generating water currents within its complex, disposable house. (After Hardy; after Lohmann.)*

the substantial locomotory capabilities possessed by most members of this class.

The most primitive (least modified) of the thaliaceans, members of the genus *Pyrosoma*, are much like colonial ascidians except that the buccal and atrial siphons are at opposite ends of the body. Like colonial ascidians, feeding is accomplished by means of cilia; the pharyngeal basket is well developed; and water is discharged through a common

exit (Fig. 22.5a). Locomotion of *Pyrosoma* spp. is achieved through ciliary activity.

The movement of other thaliaceans through the water is accomplished by closing off the buccal aperture and contracting the thick circumferential bands of muscle underlying the test. This contraction deforms the test. The resulting decrease in internal volume forces water out of the atrial siphon (since water is incompressible), and the ani-

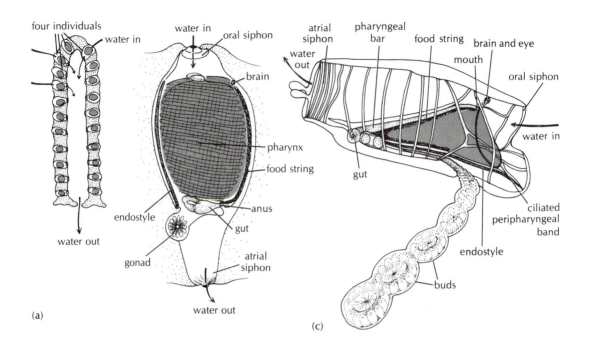

(a)

four individuals
water in
water in
oral siphon
brain
pharynx
food string
anus
endostyle
gut
gonad
atrial siphon
water out
water out

(c)

atrial siphon
pharyngeal bar
food string
brain and eye
mouth
water out
oral siphon
water in
gut
ciliated peripharyngeal band
endostyle
buds

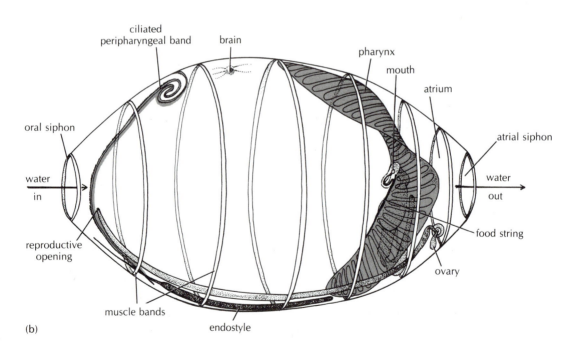

(b)

ciliated peripharyngeal band
brain
pharynx
mouth
atrium
oral siphon
water in
atrial siphon
water out
reproductive opening
food string
ovary
muscle bands
endostyle

FIGURE 22.5. *(a)* Pyrosoma atlanticum. *The general shape of a typical colony is shown to the left. The general morphology of an individual member of such a colony is shown to the right. Pyrosome anatomy is much like that of ascidians, except that the atrial and oral siphons are directly opposite each other. Shaded areas indicate sheets of mucus.* (From Berrill, 1950. The Tunicata. Ray Society, London.) *(b) Diagrammatic illustration of a doliolid* (Doliolina intermedia). *Note the conspicuous rings of musculature that generate locomotory currents. Feeding currents are produced by cilia.* (From Tokioka and Berner, 1958. Pacific Sci., 12: 317.) *(c) Diagrammatic illustration of a salp,* Cyclosalpa affinis. *The pharyngeal basket is reduced to a thin bar. The cilia on the pharyngeal bar function only in compacting the mucous net into a food string; locomotion and feeding are accomplished through contraction of the well-developed muscular bands.* (From Berrill, 1950.)

mal is jet-propelled forward as a consequence. Next, the aperture to the atrial siphon is closed while the buccal aperture is opened, and the circular muscles relax. The volume of the animal increases as the elastic test quickly regains its resting shape. Water is thus drawn into the buccal siphon and the animal is pulled forward as it prepares for the next power stroke. Note that the bands of circular muscle are antagonized by the test itself, rather than by longitudinal musculature.

In doliolids, the pharyngeal basket is reduced to a flattened plate, although cilia and stigmata are still conspicuous (Fig. 22.5b). The details of feeding have yet to be documented for these animals, but pharyngeal cilia seem to play a major role in food collection. Among the salps, however, the pharyngeal basket is reduced to a slender **branchial bar.** Water flows through a mucous bag suspended between the endostyle, branchial bar, and peripharyngeal bands; stigmata are absent (Fig. 22.5c). In salps, mus-

cular contractions thus play the dominant role in both feeding and locomotion (Fig. 22.6). The evolution of thaliaceans from ascidians appears to be a story of decreasing reliance on cilia and increasing development of musculature.

Although most species of thaliacean occur as individuals and rarely exceed several inches in length, species in the genus Pyrosoma are colonial and may attain lengths of several meters. Unfortunately for those of us living in the temperate zone, most thaliaceans occur only in warm waters.[4]

OTHER FEATURES OF UROCHORDATE BIOLOGY

1. Reproduction

Fertilization is exclusively external, occurring in the surrounding sea water, for all spe-

---

4 See Topics for Discussion, No. 5.

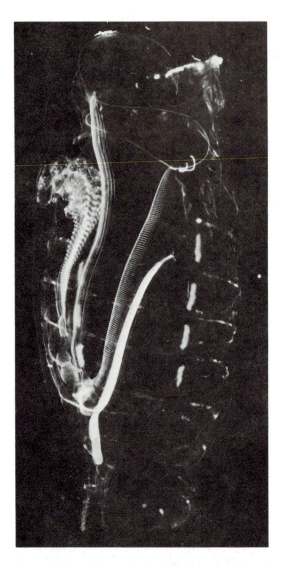

FIGURE 22.6. *Salp photographed* in situ *in the open ocean. (Courtesy of G.R. Harbison and J. Carleton.)*

hibit asexual reproduction by budding, interspersed with bouts of sexual reproduction. Most ascidians are simultaneous hermaphrodites. Larvaceans are also hermaphroditic, but reproduction is exclusively sexual. Since asexual reproduction is not found within this class, larvaceans are always solitary (i.e., there are no colonial species).

Reproduction of thaliaceans is both sexual and asexual, as with ascidians, but among the Thaliacea the two forms of reproduction are allocated between two types of individuals. The union of sperm and egg gives rise to an asexually reproducing morph called the **oozooid.** The oozooid produces no gametes. Instead, individuals called **blastozooids** are budded off from the oozooid (Fig. 22.7). These blastozooids develop functional gonads, and their gametes ultimately give rise to the next generation of oozooids.

2. *Excretory and Nervous Systems*

Specialized excretory organs are rarely found among the Urochordata. In most species, wastes are apparently removed by diffusion across general body surfaces and through the activities of scavenging amoebocytes. In several species within the ascidian genus *Molgula*, however, excretory organs have been described.[5] These ductless, blind-ending **renal sacs** accumulate nitrogenous wastes and deposit them as solid concretions composed primarily of uric acid and related compounds. These concretions accumulate within the renal sacs throughout the life of the ascidian.

cies. Gametes are discharged through the atrial siphon. In other respects, the reproductive biology differs markedly among the three classes. Ascidians commonly ex-

5 See Topics for Discussion, No. 3.

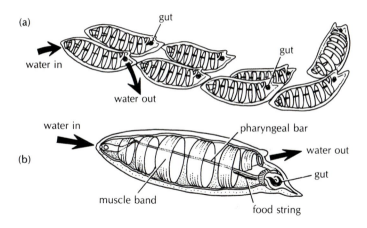

(a)

gut

water in

gut

water out

water in

pharyngeal bar

water out

(b)

gut

muscle band

food string

FIGURE 22.7. *In the life cycle of many salps, chains of blastozooids (a) are budded off from a solitary oozoid (b). The blastozooids are released from the parent while still aggregated. After swimming for a time as a group, the individual blastozooids separate and reach sexual maturity. Note that the blastozooids are produced asexually from the oozoid, so that all of the individuals in an aggregate are genetically identical with each other and with the parental oozoid. (After Hardy.)*

A conspicuous cerebral ganglion is located in the body wall, and innervates the muscles and sensory cells. Its role in the biology of ascidians seems to be minimal, since surgical removal of the cerebral ganglion causes little apparent change in activity. The brain's involvement in coordinating the activities of larvaceans and thaliaceans is likely to be much greater, since muscular activity plays a greater role in the lives of these animals. In addition, salps have a prominent, horseshoe-shaped photoreceptor associated with the cerebral ganglion.

◄ **T A X O N O M I C   S U M M A R Y**

PHYLUM CHORDATA
SUBPHYLUM UROCHORDATA
CLASS ASCIDIACEA
CLASS LARVACEA
CLASS THALIACEA
ORDER PYROSOMATIDA
ORDER DOLIOLIDA
ORDER SALPIDA

TOPICS FOR FURTHER DISCUSSION AND INVESTIGATION

1. Investigate the significance of the heart-beat reversal in urochordates, and the mechanism by which it is accomplished.

Herron, A.C., 1975. Advantages of heart reversal in pelagic tunicates. *J. Marine Biol. Assoc. U.K.*, 55: 959.

Jones J.C., 1971. On the heart of the orange tunicate, *Ecteinascidia turbinata* Herdman. *Biol. Bull.*, 141: 130.

Kriebel, M.E., 1968. Studies on cardiovascular physiology of tunicates. *Biol. Bull.*, 134: 434.

Ponec, R.J., 1982. Natural heartbeat patterns of six ascidians and environmental effects on cardiac function in *Clavelina huntsmani. Comp. Biochem. Physiol.*, 72A: 455.

2. Investigate the adaptive significance of vanadium and acid accumulation in the tests of ascidians.

Stoecker, D., 1978. Resistance of a tunicate to fouling. *Biol. Bull.*, 155: 615.

3. Discuss the evidence for functional excretory systems among ascidians.

Das, S.M., 1948. The physiology of excretion in *Molgula* (Tunicata; Ascidiacea). *Biol. Bull.*, 95: 307.

Heron, A., 1976. A new type of excretory mechanism in tunicates. *Marine Biol.*, 36: 191.

4. Discuss the similarities and differences in the feeding mechanisms of ascidians and bivalves.

MacGinitie, G.E., 1939. The method of feeding of tunicates. *Biol. Bull.*, 77: 443.

Young, C.M., and L.F. Braithwaite, 1980. Orientation and current-induced flow in the stalked ascidian *Styela montereyensis. Biol. Bull.*, 159: 428.

5. Discuss the potential ecological significance of planktonic urochordates in the open ocean.

Bruland, K.W., and M.W. Silver, 1981. Sinking rates of fecal pellets from gelatinous zooplankton (salps, pteropods, doliolids). *Marine Biol.* 63: 295.

Harbison, G.R., and R.W. Gilmer, 1976. The feeding rates of the pelagic tunicate *Pegea confederata* and two other salps. *Limnol. Oceanogr.*, 21: 517.

Madin, L.P., 1974. Field observations on the feeding behavior of salps (Tunicata: Thaliacea). *Marine Biol.*, 25: 143.

Wiebe, P.H., L.P. Madin, L.R. Haury, G.R. Harbison, and L.M. Philbin. 1979. Diel vertical migration by *Salpa aspera* and its potential for large-scale particulate organic matter transport to the deep-sea. *Marine Biol.* 53: 249.

6. Compare the locomotory mechanism of ascidian tadpoles with that of nematodes.

7. Compare the reproductive biology of thaliaceans with that of the scyphozoans.

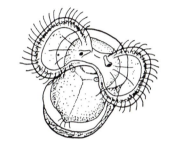

INVERTEBRATE
REPRODUCTION AND
DEVELOPMENT—
AN OVERVIEW

INTRODUCTION

The continued existence of a species depends upon the ability of individuals of that species to reproduce. Nearly every behavioral, morphological, or physiological adaptation of a species may be presumed to contribute to its reproductive success, either directly or indirectly. In a sense, then, all organisms live to reproduce. Moreover, no matter how the process of reproduction begins or proceeds, differentiation (genetically controlled specialization of cells) is always involved; much of what we presently understand about the control of gene expression comes from the study of invertebrate development. The topic of reproduction and development thus seems an ideal one to unite all of the phyla in consideration of a single aspect of invertebrate biology. As you read through this chapter, I hope you will be able to take pride in your recognition of the many terms that were new to you when you began this book.

Invertebrates show a great diversity of reproductive and developmental patterns, a diversity that far surpasses that encountered among the vertebrates. Most vertebrates fertilize internally and show some degree of parental care for the developing young. All vertebrates are deuterostomes, and thus cleavage is basically radial and indeterminate, with a coelom forming by enterocoely. Variations on the basic deuterostome theme occur, of course, largely related to differing amounts of yolk being present in the eggs. The diversity of developmental patterns observed among invertebrates is much more than variations on a theme. Among invertebrates, we encounter radical differences in:

(1) expression of sexuality;
(2) site of fertilization (if present);
(3) pattern of cell division;
(4) stage at which cell fates become determined;
(5) number of distinct tissue layers formed;
(6) mechanism through which mesoderm (if any) is formed;
(7) extent to which a body cavity develops;
(8) mechanism through which a body cavity develops; and
(9) origin of the mouth and anus (when present).

In this chapter, we will survey the patterns of reproduction and development encountered among invertebrates, identify some of the groups most closely associated with these different patterns, and consider the ecological significance, i.e., the likely adaptive benefits, of the various patterns discussed. A summary of cleavage patterns, modes of coelom formation, and other features of early metazoan embryology has already been given (Chapter 2). The goal of the present chapter is to put reproduction and development into an ecological context.

ASEXUAL REPRODUCTION

Reproduction among invertebrates may be either sexual or asexual. Sexual reproduction always involves the union of genetic material contributed by two genomes. Asexual reproduction, on the other hand, can be simply defined as reproduction in the absence of fertilization (i.e., without a union of gametes). The timing of both sexual and asexual reproductive events is controlled by both environmental and internal factors.[1]

Asexual reproduction is often a process of exact replication; in such instances, barring mutation, asexually produced offspring are genetically identical to the progenitor. This form of asexual reproduction (**ameiotic;** i.e., without meiosis) can add no genetic diversity to a population. On the other hand, through asexual reproduction a single individual can contribute to a potentially rapid increase in population size, excluding potential competitors and flooding the population with a particularly successful genotype.

Asexual reproduction need not involve egg production by a female. In sponges (porif-

---

1 See Topics for Discussion, Nos. 1, 2.

erans), hydrozoans, scyphozoans, bryozoans, thaliaceans, and some ascidians and protozoans, for example, asexual reproduction is accomplished through the budding of new individuals from pre-existing individuals (Fig. 23.1). Among the Protozoa, replication is often achieved through binary fission. In the Trematoda, asexual reproduction takes the form of ameiotic replication of larval stages. In some other groups, such as the Anthozoa, Ctenophora, Turbellaria, Rhynchocoela, Polychaeta, Asteroidea, and Ophiuroidea, body parts may be detached from the adult and left behind to regenerate into new, morphologically complete individuals (Fig. 23.2).

Egg production is intimately involved in the ameiotic, asexual reproduction of many other invertebrate species. Among selected arthropods and rotifers, asexual reproduction takes the form of **parthenogenesis,** in which eggs develop to adulthood in the absence of fertilization. As simple as this sounds, there can be unusual complications. In some mites and ticks (Arachnida), for example, the females cannot oviposit unless they first mate with a male, even though the eggs never become fertilized. Something similar occurs in many beetles, except that males do not exist in some of these species. In such cases, the females mate with males of an allied species; although no union of gametes occurs, development of the eggs will not occur in the absence of contact with sperm.

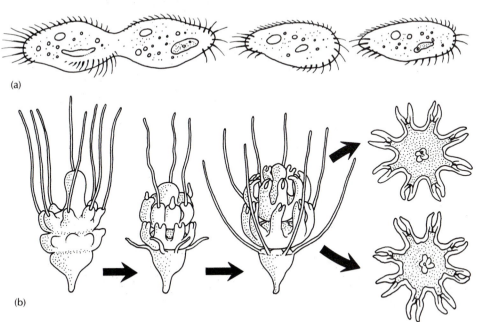

(a)

(b)

FIGURE 23.1.   (a) Binary fission in a ciliated protozoan. (b) Strobilation in the scyphozoan Stomolophus meleagris, culminating in the release of ephyrae. (From Calder, 1982. Biol. Bull., 162: 149.)

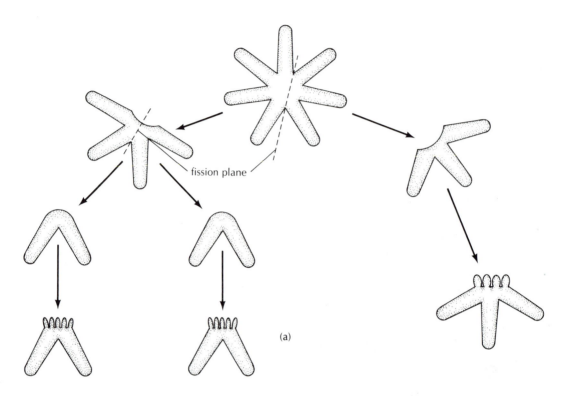

fission plane

(a)

number of individuals examined each month

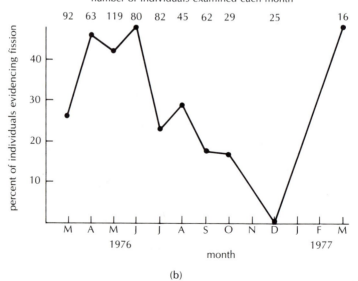

(b)

FIGURE 23.2. *(a) Asexual reproduction in the asteroid* Nepanthia belcheri. *This species routinely undergoes fission, in which six- or seven-armed adults divide to form primarily two- or three-armed individuals, which then proceed to regenerate the full complement of arms. (b) Percent frequency of fission varies over the course of one year between 0% and nearly 50% of the population. The number of individuals examined each month is shown at the top of the graph.* (From Ottesen and Lucas, 1982. Marine Biol. 69: 223.)

In other groups of invertebrates, asexual reproduction may involve meiosis, so that pairing and segregation of chromosomes occurs and new genetic combinations can be generated despite the lack of genetic input by a second individual. This occurs in some protozoans and nematodes, both parasitic and free-living, but is most commonly encountered among arthropods, especially insects and arachnids. As seen in Table 23.1, asexual reproduction is quite commonly encountered among the invertebrates. In fact, reproduction without fertilization is the primary reproductive mode in many species. Note that with few exceptions, asexual reproduction requires the presence of only a single individual.

TABLE 23.1. Forms of Sexuality Encountered among Invertebrates

TERM	FORM
1. Dioecious	♂ or ♀
2. Simultaneous hermaphrodite	⚥
3. Sequential hermaphrodite	
(a) protandric	♂ → ♀
(b) protogynous (relatively rare)	♀ → ♂

SEXUAL REPRODUCTION

Patterns of Sexuality

Although many invertebrates reproduce asexually, sexual reproduction is quite common; the fusion of haploid gametes is required for its accomplishment. Two individuals are usually involved in bringing this about. Moreover, the genetic composition of the offspring is always dissimilar to that of either parent. The two parents are usually of different sexes, in which case the species is said to be **dioecious** or **gonochoristic**. Alternatively, a single individual may be both male and female, either simultaneously (**simultaneous hermaphroditism**) or in sequence (**sequential hermaphroditism**).

Hermaphroditism is commonly encountered among invertebrates, as shown in Table 23.2. The East Coast oyster, *Crassostrea virginica*, is a good example of a species exhibiting sequential hermaphroditism. The young oyster matures as a male, later becomes a female, and may change sex every few years thereafter. Most sequential hermaphrodites change sex only once, and usually change from male to female. This is

TABLE 23.2. Summary of Invertebrate Reproductive Patterns. Present = +; Absent = −; ? = unknown; −? = probable absence; +? = probable presence; internal fertilization = within the body of the animal, but not necessarily within the reproductive tract.

TAXONOMIC GROUP	REPRODUCTIVE MODE		SPERMATO-PHORE	SEXUALITY		REPRE-SENTATIVE LARVAL STAGE	FERTILIZATION	
	ASEXUAL	SEXUAL		DIOE-CIOUS	HERMAPHRO-DITIC		INTERNAL	EXTERNAL
Protozoa	+	+	−	have mating strains		none	+	−
Porifera	+	+	−	+(rare)	+	parenchymula	+	−
Cnidaria								
Scyphozoa	+	+	−	+	+(rare)	planula	+(rare)	+
Hydrozoa	+	+	−	+	+(rare)	planula	+(rare)	+
Anthozoa	+	+	−	+	+(rare)	planula	+	+
Ctenophora	+(rare)	+	−	−	+	cydippid	+(rare)	+
Platyhelminthes								
Turbellaria	+	+	−	+(rare)	+	Müllers	+	−
Trematoda	+	+	−	−	+	miracidium, cercaria	+	−
Cestoda	+	+	−	−	+	none	+	−
Rhynchocoela	+	+	−	+	+(rare)	pilidium	+(rare)	+
Nematoda	+	+	−	+	+(rare)	none	+	−
Rotifera								
Seisonidea	−	+	−	+	−	none	+	−
Bdelloidea	+	−	−	+	−	none	−	−
Monogononta	+	+	−	+	−	none	+	−
Annelida								
Polychaeta	+	+	+	+	+(rare)	trochophore	−	+
Oligochaeta	+	+	−	−	+	none	−	+
Hirudinea	−	+	+	−	+	none	+	−
Mollusca								
Gastropoda	+(rare)	+	+	+	+	veliger	+	+(rare)
Bivalvia	−	+	−	+	+	trocophore, veliger	+(rare)	+
Polyplacophora	−	+	−	+	+(rare)	trochophore	+(rare)	+

					Larval type			
Cephalopoda	—	+	+	+	none	+(rare)	+	—
Scaphopoda	—	+	+	+	trochophore	—	—	+
Aplacophora	—	+	+	—?		+	+	+
Monoplacophora	—	+	+	—	veliger?	—	—?	+?
Arthropoda								
Merostomata	—	+	+	—	trilobite	—	+	+
Arachnida	—	+	+	+	none	—	+	—
Chilopoda	—	+	+	+	none	—	+	—
Diplopoda	—	+	+	+	none	—	+	—
Insecta	+	+	+	+	larva, pupa	—	+	—
Crustacea	+	+	+	+	nauplius, cyprid, zoea, megalopa	+(rare)	+	+(rare)
Sipuncula	—	+	+	—	trochophore, pelagosphera	+(rare)	—	+
Echiura	—	+	+	—	trochophore	—	+(rare)	+
Pogonophora	—	+	+	+	?	—	?	?
Tardigrada	+	+	+	—	none	—	+	—
Onychophora	—	+	+	+	none	—	+	+(rare)
Bryozoa	+	+	+(rare)	—	coronate, cyphonautes	+	+	+(rare)
Phoronida	+	+	+(rare)	—	actinotroch	+(rare)	+	+(rare)
Brachiopoda	—	+	+	—	nameless	+(rare)	+	+(rare)
Chaetognatha	—	+	—	+	none	+	+	—
Echinodermata								
Asteroidea	+	+	+	—	bipinnaria, brachiolaria	+(rare)	+	+
Ophiuroidea	+	+	+	—	ophiopluteus	+	+	+
Echinoidea	—	+	+	—	echinopluteus	+	+	+
Holothuroidea	—	+	+	—	auricularia	+(rare)	—	+
Crinoidea	—	+	+	—	doliolaria	—	+	+
Urochordata								
Ascidiacea	+	+	+(rare)	—	tadpole	+	+	+
Larvacea	—	+	+(rare)	—	none	+	—?	+?
Thaliacea	+	+	—	—	none	—	+	—
Hemichordata								
Enteropneusta	+(rare)	+	+	—	tornaria	—	+	+

known as **protandric hermaphroditism,** or **protandry** (*prot* = G: first; *andros* = G: male).

In contrast to species that change sex as they age, many invertebrates, including most ctenophores and cestodes, are simultaneous hermaphrodites (Fig. 23.3). Self-fertilization is rare among simultaneous hermaphrodites, although it can occur, as in the Cestoda. An advantage of simultaneous hermaphroditism is that a meeting between any two mature, consenting individuals can result in a suc-

cessful mating. This is especially advantageous in sessile animals, such as barnacles (Crustacea, Cirripedia). Indeed, the benefits of simultaneous hermaphroditism are so conspicuous that one wonders why such reproductive patterns are so rare among vertebrates. Presumably, the reproductive systems and behaviors of most advanced vertebrates have become so complex and specialized that two-sexed individuals are simply not feasible. This is perhaps unfortunate in some respects; many of the in-

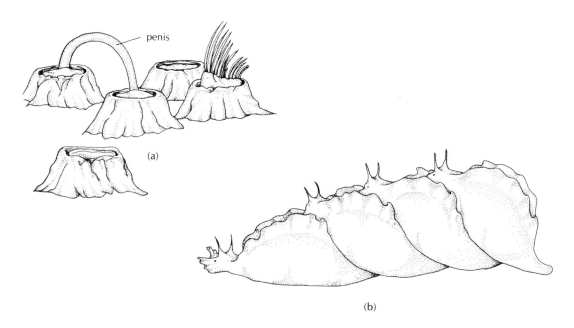

FIGURE 23.3.   *(a) Copulation among the Cirripedia. Barnacles are simultaneous hermaphrodites, a decided advantage for individuals that are incapable of locomotion as adults; any two adjacent individuals are potential mates. (After Barnes and Hughes.)*
(b) Each sea hare, Aplysia brasiliana (Gastropoda: Opisthobranchia), bears a penis at the right side of the head and a vaginal opening posteriorly. Mutual insemination is common. The individuals in the middle of the chain illustrated are functioning simultaneously as males and females. (From Purves and Orians, Life: The Science of Biology. *Sinauer Associates/Willard Grant Press; after W. Aspey.)*

equalities commonly encountered among human societies would be unlikely in a society of simultaneous hermaphrodites.

The advantages of sequential hermaphroditism are less evident. Age-dependent sex changes can deter self-fertilization (a rather extreme form of in-breeding) in hermaphroditic species. In addition, it has been suggested that it may be energetically more efficient to be one sex when small and the other sex when larger. Certainly it is true that a single spermatozoon requires less structural material and nutrients than does a single ovum and is therefore less costly to produce. The total cost of reproduction of a particular sex will, however, depend upon the total number of gametes produced, the cost of each individual gamete, and the amount of energy expended in obtaining a mate and protecting the young; maleness may or may not be cheapness, depending upon the species.

Gamete Diversity

Invertebrate gametes show considerable diversity of structure and, often, of function. Some invertebrates produce a percentage of eggs that are incapable of being fertilized and/or of sustaining development following fertilization. These **nurse eggs** are ultimately consumed by neighboring embryos (Fig. 23.4). Nurse eggs are especially common among the Gastropoda. Sperm also show a high degree of functional diversity. Many invertebrate species produce only a percentage of normal sperm; i.e., sperm that have a haploid DNA content and are capable of fertilizing an egg and promoting subsequent de-

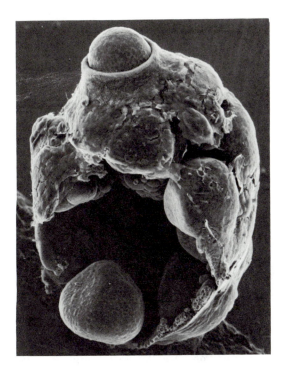

FIGURE 23.4. *Ingestion of nurse eggs during prosobranch gastropod development. The marine species* Searlesia dira *is shown in the process of ingesting a single nurse egg. A portion of the body wall has been torn away, revealing a number of previously ingested nurse eggs. Each nurse egg is approximately 230 μm in diameter. (Courtesy of Dr. Brian Rivest, from Rivest, 1983. J. Exp. Marine Biol. Ecol., 69: 217.)*

velopment of the embryo. These are **eupyrene** sperm. The other sperm can play no direct role in development because they have either an excess number of chromosomes or too few. In the extreme case, the abnormal sperm are without chromosomes entirely; i.e., they are **apyrene**. Apyrene sperm production is especially common among gastropods and insects (Fig. 23.5). The functional significance of apyrene and

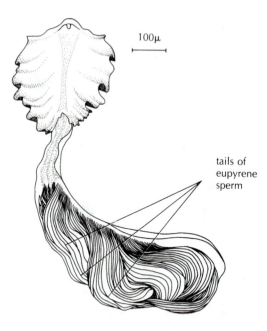

100μ

tails of
eupyrene
sperm

FIGURE 23.5. *A giant, apyrene sperm of the marine gastropod* Cerithiopsis tubercularis, *with thousands of eupyrene sperm embedded in its tail. This form of apyrene sperm, called spermatozeugmata, may function in the transport of eupyrene sperm to the egg. (From Fretter and Graham, 1976. A Functional Anatomy of Invertebrates. Academic Press.)*

other atypical sperm remains speculative at the present time.[2]

Considerable morphological diversity exists among even normal invertebrate sperm (Fig. 23.6), and the arrangement of microtubules within the axoneme (see Chapter 3, p. 35) often departs radically from the usual

---

2 See Topics for Discussion, No. 6.

9 + 2 arrangement (Fig. 23.7). Indeed, the sperm of some species lack flagella entirely. The sperm of some species are incapable of any movement, while the sperm of others may move in amoeboid fashion. Aberrant sperm are encountered especially often among the Arthropoda.

Getting the Gametes Together

All sexual development begins with the fertilization of a haploid egg; the trick, then, is to get the eggs and sperm together so that the union of gametes can occur. Invertebrates demonstrate a variety of ways of bringing this union about. On land and in fresh water, fertilization of the egg is, with few exceptions, internal, for reasons presented in Chapter 1. These environments are generally too dry or osmotically stressful for survival of exposed gametes. Internal fertilization may be accomplished in several ways. Males of many invertebrate species are equipped with a penis, through which sperm are transferred directly into the genital opening of the female. In the case of hypodermic impregnation, as encountered among some turbellarians, leeches, gastropods, and rotifers, sperm are forcefully injected through the body wall of the female. In both cases, sperm transfer is said to be **direct**.

Males of other species lack any such copulatory organs, and yet internal fertilization may still occur. Several means of achieving such **indirect sperm transfer** are encoun-

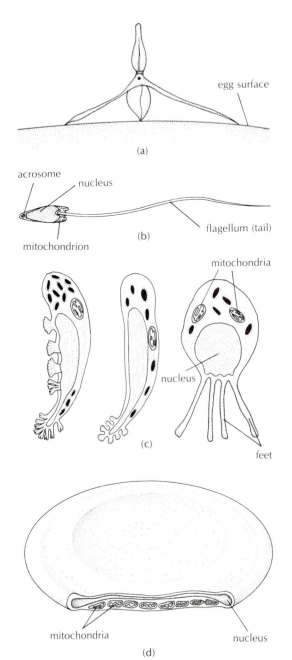

(a)

egg surface

acrosome

nucleus

flagellum (tail)

(b)

mitochondrion

mitochondria

nucleus

(c)

feet

mitochondria

nucleus

(d)

FIGURE 23.6. *Gamete diversity. (a) The tripod-like sperm of the crustacean* Galathea sp. *The sperm is about to penetrate the egg. (From Kume and Dan, 1968.* Invertebrate Embryology. *National Science Foundation; after Kortzoff.) (b) Sperm of the echinoid* Strongylocentrotus purpuratus *(the purple sea urchin). (Modified from several sources.) (c) The aflagellate sperm of gnathostomulid worms (interstitial acoelomates). These sperm use the small feet-like processes to move. (From Bacetti and Afzelius, 1976.* Biology of the Sperm Cell. *Monographs in Developmental Biology, No. 10; S. Karger; after Graebner and Adam.) (d) The disc-shaped sperm of the insect* Eosentomon transitorium. *(From Bacetti and Afzelius; after Bacetti et al.)*

tered among invertebrates.[3] Commonly, the sperm are packaged in containers of varying complexity. These sperm-filled containers, called **spermatophores** (literally, "sperm carriers"), are secreted by specialized glands found only in the male. Among terrestrial invertebrates, spermatophores are typically employed by pulmonate gastropods, onychophorans, and terrestrial arthropods, including the insects, arachnids, centipedes, and millipedes (Fig. 23.8a, b). Transfer of the spermatophore from the male to the female takes place in diverse ways. In some pseudoscorpions (Arachnida), for example, no actual mating occurs. Males may deposit spermatophores onto suitable substrates without a female being present; the sperm capsules are located by the female either chemotactically or, in a few species, by following

3 See Topics for Discussion, No. 3.

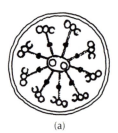 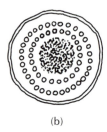 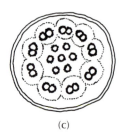 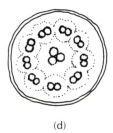

(a)   (b)   (c)   (d)

FIGURE 23.7. (a) Cross section through the sperm tail of a sea urchin. Note the 9 + 2 arrangement of microtubules in the axoneme. (b) Cross section through the axoneme of an insect sperm, Parlatoria oleae. (c) Axoneme of a caddisfly sperm, Polycentropus sp. The microtubules have a 9 + 7 arrangement. (d) 9 + 3 arrangement of microtubules in the axoneme of spider sperm, Pholeus phalangioides. (All based on Baccetti and Afzelius, 1976. Biology of the Sperm Cell. Monographs in Developmental Biology, No. 10. S. Karger.)

silk threads secreted by the male. Once the female has located a spermatophore, she inserts it into her genital opening, discharging the sperm from the container. This mechanism seems to be admirably suited for internal fertilization in a species in which the proximity of one individual to another frequently prompts physical attack and even cannibalism.

The pseudoscorpion example to the contrary, internal fertilization generally requires a high degree of cooperation between pairs of individuals, and is often preceded by elaborate courtship displays. This is typically true when fertilization is accomplished by copulation, but is also common when fertilization is achieved through use of spermatophores. In what appear to be the more highly evolved

pseudoscorpions, for example, males again deposit spermatophores on the ground, but only in the presence of females and only after a complex mating dance. The male then physically guides the female and positions her over the stalk of the attached spermatophore, whereupon she takes up the sperm packet through her genital opening. The two partners then quickly separate from each other. The female removes the capsule a short time later, after it has been emptied by osmotically generated pressures and the sperm has been safely stored within her.

This mode of sperm transfer closely resembles that observed in true scorpions. Here, too, actual deposition of the stalked spermatophore takes place only in the presence of the female and is preceded by an intricate

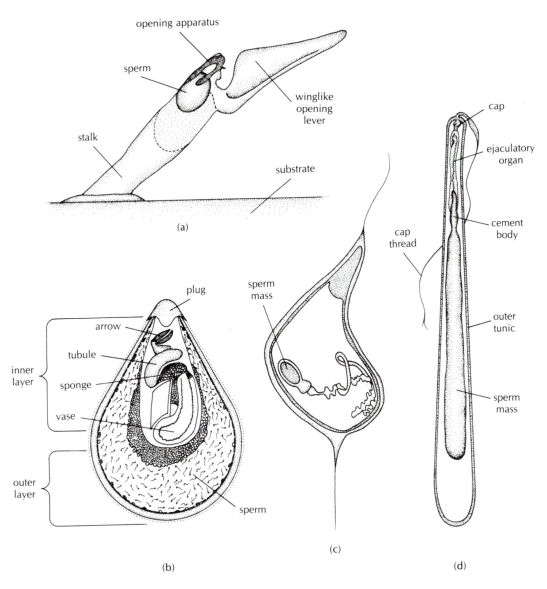

FIGURE 23.8. (a) The spermatophore of a scorpion. (From Barnes, 1980.) Invertebrate Zoology, 4th ed. Saunders; after Angermann.) (b) Longitudinal section through the spermatophore of a tick (Arachnida). After transfer to the female, the outer layers of the spermatophore elongate over a one-minute period. The tip of the inner section of the spermatophore then opens, and within one additional second, the arrow, tubule, sponge, and vase are shot out from the spermatophore, discharging sperm. (From Feldman-Muhsam. 1983. J. Insect Physiol. 29: 449.) (c) Spermatophore of a vermetid gastropod, tentatively identified as Dendropoma platypus. (From Hadfield and Hopper, 1980. Marine Biol., 57: 315.) (d) A cephalopod spermatophore. (From Brown, 1950. Selected Invertebrate Types. John Wiley & Sons.)

dance, during which time the male searches for a suitable substrate on which to cement the sperm-filled container. The spermatophore of scorpions is quite complex, possessing a mechanically operated sperm-ejection lever (Fig. 23.8a). Again, the male positions the female so that her genital opening is over the tip of the capsule (Fig 23.9). The capsule is then inserted into the genital opening just far enough to operate the lever of the spermatophore; the ejected sperm are then taken up and stored for later use.

In some other arachnids, the exact functional significance of the spermatophore is made dramatically apparent. The males of some species forcefully subdue a female, force her genital pore open, deposit a spermatophore on the ground, pick it up with the chelicerae, force the spermatophore into the genital opening, close the opening, and leave. Spermatophore transfer in centipedes and millipedes follows a somewhat similar script, except that the female voluntarily picks up and inserts the sperm packet following an often elaborate mating dance. Clearly, the spermatophore is serving as a functional substitute for a copulatory organ in transporting the spermatozoa to the female.

Spermatophores are also used by many marine and some freshwater invertebrates. Spermatophores are known to occur in monogonont rotifers, polychaetes, oligochaetes, leeches, gastropods (Fig. 23.8c), cephalopods, crustaceans (Fig. 23.10), phoronids, pogonophorans, and chaetognaths. In some species, the spermatophores are simply discharged into the sea and reach a female by chance. Species employing this mode of sperm transfer always live in close association, i.e., communally, so the loss of sperm is not so great as might be supposed. Floating spermatophores have been reported among the Gastropoda, Polychaeta, Phoronida, and Pogonophora. More commonly, spermatophores are delivered by the male directly.

Among the cephalopods, the complexity of the spermatophore, and of its mode of transfer to the female, rivals that encountered among terrestrial invertebrates. Males often use their chromatophores to perform species-specific color displays for the female as a prelude to the transfer of sperm. The spermatophore is a large cylindrical mass of sperm, incorporating a complex osmotically or mechanically activated sperm discharge mechanism (Fig. 23.8d). Large numbers of spermatophores are stored in a pouch, called **Needham's sac,** opening into the mantle

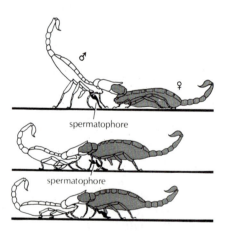

FIGURE 23.9. *Exchange of spermatophores in scorpions. The male is shown on the left. After the spermatophore is deposited by the male, the female is guided over it. (From Angermann, 1957. Z. Fur Tierpsychol. 14: 276.)*

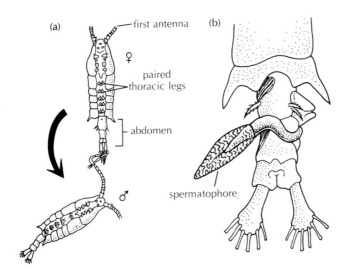

FIGURE 23.10. *Spermatophore exchange in a marine copepod,* Centropages typicus. *(a) The male grabs the female, using the hinged right first antenna. The male then swings the female around and grasps her around the abdomen, using his clawed right fifth thoracic appendage. (b) The spermatophore is shown in position after its transfer to the female. (From Blades, 1977. Marine Biol., 40: 57.)*

cavity. The spermatophores of some species attain lengths of one meter, and contain up to approximately 10^{10} sperm! Typically, at the appropriate moment, the male grabs one or more spermatophores from Needham's sac and inserts them into the mantle cavity of the female, adjacent to the genital opening. The organ used to insert the spermatophores is a modified arm called the **hectocotylus** (Fig. 23.11). In some cephalopod species, the hectocotylus breaks off from the male following its insertion into the mantle cavity of the female. Before the reproductive biology of these species was understood, the disconnected hectocotylus was thought to be a parasitic worm, a not unreasonable assumption.

The evolutionary movement of invertebrates to land and fresh water from the sea may well have been accompanied not by the development of entirely new systems of fertilization, but rather by the modification of already existing ones. Indeed, reproduction through the use of spermatophores or through copulation may be considered a preadaptation for life on land; development of spermatophores may well have been one of the evolutionary adaptations that made the transition from salt water to land or fresh water possible.

In marine animals, internal fertilization may be accomplished in the absence of any form of physical copulation or sophisticated sperm packaging. Because the concentration

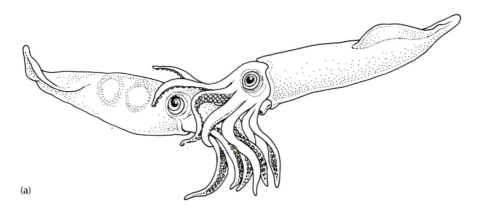

(a)

FIGURE 23.11. *(a) Mating in the squid* Loligo. *(After Barnes; after R.F. Sisson.) (b)* Octopus lentus, *showing hectocotyl arm. The end of the arm forms a broad, cup-like depression that holds the spermatophores after the male has removed them from the mantle cavity. (From Meglitsch, 1972.* Invertebrate Zoology, *2nd ed. Oxford University Press; after Verrill.)*

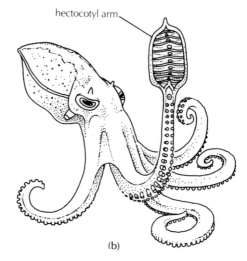

hectocotyl arm

(b)

of dissolved salts in sea water closely approximates that of most cells and tissues, sperm can be freely discharged into the sea water and transported to the female by water currents. Those species of bryozoans, echinoderms, bivalves, sponges, and cnidarians that have internal fertilization commonly employ this mechanism.

Once an egg has been fertilized internally, the embryos may develop within the body of the female until released as miniatures of the adult—as, for example, in some brooding bivalves and ophiuroids. Alternately, the fertilized eggs may be packaged in groups within egg capsules or egg masses, which are then either protected by the female or affixed to, or buried within, a substrate and abandoned. These encapsulating structures are especially complex among the Gastropoda (Fig. 23.12).

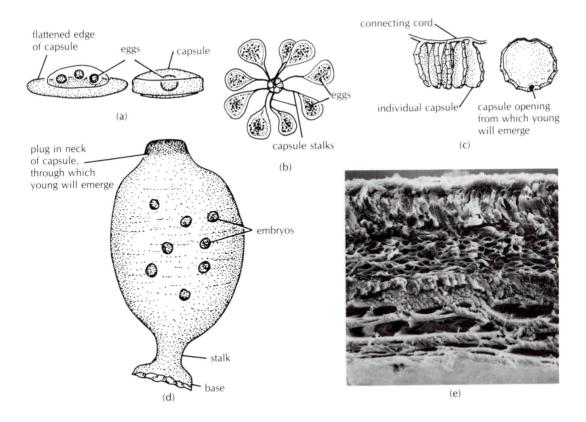

FIGURE 23.12. *Representative gastropod egg capsules of marine prosobranch gastropods. (a) Planktonic egg capsules of two periwinkles,* Littorina littorea *and* Tectarius muricatus. *(b) Cluster of egg capsules of* Turritella. *(c) Necklace of egg capsules deposited by the whelk* Busycon carica. *Each capsule is about the size of a quarter. (d) Egg capsule of the oyster drill,* Urosalpinx cinerea. *(e) Cross section of the egg capsule wall of* Nucella (= Thais) lima, *showing the complex, multilayered construction. The wall is approximately 55 μm thick. [(a–c) from Abbott et al., 1968. Seashells of North America. Western Publ. Co. (d) From Purves and Orians, 1982. Life: The Science of Biology. Sinauer Associates/ Willard Grant Press. (e) From Pechenik, unpublished photo.]*

Among marine invertebrates, union of gametes may also be accomplished in the absence of the structurally and behaviorally complex mechanisms generally associated with internal fertilization. In the ocean, fertilization may be achieved by the coordinated release of eggs and sperm into the surrounding sea water. **External fertilization** is commonly encountered in the marine environment, as seen in Table 23.2.

Larval Forms

The product of an external fertilization generally develops into a free-living, swimming larva, an individual that grows and differentiates entirely in the water, as a member of the plankton. Larval forms are also frequently produced by species that have internal fertilization, the larvae either emerging from the females after a period of brooding, or from egg capsules or egg masses. The larvae of most invertebrate species are ciliated, the cilia serving for locomotion and, in species with feeding larvae, for food collection as well. External ciliation is incompatible with the chitinous exoskeleton of larval arthropods; among such larvae, locomotion and food collection must be achieved using specialized appendages (Table 23.3). External ciliation is also lacking during the development of nematodes, which, you will recall, are also enclosed within a complex cuticle. Urochordates and chaetognaths constitute the final exceptions to the rule of ciliated larval stages amongst marine invertebrates.

The adaptive benefits of larval forms in the marine environment are easily imagined for species that are slow-moving or sessile as adults: genetic exchange between geographically separated populations of the same species; rapid recolonization of areas following local extinctions; lack of direct competition with adults for food or space. The latter benefit is also of particular significance for terrestrial insects.

Through the course of evolution by natural selection, larval forms have become increasingly well-adapted to their own niches, and may little resemble the adults of their own species, either morphologically or physiologically (Fig. 23.13). Larvae and adults may be considered as ecologically distinct organisms—often exploiting entirely different habitats, life-styles, and food sources—that just happen to have a genome in common. This latter point is crucial. Despite their ecological dissimilarity, the success of the one form in its stage of the life cycle determines the very existence of the other. Reproductive patterns comprising two or more ecologically distinct phases are termed **complex life cycles.**

The transition between phases of a complex life history often takes the form of an abrupt morphological, physiological, and ecological revolution termed a **metamorphosis** (Fig. 23.14, 18.14). The greater the degree of difference between the adult and larval life-styles, and the greater the degree of adaptation to those different life-styles, the more dramatic the metamorphosis.[4] Complex life cycles are believed to be the original condition for marine invertebrates. Such life cycles seem to have been generally lost in association with the invasion of land and fresh water, but have been re-evolved in at least one group, the Insecta. The percentage of insect species exhibiting **holometabolous development** (i.e., development involving a conspicuous metamorphosis) has increased from about 10% (325 million years ago), to about 63% (200 million years ago), to about 90% (presently). The adaptive benefits of complex life histories must indeed be considerable.

---

4 See Topics for Discussion, No. 5.

TABLE 23.3 Representative Larval Invertebrates

PHYLUM	CHARACTERISTIC LARVA	ADULT	PAGE REF.
Porifera	amphiblastula	sponge	82
Cnidaria (= Coelenterata)	strobilating scyphistoma / ephyra	jellyfish (Scyphozoa)	94
	planula	hydroid (Hydrozoa)	97
		anemone (Anthozoa)	103
Platyhelminthes	Müller's larva	flatworm (Turbellaria)	138

475

TABLE 23.3 (continued)

PHYLUM	CHARACTERISTIC LARVA			ADULT	PAGE REF.
Platyhelminthes (continued)					145

miracidium → redia → cercaria → fluke (Trematoda)

| | | | | | |
| Rhyncocoela (= Nemertea) | | | | | 167 |

pilidium

ribbon worm

| Annelida | | | | | 221 |

setigerous larva · polychaete worm (Polychaeta)

trochophore

| Sipuncula | | | | | 240 |

pelagosphaera

sipunculan

TABLE 23.3 (continued)

PHYLUM	CHARACTERISTIC LARVA		ADULT	PAGE REF.
Mollusca				295

chiton (Polyplacophora)

trochophore

scaphopod (Scaphopoda)

295

veliger

snail (Gastropoda)

295

veliger

clam (Bivalvia)

295

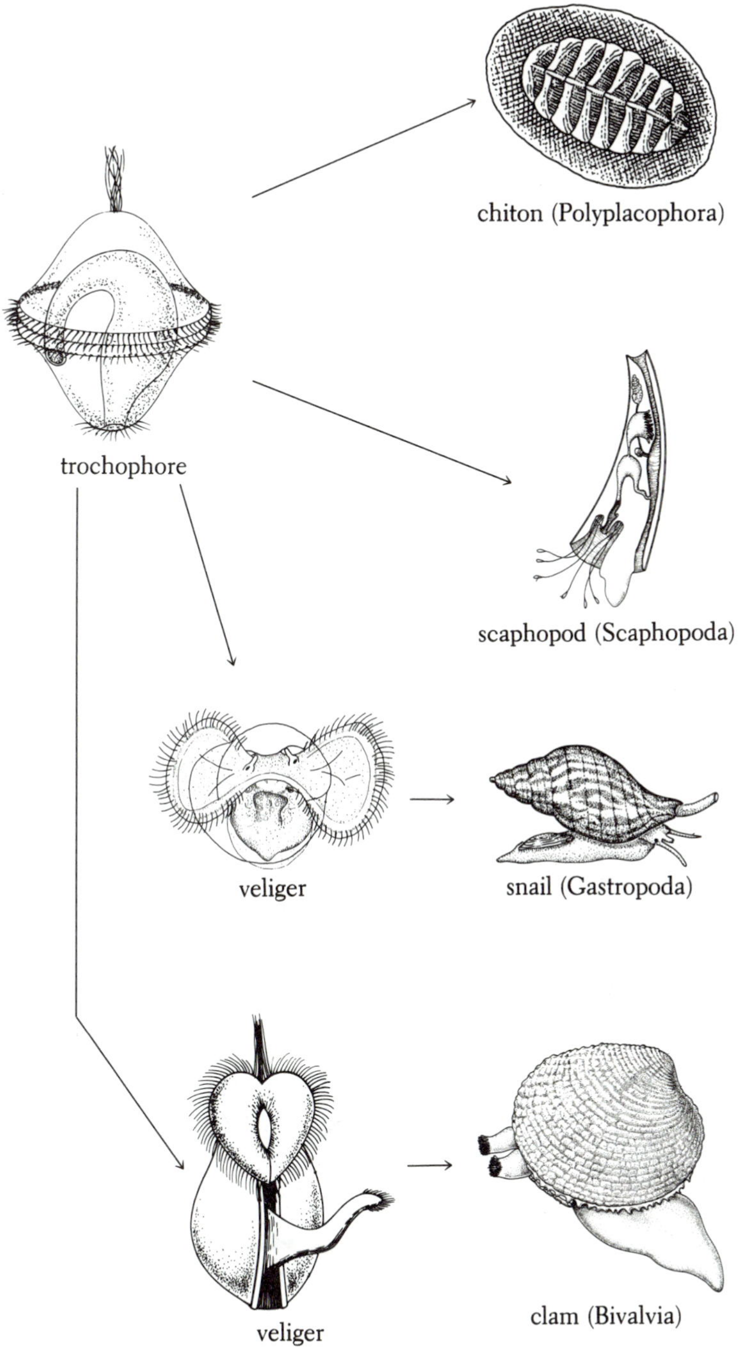

477

TABLE 23.3 (continued)

PHYLUM	CHARACTERISTIC LARVA		ADULT	PAGE REF.
Arthropoda				345

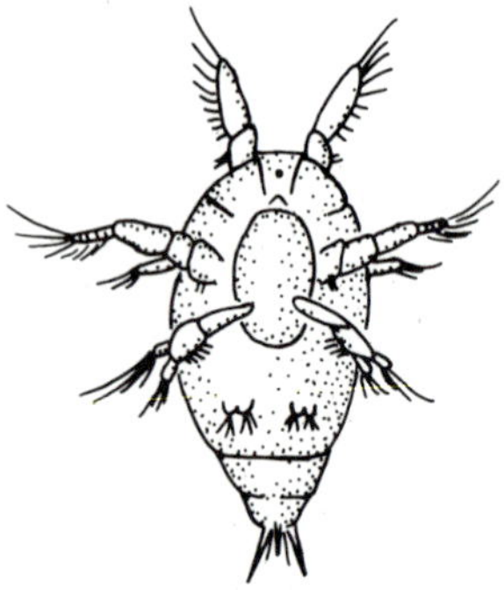

nauplius

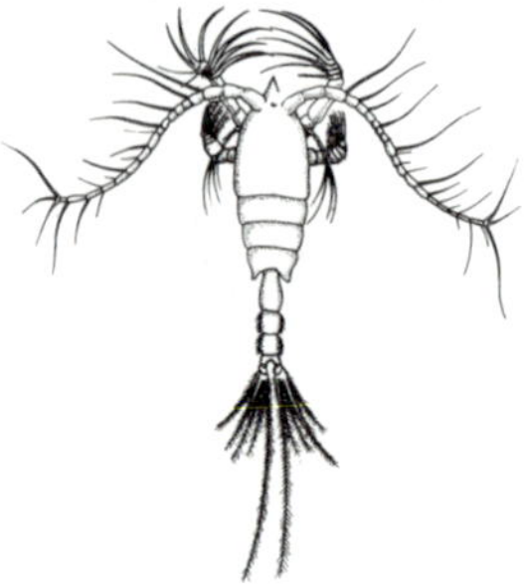

copepod
(Crustacea, Copepoda)

345

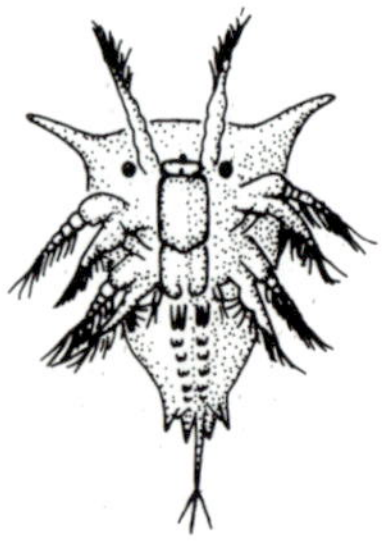

nauplius

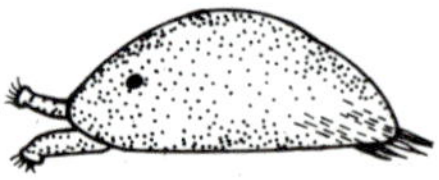

cyprid

barnacle
(Crustacea, Cirripedia)

345

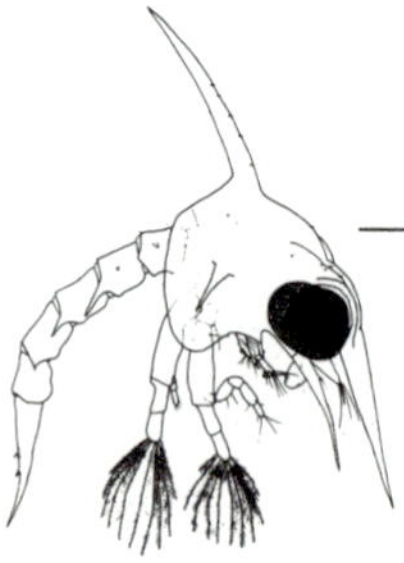

zoea

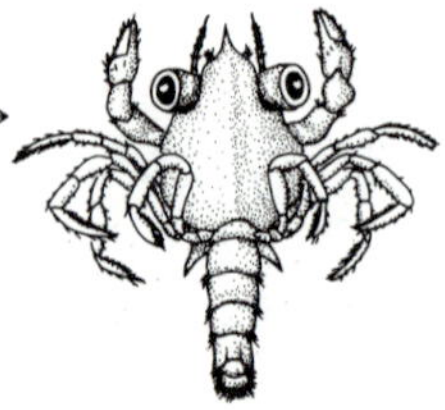

megalopa

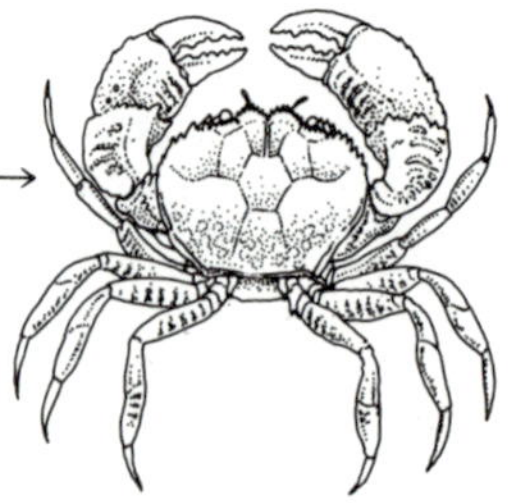

crab
(Crustacea, Malacostraca,
Decapoda)

TABLE 23.3 (continued)

PHYLUM	CHARACTERISTIC LARVA		ADULT	PAGE REF.
Arthropoda (continued)				347

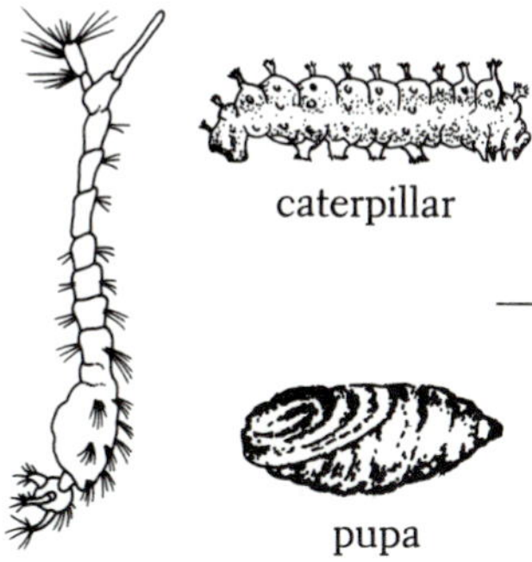

nymph

caterpillar

pupa

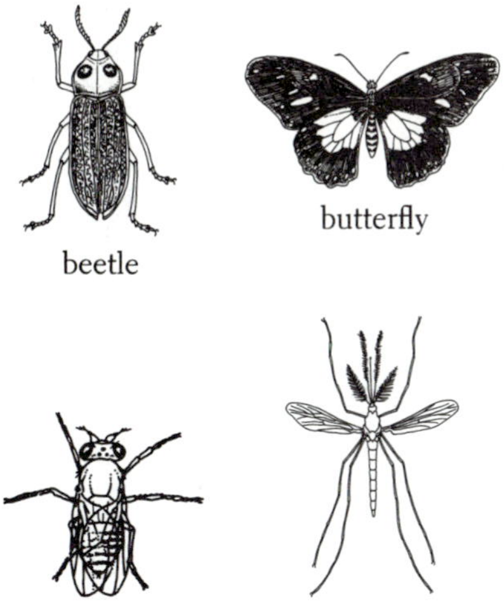

beetle

butterfly

fly

midge

(Insecta)

Bryozoa 393

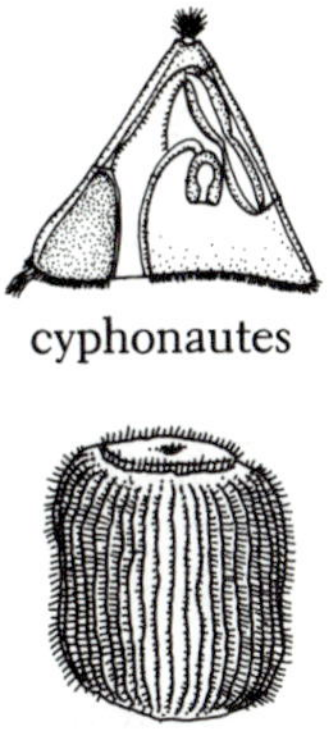

cyphonautes

coronate larva

bryozoans

Phoronida 393

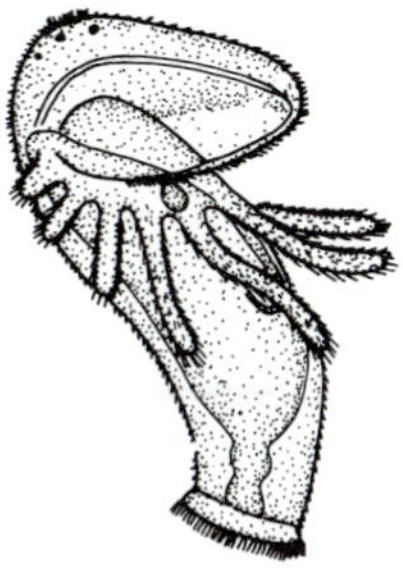

actinotroch

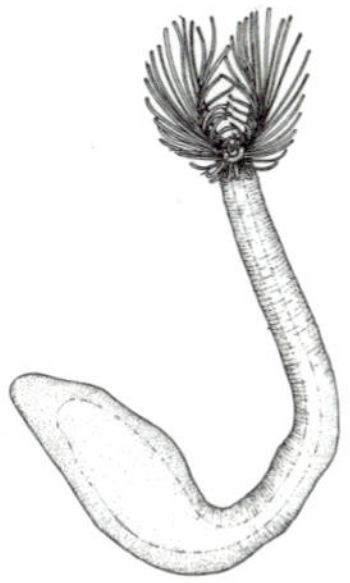

phoronid

479

TABLE 23.3 (continued)

PHYLUM	CHARACTERISTIC LARVA	ADULT	PAGE REF.
Brachiopoda	brachiopod larva	lampshell	393
Echinodermata	doliolaria	sea lily (Crinoidea)	423
	bipinnaria brachiolaria	sea star (Asteroidea)	423
	echinopluteus	sand dollar (Echinoidea) sea urchin	423

TABLE 23.3 (continued)

PHYLUM	CHARACTERISTIC LARVA	ADULT	PAGE REF.
Echinodermata (continued)	ophiopluteus	brittle star (Ophiuroidea)	423
	auricularia	sea cucumber (Holothuria)	423
Hemichordata	tornaria	acorn worm	440
Chordata, Urochordata	tadpole	tunicate, sea squirt (Ascidiacea)	449

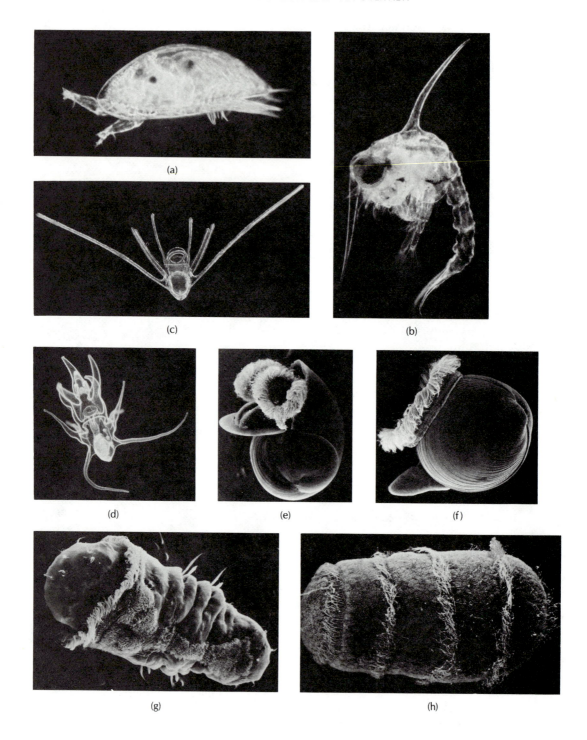

(a)

(c)

(b)

(d)

(e)

(f)

(g)

(h)

FIGURE 23.13. *Invertebrate larvae. (a) Barnacle cyprid. (b) Zoea of a decapod crustacean. (c) Ophiopluteus of Ophiothrix fragilis, an ophiuroid. (d) Brachiolaria of the asteroid Asterias vulgaris. (All courtesy of D.P. Wilson/Eric and David Hosking.) (e) Veliger of the opisthobranch Rostanga pulchra. (Courtesy of F.S. Chia, from Chia, 1978. Marine Biol., 46: 109.) (f) Veliger of the bivalve* Lyrodus pedicellatus, *showing shell, foot, and ciliated velum. Shell length is about 420 μm. (Courtesy of C. Bradford Calloway.) (g) Advanced polychaete trochophore larva. Note that setae have already developed on two of the segments. (Courtesy of F.S. McEuen, from McEuen, 1983. Marine Biol., 76: 301.) (h) Doliolaria of the crinoid* Florometra serratissima. *(Courtesy of Philip V. Mladenov, from Mladenov, P.V., and F.S. Chia, 1983. Marine Biol., 73: 309.)*

The larvae of some invertebrates are **lecithotrophic** (*lecitho* = G: yolk; *trophy* = G: feeding); i.e., they subsist on nutrient reserves supplied to the egg by the parent and are independent of the outside world for food. Lecithotrophic development is commonly encountered among most groups of invertebrates, particularly among marine invertebrates living at high latitudes or in very deep water. Lecithotrophy is much less common in shallow water environments of the tropics and subtropics. Here, the larval stages typically develop functional guts and feed upon other members of the plankton, both plant (phytoplankton) and animal (zooplankton). Such species are said to be **planktotrophic.**

DISPERSAL AS A COMPONENT OF THE LIFE-HISTORY PATTERN

Most freshwater, marine, or terrestrial animals have a dispersal stage at some point in their life histories. For many marine invertebrates, which exploit the properties of sea water by living a sedentary or even a sessile adult existence, dispersal is accomplished by a planktonic larval stage, as discussed above. How much dispersal occurs depends upon how long the larval form can be maintained, and on the speed and direction of water currents in which the larvae live (Fig. 23.15).

Freshwater invertebrates must deal with an often ephemeral habitat. Some groups, such as the sponges, tardigrades, bryozoans, and a number of crustaceans and rotifers, often avoid the need to disperse away from unfavorable conditions by forming resistant stages of arrested development. Such stages take the form of gemmules in sponges, resting eggs in crustaceans and rotifers, and statoblasts in bryozoans. Rotifers, tardigrades, and many protozoans may also enter into a cryptobiotic state during periods of dehydration. When environmental conditions improve, these various resting stages become revitalized. A considerable amount of windborne dispersal to new habitats may also occur for these different diapause forms.

For terrestrial invertebrates, a sedentary adult existence is rare, due to the air being

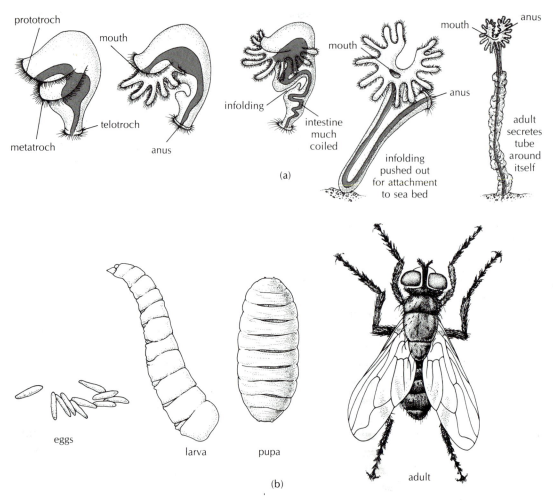

FIGURE 23.14. *Examples of invertebrate metamorphosis.*
(a) Metamorphosis of an actinotroch larva into an adult phoronid.
Note the dramatic reorientation of the digestive tract. (From Hardy,
1965. The Open Sea. Houghton Mifflin.) (b) Stages in the
development of the common housefly. This animal undergoes two
distinct metamorphoses: one from the larval to the pupal stage, and
another from the pupa to the adult. (After Engelmann and Hegner;
after Packard.)

dry and of low density. Not surprisingly then, dispersal of terrestrial species is generally achieved by the adults, and the developing larvae are the stay-at-homes. It is interesting that the major exceptions to this generalization are found among the "suspension-feeding" arachnids, i.e., the spiders. In many species, shortly after the young spiders emerge from the silken cocoon built for them by the mother, they climb to the top of the nearest twig or blade of grass and allow themselves to be taken by the wind. This is termed

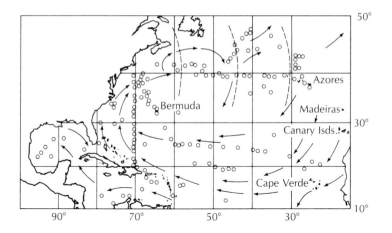

FIGURE 23.15. *Dispersal of larvae of the tropical prosobranch* gastropod Cymatium parthenopeum *throughout the Atlantic Ocean. Circles indicate stations at which larvae of this species were collected. Arrows indicate surface circulation patterns. The adults are restricted to shallow-water habitats on the east coast of the U.S. and the west coast of Europe (From Pechenik et al., 1984. Science, 224: 1097.)*

ballooning. Air currents may disperse baby spiders for hundreds of miles (Fig. 23.16).

The degree to which the dispersal phase selects the appropriate habitat for the sedentary phase obviously determines the probability that a given individual will indeed survive to reproductive maturity. Among marine invertebrates with free-living larvae, there has undoubtedly been extreme selective pressure acting against random metamorphosis to adult form and habitat. Instead, the dispersing larvae of most species are highly selective about where they will metamorphose, and can delay their metamorphosis for varying periods of time (up to many months for some species of molluscs, echinoderms, and arthropods) if they fail to encounter the appropriate cues. The cues

that trigger metamorphosis are often associated with some component of the adult environment: the adult food source or prey species, for example, or, quite commonly, adults of the same species (Table 23.4).[5]

Among the terrestrial insects, the larval stages are the sedentary eating machines and the adults are the dispersal stages. Not surprisingly, the larvae of many species have become highly adapted for life on specific hosts and the adults, in turn, have often evolved adaptations promoting the placement of eggs in the habitats supporting the best larval growth and survival.

Even though much of the existing di-

---

5 See Topics for Discussion, No. 4.

FIGURE 23.16.   Young spiders ballooning.
*... A warm draft of rising air blew softly through the barn cellar.
The air smelled of the damp earth, of the spruce woods, of the sweet
springtime. The baby spiders felt the warm updraft. One spider
climbed to the top of the fence. Then it did something that came as
a great surprise to Wilbur. The spider stood on its head, pointed its
spinnerets into the air, and let loose a cloud of fine silk. The silk
formed a balloon. As Wilbur watched, the spider let go of the fence
and rose into the air.*

"Good-bye!" it said, as it sailed through the doorway.

*"Wait a minute!" screamed Wilbur. "Where do you think you're
going?"*

*But the spider was already out of sight. . . . The air was soon
filled with tiny balloons, each balloon carrying a spider.*

*Wilbur was frantic. Charlotte's babies were disappearing at a great
rate. (Illustration and text from* Charlotte's Web, *by E.B. White,
pictures by Garth Williams. Copyright, 1952, by E.B. White; text
copyright renewed © 1980 by E.B. White; illustrations copyright
renewed © 1980 by Garth Williams. Reprinted by permission of
Harper & Row, Publishers, Inc.)*

versity of reproductive pattern has not found its way into the preceding discussion, it should be clear that invertebrates show a dazzling variety of ways to reproduce. Many of the patterns that seem strange to us— external fertilization, simultaneous and sequential hermaphroditism, and parthenogenesis, for example—are actually very common reproductive modes. We might ask, why are there so many different ways to re-

TABLE 23.4. Gregarious metamorphosis by barnacle larvae. Groups of twelve cyprid larvae of *Balanus balanoides* were placed in each of ten dishes of sea water (controls) or in dishes containing sea water plus the adults and shells of *B. balanoides* or of two other barnacle species, *B. crenatus* or *Elminius modestus*. The number of metamorphosed individuals was assessed after 24 hours. (Data from Knight-Jones, 1953. *J. Exp. Biol.* 30: 584.)

	CONTROL	*B. BALANOIDES*	*B. CRENATUS*	*E. MODESTUS*
Total no. larvae in experiment	120	120	120	120
Average no. metamorphosing per dish (± one standard error about the mean)	0.4 ± 0.31	9.5 ± 0.81	3.7 ± 0.93	1.6 ± 0.85
Average percent metamorphosed (± one s.e. about the mean)	3.3 ± 2.5%	79.2 ± 6.7%	30.8 ± 7.7%	13.3 ± 7.0%

produce? The question is fun to think about, but, unfortunately, very difficult to answer. For one thing, the life histories of most species have yet to be documented. For another, the adaptive significance of the different patterns is difficult to demonstrate convincingly. Moreover, the selective pressures responsible for presently observed reproductive patterns may be quite different today from what they were hundreds of thousands of years ago. Lastly, the role of historical accident in shaping the reproductive patterns of various groups of invertebrates is hidden among evolutionary interrelationships that are often unclear, and which may never be satisfactorily unravelled.

TOPICS FOR FURTHER DISCUSSION AND INVESTIGATION

1. What are the roles of temperature and light in regulating the cycles of reproductive activity of invertebrates?

Davison, J., 1976 *Hydra hymanae*: regulation of the life cycle by time and temperature. *Science*, 194: 618.

De March, B.G., 1977. The effects of photoperiod and temperature on the induction and termination of reproductive resting stage in the freshwater amphipod *Hyallela azteca* (Sanssure). *Can. J. Zool.*, 55: 1595.

Fell, P. E., 1974. Diapause in the gemmules of the marine sponge, *Haliclona loosanoffi*, with a note on the gemmules of *Haliclona oculata*. *Biol. Bull.*, 147: 333.

Hayes, J.L., 1982. Diapause and diapause dynamics of *Colias alexandra* (Lepidoptera: Pieridae). *Oecologia (Berl.)*, 53: 317.

Landry, M.R., 1975. Dark inhibition of egg hatching of the marine copepod *Acartia clausi* Giesbr. *J. Exp. Marine Biol. Ecol.*, 20: 43.

Marcus, N.H., 1982. Photoperiodic and temperature regulation of diapause in *Labidocera aestiva* (Copepoda: Calanoida). *Biol. Bull.*, 162: 45.

Pearse, J.S., and D.J. Eernisse, 1982. Photoperiodic regulation of gametogenesis and gonadal growth in the sea star *Pisaster ochraceus*. *Marine Biol.*, 67: 121.

Rokop, F.J., 1974. Reproductive patterns in the deep-sea benthos. *Science*, 186: 743.

Rose, S.M., 1939. Embryonic induction in *Ascidia*. *Biol. Bull.*, 77: 216.

Stross, R.G., and J.C. Hill, 1965. Diapause induction in *Daphnia* requires two stimuli. *Science*, 150:1462.

Vowinckel, C., 1970. The role of illumination and temperature in the control of sexual reproduction in the planarian *Dugesia tigrina* (Girarad). *Biol. Bull.*, 138: 77.

West, A.B., and C.C. Lambert, 1976. Control of spawning in the tunicate *Styela plicata* by variations in a natural light regime. *J. Exp. Zool.*, 195: 263.

2. What aspects of reproductive activity appear to be under chemical control among invertebrates?

Crisp, D.J., 1956. A substance promoting hatching and liberation of young in cirripedes. *Nature*, 178: 263.

Davis, W.J., G.J. Mpitsos, and J.M. Pinneo, 1974. The behavioral hierarchy of the mollusc *Pleurobrachaea*. II. Hormonal suppression of feeding associated with egg laying. *J. Comp. Physiol.*, 90: 225.

Engelmann, F., 1959. The control of reproduction in *Diploptera punctata* (Blattaria). *Biol. Bull.*, 116: 406.

Forward, R.B., Jr., and K.J. Lohmann, 1983. Control of egg hatching in the crab *Rhithropanopeus harrisii* (Gould). *Biol. Bull.*, 165: 154.

Golden, J.W., and D.L. Riddle, 1982. A pheromone influences larval development in the nematode *Caenorhabditis elegans*. *Science*, 218: 578.

Golding, D.W., 1967. Endocrinology, regeneration and maturation in *Nereis*. *Biol. Bull.*, 133: 567.

Kanatani, H., and M. Ohguri, 1966. Mechanism of starfish spawning. I. Distribution of active substance responsible for maturation of oocytes and shedding of gametes. *Biol. Bull.*, 131: 104.

Katona, S.K., 1973. Evidence for sex pheromones in planktonic copepods. *Limnol. Oceanogr.*, 18: 574.

Kelly, T.J., and L.M. Hunt, 1982. Endocrine influence upon the development of vitellogenic competency in *Oncopeltus fasciatus*. *J. Insect Physiol.*, 28: 935.

Kupfermann, I., 1967. Stimulus of egg laying: possible neuroendocrine function of bag cells of abdominal ganglia of *Aplysia californica*. *Nature, London*, 216: 814.

Marthy, H., J.R. Hauser, and A. Scholl, 1976. Natural tranquilizer in cephalopod eggs. *Nature*, 261: 496.

Pechenik, J.A., 1975. The escape of veligers from the egg capsules of *Nassarius obsoletus* and *Nassarius trivittatus* (Gastropoda: Prosobranchia). *Biol. Bull.*, 149: 580.

Reynolds, S.E., P.H. Taghert, and J.W. Truman, 1979. Eclosion hormone and bursicon titres and the onset of hormonal responsiveness during the last day of adult development in *Manduca sexta* (L.) *J. Exp. Biol.* 78: 77.

Shorey, H.H., and R.J. Bartel, 1970. Role of a volatile sex pheromone in stimulating male courtship behaviour in *Drosophila melanogaster*. *Anim. Behav.*, 18: 159.

Takeda, N., 1979. Induction of egg-laying by steroid hormones in slugs. *Comp. Biochem. Physiol.*, 62A: 273.

Truman, J.W., and P.G. Sokolove, 1972. Silk moth eclosion: hormonal triggering of a centrally programmed pattern of behavior. *Science*, 175: 1491.

Wigglesworth, V.B., 1934. The physiology of ecdysis in *Rhodnius prolixus* (Hemiptera). II. Factors controlling moulting and "metamorphosis." *Quart. J. Microsc. Sci.*, 77: 191.

3. Discuss the morphological and behavioral adaptations for indirect sperm transfer.

Blades, P.I., 1977. Mating behavior of *Centropages typicus* (Copepoda: Calanoida). *Marine Biol.*, 40: 57.

Feldman-Muhsam, B., 1967. Spermatophore formation and sperm transfer in ornithodoros ticks. *Science*, 156: 1252.

Legg, G., 1977. Sperm transfer and mating in *Ricinoides hanseni* (Ricinulei: Arachnida). *J. Zool., London*, 182: 51.

Reeve, M.R., and M.A. Walter, 1972. Observations and experiments on methods of fertilization in the chaetognath *Sagitta hispida*. *Biol. Bull.*, 143: 207.

Weygoldt, P., 1966. Mating behavior and spermatophore morphology in the pseudoscorpion *Dinocheirus tumidus* Banks (Cheliferinea: Chernetidae). *Biol. Bull.*, 130: 462.

4. How are the cues used by dispersal stages related to the habitat requirements of the sedentary or sessile stage in a complex life cycle?

Brewer, R.H., 1976. Larval settling behavior in *Cyanea capillata* (Cnidaria: Scyphozoa). *Biol. Bull.*, 150: 183.

Chew, F.S., 1977. Coevolution of pierid butterflies and their cruciferous food plants. II. The distribution of eggs on potential foodplants. *Evolution*, 31: 568.

Cohen, L.M., H. Neimark, and L.K. Eveland, 1980. *Schistosoma mansoni*: response of cercariae to a thermal gradient. *J. Parasitol.*, 66: 362.

Grosberg, R.K., 1981. Competitive ability influences habitat choice in marine invertebrates. *Nature*, 290: 700.

Highsmith, R.C., 1982. Induced settlement and metamorphosis of sand dollar (*Dendraster excentricus*) larvae in predator-free sites: adult sand dollar beds. *Ecology*, 63: 329.

Knight-Jones, E.W. 1953. Laboratory experiments on gregariousness during settling in *Balanus balanus* and other barnacles. *J. Exp. Biol.*, 30: 584.

Knight-Jones, E.W., 1955. The gregarious settling reaction of barnacles as a measure of systematic affinity. *Nature, London*, 175: 266.

MacInnes, A.J., 1969. Identification of chemicals triggering cercarial penetration responses of *Schistosoma mansoni*. *Nature, London*, 224: 1221.

MacInnes, A.J., W.M. Bethel, and E.M. Cornfield, 1974. Identification of chemicals of snail origin that attract *Schistosoma mansoni* miracidia. *Nature, London*, 248: 361.

Olson, R., 1983. Ascidian-*Prochloron* symbiosis: the role of larval photoadaptations in midday larval release and settlement. *Biol. Bull.*, 165: 221.

Scheltema, R.S., 1961. Metamorphosis of the veliger larvae of *Nassarius obsoletus* (Gastropoda) in response to bottom sediment. *Biol. Bull.*, 120: 92.

Spight, T.M., 1977. Do intertidal snails spawn in the right places? *Evolution*, 31: 682.

Wallace, R.L., 1978. Substrate selection by larvae of the sessile rotifer *Ptygura beauchampi*. *Ecology*, 59: 221.

Williams, K.S., 1983. The coevolution of *Euphydryas chalcedona* butterflies and their larval host plants. III. Oviposition behavior and host plant quality. *Oecologia (Berl.)*, 56: 336.

Wood, E.M., 1974. Some mechanisms involved in host recognition and attachment of the glochidium larva of *Anodonta cygnea* (Mollusca: Bivalvia). *J. Zool.*, 173: 15.

5. Describe the morphological changes that take place during metamorphosis in one group of marine invertebrates.

Atkins, D., 1955. The cyphonautes larvae of the Plymouth area and the metamorphosis of *Membranipora membranacea* (L.). *J. Marine Biol. Assoc. U.K.*, 34: 441.

Berrill, N.J., 1947. Metamorphosis in ascidians. *J. Morphol.*, 81: 249.

Bickell, L.R., and S.C. Kempf, 1983. Larval and metamorphic morphogenesis in the nudibranch *Melibe leonina* (Mollusca: Opisthobranchia). *Biol. Bull.*, 165: 119.

Bonar, D.B., and M.G. Hadfield, 1974. Metamorphosis of the marine gastropod *Phestilla sibogae* Bergh (Nudibranchia: Aeolidacea). I. Light and electron microscopic analysis of larval and metamorphic stages. *J. Exp. Marine Biol. Ecol.* 16: 227.

Cameron, R.A., and R.T. Hinegardner, 1978. Early events in sea urchin metamorphosis, description and analysis. *J. Morphol.*, 157: 21.

Cloney, R.A., 1977. Larval adhesive organs and metamorphosis in ascidians. *Cell Tissue Res.*, 183: 423.

Cole, H.A., 1938. The fate of the larval organs in the metamorphosis of *Ostrea edulis*. *J. Marine Biol. Assoc. U.K.*, 22: 469.

Dean, D., 1965. On the reproduction and larval development of *Streblospio benedicti* Webster. *Biol. Bull.*, 127: 67.

Factor, J. R., 1981. Development and metamorphosis of the digestive system of larval lobsters, *Homarus americanus* (Decapoda: Nephropidae). *J. Morphol.*, 169: 225.

Lang, W.H., 1976. The larval development and metamorphosis of the pedunculate barnacle *Octolasmis mülleri* (Coker, 1902) reared in the laboratory. *Biol. Bull.*, 150: 255.

Reed, C.G., and R.M. Woollacott, 1982. Mechanisms of rapid morphogenetic movements in the metamorphosis of the bryozoan *Bugula neritina* (Cheilostomata, Cellularioidea). I. Attachment to the subtratum. *J. Morphol.*, 172: 335.

Rivest, B., 1978. Development of the eolid nudibranch *Cuthona nana* (Alder and Hancock, 1842), and its relationship with a hydroid and hermit crab. *Biol. Bull.*, 154: 157.

6. Discuss the possible adaptive significance of the apyrene sperm of insects, and suggest ways to test the various hypotheses.

Silberglied, R.E., J.G. Shepherd, and J.L. Dickinson, 1984. Eunuchs: the role of apyrene sperm in Lepidoptera? *Amer. Nat.*, 123: 255.

7. What impact do environmental pollutants have on the reproductive biology of marine invertebrates?

Bellam, G., D.J. Reish, and J.P. Foret, 1972. The sublethal effects of a detergent on the reproduction, development, and settlement in the polychaetous annelid *Capitella capitata*. *Marine Biol.*, 14: 183.

Bigford, T.E., 1977. Effects of oil on behavioral responses to light, pressure and gravity in larvae of the rock crab *Cancer irroratus*. *Marine Biol.*, 43: 137.

Calabrese, A.J., J. R. MacInnes, D.A. Nelson, and J.E. Miller, 1977. Survival and growth of bivalve larvae under heavy-metal stress. *Marine Biol.*, 41: 179.

Dafni, J., 1980. Abnormal growth patterns in the sea urchin *Tripneustes* cf. *gratilla* (L.) under pollution (Echinodermata, Echinoidea). *J. Exp. Marine Biol. Ecol.*, 47: 259.

Epifanio, C.E., 1971. Effects of dieldrin in sea water on the development of two species of crab larvae, *Leptodius floridanus* and *Panopeus herbstii*. *Marine Biol.*, 11: 356.

Heslinga, G.A., 1976. Effects of copper on the coral-reef echinoid *Echinometra mathaei*. *Marine Biol.*, 35: 155.

Kobayashi, N., 1980. Comparative sensitivity of various developmental stages of sea urchins to some chemicals. *Marine Biol.*, 58: 163.

Muchmore, D., and D. Epel, 1977. The effects of chlorination of wastewater on fertilization in some marine invertebrates. *Marine Biol.*, 19: 93.

GLOSSARY OF FREQUENTLY USED TERMS

aboral: the part of the body farthest from the mouth.

acoelomate: lacking a body cavity between the gut and the outer body wall musculature.

annulation: external division of a worm-shaped body into a series of conspicuous rings.

archenteron: a cavity that eventually becomes the digestive tract of the adult or larva, formed during the development of an embryo.

asexual reproduction: reproduction that does not involve the fusion of gametes; reproduction without fertilization.

benthic: living on or within a substrate.

benthos: the animals and plants living on or within a substrate.

binary fission: asexual division of one organism into two nearly identical organisms.

bioluminescence: biochemical production of light by living organisms.

biramous: two-branched.

blastocoel: the internal cavity commonly formed by cell division early in embryonic development, prior to gastrulation.

brooding: parental care of developing young.

budding: a form of asexual reproduction in which new individuals develop from a portion of the parent, as in many protozoans, cnidarians, and polychaetes.

cephalization: concentration of nervous and sensory systems in the anterior part of the body, which becomes known as the head.

chromatophore: a pigment-containing cell that can be used by an animal to vary its external coloration.

cilium: a threadlike locomotory organelle containing a highly organized array of microtubules; shorter than a flagellum.

cirri: (1) among ciliated protozoans, a group of cilia that function as a single unit; (2) among barnacles, the thoracic appendages, which are modified for food collection; (3) among crinoids, the prehensile appendages located aborally, used for walking and for clinging to solid substrates.

coelom: an internal body cavity lying between the gut and the outer body wall musculature, and which is lined with derivatives of the mesoderm.

colony: a group of genetically identical individuals formed asexually from a single colonizing individual.

conjugation: a temporary physical association in which genetic material is exchanged between two ciliate protozoans.

convergent evolution: the process whereby similar characteristics are independently evolved by different groups of organisms in response to similar selective pressures.

cuticle: a noncellular, secreted body covering.

cyst: a secreted covering that protects many small invertebrates, including some protozoans, rotifers and nematodes, from environmental stresses such as desiccation and overcrowding.

deposit-feeding: ingesting substrate (sand, soil, mud) and extracting the organic fraction.

desiccation: dehydration.

dioecious: characterized by having separate sexes; i.e., an individual is either male or female, but never both.

diploblastic: possessing only two distinct tissue layers during embryonic development.

ectoderm: an embryonic tissue layer; forms epidermal, nervous, and sensory organs and tissues.

encystment: the secretion of a protective outer covering that permits some small invertebrates to withstand exposure to extreme environmental stresses such as desiccation and overcrowding.

endoderm: an embryonic tissue layer; forms wall of digestive system.

enterocoely: formation of a coelom through out-pocketing of the inner portion of the archenteron in some animals.

estuary: a partially enclosed body of water influenced by both tidal forces and by freshwater input from the land.

eutely: species-specific constancy of cell numbers or nuclei; growth occurs through increased cell size rather than increased cell number.

exoskeleton: a system of external levers and joints that permits pairs of muscles to act against, or antagonize each other; the exoskeleton is also protective.

filiform: threadlike.

filter-feeder: an organism that filters food particles from the surrounding medium.

flagellum: a threadlike locomotory organelle containing a highly organized array of microtubules; longer than a cilium.

flame cell: flagellated cells associated with protonephridia, as in flatworms, rotifers, and some polychaetes.

gametes: the sex cells involved in fertilization.

gastrovascular canals: fluid-filled canals opening at the mouth of cnidarians and ctenophores, and which function in gas exchange and in the distribution of nutrients.

gastrulation: creation of a new tissue layer by the movement of cells in the early embryo (blastula).

gill: a structure specialized for gas exchange in aquatic animals.

hermaphrodite: a single individual that functions as both male and female, either simultaneously or in sequence.

hydrostatic skeleton: a constant-volume, fluid-filled cavity that permits mutual antagonism of muscle pairs.

interstitial: living in the spaces between sand grains.

intertidal: living in the area between high and low tides, and thus alternately exposed to the air and to the sea.

larva: a free-living developmental stage in the life-history of many invertebrate species.

mesenteries: (1) infoldings of gastroderm and mesoglea extending into the gastrovascular cavity of cnidarians; (2) sheets of peritoneum from which the digestive tract is suspended in coelomates.

mesoderm: an embryonic tissue layer that gives rise to muscles, gonads, and other organs of the adult.

mesoglea: gelatinous layer found between the epidermis and the gastrodermis of cnidarians.

mesohyl: nonliving, middle gelatinous layer of sponges; living cells are often found within it.

metamerism: serial repetition of organs and tissues, including the body wall, sensory systems, and musculature.

metamorphosis: a dramatic transformation of morphology and function occurring over a short period of time during development.

metanephridium: an organ open to the body cavity through a ciliated funnel (nephrostome) and involved in excretion or in the regulation of water balance or salt content.

metazoan: a multicellular animal.

microtubules: tubulin-containing cylinders characteristic of cilia and flagella.

monoecious: characterized by the presence of both sexes in a single individual, either in sequence or sequentially; hermaphroditic.

nematocyst: cnidarian cell type that explosively emits long threads specialized for defense and food capture.

ocellus: a simple, pigment-containing photoreceptor found in a variety of unrelated invertebrates.

osmosis: diffusion of water across a semipermeable membrane (permeable to water but not to solute) along a concentration gradient.

parasitism: a close association between species in which one member benefits at the expense of the partner.

parthenogenesis: development of an unfertilized egg into a functional adult.

pelagic: aquatic, living between the water surface and the bottom.

peristalsis: progressive waves of muscular contraction passing down the length of an organism or organ system.

peritoneum: the mesodermal lining of the body cavity of coelomates.

phagocytosis: the process in which food particles are surrounded by cell membrane and incorporated into the cell cytoplasm, forming a food vacuole.

plankton: animals (zooplankton) and plants (phytoplankton) that have only limited locomotory capabilities and are therefore distributed by currents and other water movements.

pre-adaptation: a trait that is adaptive only in a new set of physical or biological circumstances.

primitive: characteristics believed to closely resemble those of the ancestral form; possessing such characteristics.

proboscis: a tubular extension at the anterior of an animal, generally used for locomotion or food collection; may or may not be directly connected to the gut.

protandric hermaphroditism: a single individual functions as male and then female in sequence.

pseudocoel: an internal body cavity lying between the outer body wall musculature and the gut, generally formed by persistence of the embryonic blastocoel.

pseudopodia: amorphous protrusions of cytoplasm involved in the locomotion and feeding of amoebae and related protozoans.

radial cleavage: a form of early cell division in which all cleavage planes are perpendicular, so that daughter cells come to lie directly in line with each other.

schizocoely: coelom formation accomplished by a split in the mesoderm during embryonic development of some animals.

sedentary: bottom-dwelling and capable of only limited locomotion.

septa: peritoneal (mesodermal) sheets separating adjacent segments, as in annelids, or body divisions, as in chaetognaths.

sessile: bottom-dwelling and generally incapable of locomotion.

sexual reproduction: reproduction involving fusion of gametes.

spermatophore: a container of sperm transferred from one individual to another during mating.

spicules: calcareous or silicious formations present in the tissues of some organisms and generally serving protective or supportive functions.

spiral cleavage: pattern of cell division in which cleavage planes are at 45° to the animal–vegetal axis of the egg.

statocyst: a sense organ that informs the bearer of orientation of the body to gravity.

suspension feeder: an animal that feeds by extracting particulates from the surrounding medium; this may be accomplished by filtering or by other means.

test: any hard external covering; may be secreted by the animal or constructed from surrounding materials.

triploblastic: possessing three distinct tissue layers during embryonic development.

vector: any organism that transmits parasites from one host species to another.

vermiform: worm-shaped; i.e., soft-bodied and substantially longer than wide.

zooid: a single member of a colony.

zooplankton: the animal component of the plankton, having only limited locomotory powers.

INDEX

Boldface indicates an illustration; t indicates a table.